Lecture Notes in Electrical Engineering

Volume 1172

The book series *Lecture Notes in Electrical Engineering* (LNEE) publishes the latest developments in Electrical Engineering—quickly, informally and in high quality. While original research reported in proceedings and monographs has traditionally formed the core of LNEE, we also encourage authors to submit books devoted to supporting student education and professional training in the various fields and applications areas of electrical engineering. The series cover classical and emerging topics concerning:

- Communication Engineering, Information Theory and Networks
- Electronics Engineering and Microelectronics
- Signal, Image and Speech Processing
- Wireless and Mobile Communication
- Circuits and Systems
- Energy Systems, Power Electronics and Electrical Machines
- Electro-optical Engineering
- Instrumentation Engineering
- Avionics Engineering
- Control Systems
- Internet-of-Things and Cybersecurity
- Biomedical Devices, MEMS and NEMS

For general information about this book series, comments or suggestions, please contact leontina.dicecco@springer.com.

To submit a proposal or request further information, please contact the Publishing Editor in your country:

China

Jasmine Dou, Editor (jasmine.dou@springer.com)

India, Japan, Rest of Asia

Swati Meherishi, Editorial Director (Swati.Meherishi@springer.com)

Southeast Asia, Australia, New Zealand

Ramesh Nath Premnath, Editor (ramesh.premnath@springernature.com)

USA, Canada

Michael Luby, Senior Editor (michael.luby@springer.com)

All other Countries

Leontina Di Cecco, Senior Editor (leontina.dicecco@springer.com)

**** This series is indexed by EI Compendex and Scopus databases. ****

Sanjay Yadav · Yogendra Arya ·
Shailesh M. Pandey · Noredine Gherabi ·
Dimitrios A. Karras
Editors

Proceedings of 3rd International Conference on Artificial Intelligence, Robotics, and Communication

ICAIRC 2023

Springer

Editors
Sanjay Yadav
Formerly with CSIR-National Physical Lab
New Delhi, Delhi, India

Shailesh M. Pandey
Department of Mechanical Engineering
National Institute of Technology Patna
Patna, Bihar, India

Dimitrios A. Karras
School of Sciences
University of Athens
Athens, Greece

Yogendra Arya
Department of Electrical Engineering
J.C. Bose University of Science
and Technology
Faridabad, Haryana, India

Noredine Gherabi
National School of Applied Sciences
Sultan Moulay Slimane University
Beni-Mellal, Khouribga, Morocco

ISSN 1876-1100 ISSN 1876-1119 (electronic)
Lecture Notes in Electrical Engineering
ISBN 978-981-97-2199-3 ISBN 978-981-97-2200-6 (eBook)
https://doi.org/10.1007/978-981-97-2200-6

This Springer imprint is published by the registered company Springer Nature Singapore Pte Ltd.
The registered company address is: 152 Beach Road, #21-01/04 Gateway East, Singapore 189721, Singapore

Paper in this product is recyclable.

Preface

This book is based on a select collection of contributions from the 2023 3rd International Conference on Artificial Intelligence, Robotics, and Communication (ICAIRC 2023) held on December 22–24, 2023, at Xiamen, China. This book aims to present the essence of the academic conference and explore the importance, significance, and social impact of artificial intelligence, robotics, and communication technologies.

With the rapid development of science and technology, artificial intelligence, robotics, and communication technology are increasingly becoming an indispensable part of our lives. Innovations and advances in these areas have not only changed our daily lives, but have also had a profound impact on society, economy, and culture. This book brings together papers from leading experts and researchers from around the world who have made significant breakthroughs and achievements in cutting-edge explorations in artificial intelligence, robotics, and communication.

The content of this book covers topics on Edge Intelligence, Mobile Robotics, Robot Control, Visual Serving, Information Security, etc. Through rigorous research methods and empirical analysis, these papers demonstrate the wide application of artificial intelligence, robotics, and communication technologies in transportation, medical, industry, and other fields. They reveal the potential of these technologies and their ability to increase productivity, solve social problems, and improve people's quality of life.

We sincerely hope that this book will stimulate readers' interest in artificial intelligence, robotics, and communication technologies and promote exchanges and cooperation between academia and industry. We firmly believe that by sharing knowledge and experience, we can further advance these areas and contribute to building an intelligent, efficient, and sustainable society.

Finally, we would like to thank all the authors and reviewers who contributed to this book. Their efforts and enthusiasm made the publication of this book possible. We hope this book will inspire you and serve as a reference for your exploration in the fields related to artificial intelligence, robotics, and communication.

New Delhi, India — Sanjay Yadav
Faridabad, India — Yogendra Arya
Patna, India — Shailesh M. Pandey
Beni-Mellal, Morocco — Noredine Gherabi
Athens, Greece — Dimitrios A. Karras

Contents

About the Editors

Prof. (Dr.) Sanjay Yadav born in 1962, obtained his master degree in science (M.Sc.) in 1985 and Ph.D. degree in Physics in 1990. He is Former Chief Scientist and Head, Physico Mechanical Metrology Division of NPL, India and Former Faculty of Physical Sciences, Academy of Scientific and Innovative Research (AcSIR), HRDG, Ghaziabad. He taught Advanced Measurement Techniques and Metrology course, took practical classes and supervising graduate, master and Ph.D. students since 2011. He is working as the Editor-in-Chief (EIC) of the *MAPAN: The Journal of Metrology Society of India*. He has significantly contributed to pressure metrology, biomedical instrumentation, ultrasonic transducers and instrumentation systems. His current research interests include research and developmental activities in physico mechanical measurements; establishment, realization, maintenance and up-gradation of national pressure and vacuum standards; dissemination of national practical pressure scale to users through apex level calibration, training and consultancy services; inter-laboratory comparisons, proficiency testing programme and key comparisons, implementation of Quality System in the laboratory as per ISO/IEC 17025 standard and Finite Element Analysis (FEA) and Monte Carlo Simulations for pressure balances. He has over 450 research publications to his credit, published in the national and international journals of repute and conferences besides contributing in several books as editor and author, published by Springer and MSI as well as in drafting several project, scientific and technical reports, documents and policy papers.

Dr. Yogendra Arya is currently working as an Associate Professor with the Department of Electrical Engineering, J. C. Bose University of Science and Technology, YMCA, Faridabad, India. He received his A.M.I.E. in Electrical Engineering from The Institution of Engineers (India), in 2008, M.Tech. (Instrumentation and Control) from DCRUST, Murthal, Haryana in 2010 and a Ph.D. from Delhi Technological University, Delhi in 2018. He has published 60+ research papers in reputed International journals. He has been placed among "Top 2% of Researchers in the World" for 2019, 2020, 2021, and 2022 by Stanford University, USA and Elsevier. He has Google Scholar citations of 3500+ and h-index of 35. He has received "Best Associate Editor Award-2021, 2022, and 2023" from *Journal of Electrical Engineering and*

Technology. He is Associate Editor of a few SCIE journals. He is a regular reviewer of leading journals. He is a Fellow of IETE, SM of IEEE, an Associate Member of The Institution of Engineers (India), and a Life Member of ISTE. His major areas of research interest include instrumentation, control, AGC/LFC in smart grids, fuzzy control, energy storage systems, and renewable energy systems.

Dr. Shailesh M. Pandey's academic journey is rooted in his pursuit of excellence and dedication to his field. He completed his B.Tech. in Mechanical Engineering in the year 2007 and Qualified GATE in 2010. He holds a Master's degree (2012) and a Ph.D. (2017) in Mechanical Engineering from DTU (Delhi Technological University), a prestigious institution known for its strong engineering programs. For his M.Tech. dissertation and Doctoral research work, he had the privilege of being supervised by two esteemed professors, Prof. Qasim Murtaza and Prof. R. S. Walia from Delhi Technological University (DTU) Delhi. Currently, Dr. Pandey serves as a Training and Placement Officer at NIT Patna, where he guides and supports students in their career development, a position he has held since 2020. His area of research includes surface modification, coating, tribology, material processing, composite materials, metal matrix nanocomposites, nano-coatings, wear, deformation, and corrosion. He has guided more than 25 B.Tech. projects and 10 M.Tech. dissertations. At present, 5 Ph.D. Students are working under his supervision. Dr. Pandey is also the founder and president of the educational and skill-development NGO "Avhikalpana" and a Trust "Research and Innovation Trust".

Prof. Dr. Noredine Gherabi is a professor of computer science with industrial and academic experience. He holds a doctorate in computer science. In 2013, he worked as a professor of computer science at Mohamed Ben Abdellah University, and since 2015 has worked as a research professor at Sultan Moulay Slimane University, Morocco. Member of the International Association of Engineers (IAENG). He has several contributions to information systems namely: big data, semantic web, pattern recognition, and intelligent systems. He has papers (book chapters, international journals, and conferences/workshops), and edited books. He has served on executive and technical program committees and as a reviewer of numerous international conferences and journals, he convened and chaired more than 30 conferences and workshops.

Prof. Dr. Dimitrios A. Karras received his Diploma and M.Sc. Degree in Electrical and Electronic Engineering from the National Technical University of Athens (NTUA), Greece in 1985, and a Ph.D. Degree in Electrical Engineering, from the NTUA, Greece in 1995, with honors. From 1990 to 2004 he collaborated as a visiting professor and researcher with several universities and research institutes in Greece. Since 2004, after his election, he has been with the Sterea Hellas Institute of Technology, Automation Department, Greece as an associate professor in Intelligent Systems-Decision Making Systems, Digital Systems, Signal Processing, till 12/2018, as well as with the Hellenic Open University, Department of Informatics as a visiting professor in Communication Systems (the latter since 2002 and up to 2010). Since

1/2019 is an Associate Professor in Intelligent Systems-Decision Making Systems, Digital Systems, and Signal Processing at the National and Kapodistrian University of Athens, Greece, School of Science, Department General as well as Adjunct Associate Professor with the EPOKA university, Computer Engineering Department, Tirana. He has published more than 70 research-refereed journal papers in various areas of intelligent and distributed/multiagent systems, Decision Making, pattern recognition, image/signal processing, and neural networks as well as in bioinformatics and more than 185 research papers in International refereed scientific Conferences. His research interests span the fields of intelligent and distributed systems, Decision Making Systems, multiagent systems, pattern recognition and computational intelligence, image and signal processing and systems, biomedical systems, communications, and networking as well as security. He has served as a program committee member as well as program chair and general chair in several international workshops and conferences.

Artificial Intelligence Control and Information Detection

Generalization Method of Virtual Test Scenarios for Autonomous Vehicles

Zhe Tan, Zhimin Bai, Qiang Yang, and Xiaojuan Li

Abstract Automatic driving is an inevitable development trend in the traffic field. However, unlike traditional driving modes, lots of simulation test scenarios for autonomous vehicles are required before the operation in real environment to ensure the safety during the actual operation. At present, the actual environment testing method has high cost and low efficiency and cannot test a large number of test scenarios within a certain time period. Therefore, the paper mainly studies the generalization method of virtual test cases for autonomous vehicles. The test scenarios are divided into three types: functional scenario, logical scenario, and concrete scenario. The main indicators of the actual test environment and their hierarchical relationship are analyzed, and the indicator hierarchical model is established. Based on the hierarchical relationship of different indicators and multi-tree building method, the paper studies the generalization method of functional scenarios. Then, the distribution functions and values' range of parameters are obtained by extracting and analyzing the quantifiable parameters of function scenario, and the logical scenarios are obtained. Within the range of different indicators, the selection method of specific values is studied. By analyzing the spatial relationship between different parameters, the combination method between these parameters is determined, and the concrete scenarios that can be used in actual test cases are obtained. Finally, the feasibility of the proposed method is verified by a case study.

1 Introduction

The virtual simulation test of the scenario is important for autonomous vehicles. Before the autonomous vehicles operate in the real traffic environment, the test and evaluation for the security of the autonomous vehicles is critical. Nowadays,

Z. Tan · Z. Bai · Q. Yang
Beijing Saimo Technology Ltd. Co, Beijing, China

X. Li (✉)
College of Urban Rail Transit and Logistics, Beijing Union University, Beijing, China
e-mail: lixj@imu.edu.cn

S. Yadav et al. (eds.), *Proceedings of 3rd International Conference on Artificial Intelligence, Robotics, and Communication*, Lecture Notes in Electrical Engineering 1172, https://doi.org/10.1007/978-981-97-2200-6_1

automated or semi-automated automobile testing lines are widely used in many mass-produced automobile factories. And this will inevitably generate a lot of costs and uncertainty. Therefore, the paper proposes a generation method of virtual scenarios for autonomous vehicles.

The PEGASUS project developed a verification framework for the issuance process of autonomous vehicles. Pütz [1] introduced the database entity and procedure of the PEGASUS project. Staplin [2] developed a test scenario framework that incorporates elements of each operational components. The framework used a checklist-type approach to identify high-level scenario tests by specifying relevant tactical maneuvers. Zhou [3] elaborated and compared the latest developments of five simulation test methods for autonomous vehicles in order to test and verify the rationality and stability of algorithms and system more efficiently. Menzel [4] divided the scenarios into three categories: functional scenarios, logical scenarios, and concrete scenarios, and introduced the conversion relationship among these three types of scenarios. Wang [5] proposed a test scenario generation system and method for the autonomous vehicles. This system included some function modules such as test content generation module, test scenario base, test scenario layout module, and test route generation module. Xia [6] provided a method to generate more effective and compact test scenarios for intelligent driving systems. An index is firstly proposed to measure the complexity of scenario by AHP method. And the complicated cases are preferred to be clustered together into continuous scenarios. Ni [7] proposed an improved deep network-based scene classification method. An improved faster region with convolutional neural network features (RCNN) network is used to extract the features of representative objects in the scene. Zhu [8] proposed that the test scenario for autonomous vehicles is a combination between the driving environment and the driving situation. By analyzing virtual technology of scenarios for autonomous vehicles, the test scenarios were built based on the elements data collection. Shu [9] screened out basic test scenario groups which could cover various levels and functions of the real driving environment by determining complex scenario groups and selecting parameter combination criterion. Wei [10] combined two methods, traffic data-driven basic scene construction and mechanism and theory-based logical scenario construction, to obtain a basic scenario composed of several traffic accident feature variables. The other traffic participants combine their movement directions around the main vehicle in the basic scenario, in order to generalize more logical scenarios. Zhang [11] proposed a road testing scenario for L3-level autonomous driving highways based on the application scenarios of L3-level autonomous driving highways, including emergency response of autonomous driving vehicles, human intervention capabilities, and comprehensive driving capabilities and testing items.

The paper studies the generalization method of simulation test scenarios for autonomous vehicles. Through the abstraction of the actual scenario, the relevant indicators of the test scenario are established, and the hierarchical relationship between the indicators is analyzed. In this paper, the scenarios are divided into three levels according to the demand which are functional scenario, logical scenario, and concrete scenario. And the generation method for these different scenarios is studied.

The functional scenario is a structured informal description. The data of functional scenario is analyzed and clustered through the actual road conditions. Based on functional scenarios, the parameters' distribution and space of each scenario are analyzed, and the logical scenarios are obtained. Then, based on the logical scenarios, the specific value selection method is studied by the parameter distribution function. And the concrete scenario can be obtained by selecting and reorganizing the discrete parameter values.

2 Indicators Hierarchical Model

Each virtual test scenario has different parameters, mainly including road geometric parameters, traffic flow parameters, environmental parameters, and autonomous vehicle parameters. The paper determines the hierarchical relationship of these parameters for generating different test scenario. The indicators of each level for the different test scenario are shown in Table 1. According to the indicators hierarchical model, the tree model can be built for generating different test scenario. By searching the sequence of the multi-tree parameters, the different functional scenarios can be obtained. For the same level, the different scenarios are generated simultaneously, such as four different scenarios can be obtained in Level 3. For the different levels, the indicators can be added in the scenarios generated by the upper level, and the new scenarios can be obtained.

3 Functional Scenario Construction

The function scenarios can be generated as follows.

1. The basic scenarios are generated by Levels 1–4 for different road conditions. And the different conditions of traffic signs and lines are added to Level 5.
2. Based on the scenarios of different road conditions, the indicators of traffic flow are added to Levels 6–8. The position relationship between the two vehicles (test vehicle and the surrounding vehicle) can be confirmed by the position relationship matrix. Then the different behaviors of the surrounding vehicles are added to generate the different scenarios. Finally, the behaviors of the test vehicle are selected by the behavior correlation matrix of the surrounding vehicle and the test vehicle. And the new scenarios by considering the behavior of vehicles in the avoidance area and the behavior of test vehicles are generated.
3. Based on the above scenarios, the environmental indicators of Level 9 are added. Environmental scenarios mainly include rainy, snowy, strong wind, and foggy weather. And the new scenarios for the night driving are generated by Level 9 and Level 10.

Table 1 Indicators of each level

Level 1	Scenario	Linear geometry			
	Indicators	Curve radius, curve corner, super high, widening, slope, slope length			
Level 2	Scenario	Cross section			
	Indicators	Road width, lane width, road shoulder, road smoothness, road anti-skid performance, maximum speed limit, minimum speed limit			
Level 3	Scenario	Intersection	Flyover	Bridge	Tunnel
	Indicators	Left-turn lane, right-turn lane, signal lights, traffic police command	Ramp curve radius, ramp longitudinal slope, ramp slope length, merging zone, diverging zone	Bridge head curve, bridge slope	Tunnel length, tunnel lighting
Level 4	Scenario	Road obstacles			
	Indicators	Road traffic accidents, moving objects on the road, construction occlusion			
Level 5	Scenario	Traffic signs and lines			
	Indicators	Warning sign, prohibition sign, mandatory sign, guide sign, tourist sign, road construction safety sign, auxiliary sign, no marking, indicating marking, warning marking			
Level 6	Scenario	Types of traffic participants			
	Indicators	Motor vehicles, motorcycles, bicycles, pedestrians			
Level 7	Scenario	Test vehicle behavior		Surrounding vehicle behavior	
	Indicators	Acceleration, deceleration, synchronization, go straight, cut in, cut out		Acceleration, deceleration, synchronization, go straight, cut in, cut out	
Level 8	Scenario	The second action of the test vehicle			
	Indicators	Emergency deceleration, emergency lane change			
Level 9	Scenario	Rain	Snow	Wind	Foggy
	Indicators	Rainfall, drainage	Snowfall, snowmelt degree	Wind force, wind speed, wind direction	Visibility, tire slip degree
Level 10	Scenario	Night driving			
	Indicators	Light intensity			

The specific algorithm steps of the functional scenario are as follows:

Step 1: Define parameter x_{ij}^{d} as 0–1 variable,

$$x_{ij}^{d} = \begin{cases} 1 & x_{ij} \text{ exist or function in the scene } d \\ 0 & or \end{cases}$$

Step 2: The indicators relationship set of the functional scenario is built based on the principle of multi-tree construction as follows.

Step 2.1: Define the index x_{ij} of each node, i is the Number of the level and the j is the Number of the indicators for each level.
Step 2.2: Add a domain for recording node names.
Step 2.3: Increase the array used to store the address of the child node.
Step 2.4: Read the data file and build a multi-tree.

Step 2.4.1: Reading start from the index with $i = 0$ and $j = 1$, name the index of this layer as the j-th layer index, and name the index of this layer as the parent node. Then read the next level child nodes of the j-th level indicator in turn.
Step 2.4.2: Determine the functional variables of the sub-node x_{ij} of the j-th level index.
Step 2.4.3: Let $:= j + 1$.
Step 2.4.4: Read each node index x_{ij} of the j-th level in turn, and assign each node index as a child node to its corresponding parent node in the j-1th level index.
Step 2.4.5: Judge whether the j-th layer is a bottom node, if it is, turn to Step 2.4.6, otherwise turn to Step 2.4.3.
Step 2.4.6: Let $i = i + 1$ ($i < = 10$) and turn to the Step 2.4.1.

Step 3: Based on the deep search algorithm for the multi-tree, by judging the value of the function variable, the function scenarios set is established.

Step 3.1: Define an empty set of functional scenario Z.
Step 3.2: Visit the vertex node of the multi-tree x_{01}.
Step 3.3: Starting from the currently specified starting vertex, if there is any adjacent vertex of the currently visited vertex that has not been visited, choose one to visit; otherwise, return to the most recently visited vertex until there is a path connected to x_{01} in the multi-tree. The nodes are all visited. And in the sequential access, the functional variables of $x_{ij} = 1$ are included in the variable set Z.
Step 3.4: Judge whether there are any nodes in the multi-tree that have not been visited at this time. If yes, choose one of the unvisited vertices as the starting vertex and visit it, turn to Step3.3, otherwise, turn to Step4 when the traversal ends.

Step 4: Update the functional scenario set Z.

4 Logic Scenarios Construction

By analyzing the specific parameters that can be collected from the actual environment, the distribution range of each parameter can be determined. And the distribution of each key parameter is calculated. Determine the value a% according to the value range of each parameter when generating a concrete scenario. That is the proportion of the statistical values of parameter needs more than a%. Based on the distribution, the continuous rate of change can be calculated and parameter range can be obtained.

Specific steps of logical scenario generation as follows:

Step 1: Select a certain functional scenario to generalize the logical scenarios.
Step 2: Extract and classify the relevant parameters of the functional scenario.

> **Step 2.1**: Create two empty sets, set M and set N.
> **Step 2.2**: Classify the extracted relevant parameters into two categories: basic function parameters that are not generalizable and variable parameters that can be generalized. List the non-generalizable basic function parameters into the set M, and list the generalizable variable parameters into the set N.

Step 3: Extract the parameter value of each generalizable variable parameter in the set N in the real environment in turn, and determine the range of each optional parameter set.

> **Step 3.1**: The parameters in the set N are marked as i, and x_{i} represents the i-th parameter ($i = 1 \cdot \cdots \cdot n$).
> **Step 3.2**: Set: $i = 0$.
> **Step 3.3**: Let: $i = i + 1$.
> **Step 3.4**: Extract the parameter value of parameter xi in the real environment.
> **Step 3.5**: Perform numerical fitting on the obtained parameter values in the real environment to obtain the probability density function of the normal distribution $f(x)$.
> **Step 3.6**: Determine the acceptable probability section of the parameter according to the parameter value of the probability distribution function had been obtained.
> **Step 3.7**: Judge whether $i = n$, if it is, turn to Step4, otherwise, turn to Step3.3.

Step 4: According to the determined specific values of the parameters in the set N and the basic function parameter values in the set M, the logical scenarios can be obtained.

5 Concrete Scenarios Construction

According to the parameter distribution range analyzed above, the parameter change trend function and parameter probability are established. Let x_i represent the i-th parameter and $f(x_i)$ represent the probability density function of parameter x_i. The

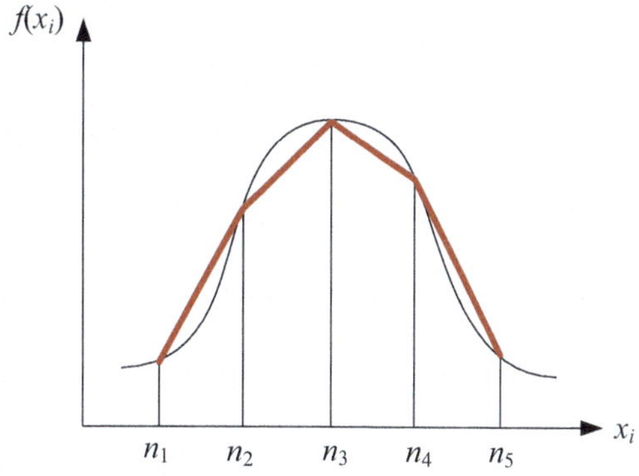

Fig. 1 Example of specific value selection method

density function can be extracted by the parameter extraction method in the logical scenario.

As shown in Fig. 1, take the normal distribution as an example. For the parameter distribution interval, it is equally divided into N parts. It is assumed that the selected number of parameters in each part obeys a uniform distribution. The number of specific parameters selected in the j-th part obeys the uniform distribution of U_j as Formula (1).

$$N_j = U_j(f(x_i)_{n_j}, f(x_i)_{n_{j+1}}), j = 1...N - 1 \tag{1}$$

According to the number of values and value ranges of each part, the specific values of the parameters are selected on average. And the concrete scenario can be obtained.

The interval section expression mode $[n_j, n_{j+1})(j = 1, 2, 3 \cdots\cdots)$ represents the value ranges of the j-th part. m_j is the number of values for the j-th part. The a_j is the value set of the j-th part represented by Formula (2). Set A is the total value set represented by Formula (3). If the value produces a decimal when dividing an integer, the decimal is rounded down to an integer.

$$a_j = \left\{n_j, n_j + \frac{n_{j+1} - n_j}{m_j}, n_j + \frac{n_{j+1} - n_j}{m_j} * 2, n_j + \frac{n_{j+1} - n_j}{m_j} * 3, \cdots, \frac{n_{j+1} - n_j}{m_j} * (m_j - 1)\right\} \tag{2}$$

$$A = \{a_1, a_2, a_3, \cdots\} \tag{3}$$

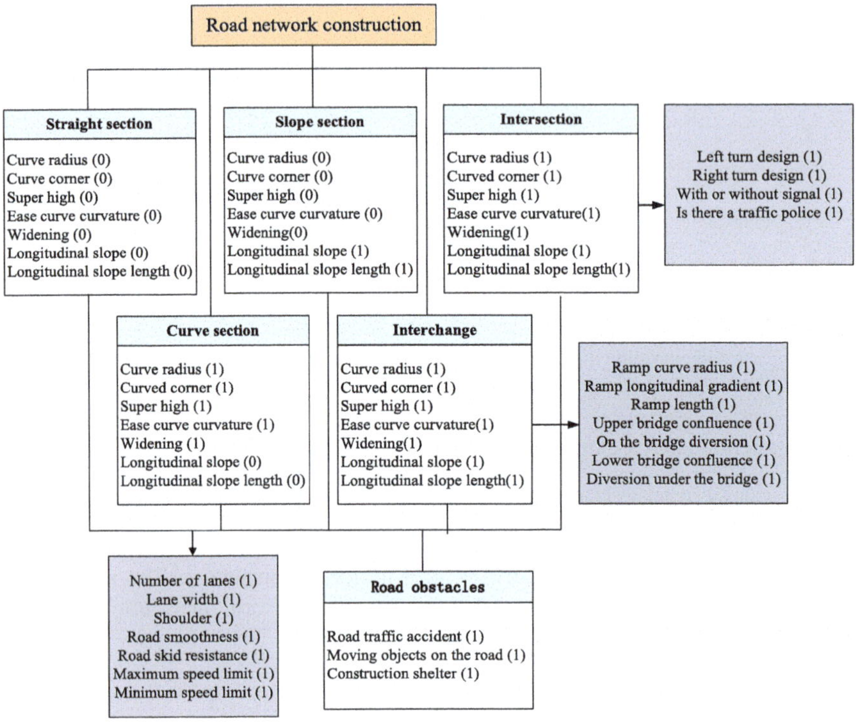

Fig. 2 Different road conditions for the functional scenarios

6 Case Study

6.1 *Functional Scenario Generation*

The functional scenario of different road conditions, traffic flow conditions, and environmental conditions can be generated by the methods shown in Figs. 2, 3, and 4.

6.2 *Examples of Logical Scenarios Generalization*

According to the functional scenario, the parameter value analysis is performed for the parameter whose value is 1 for each functional scenario. The generation method of logic scenario is analyzed by the traffic part of one functional scenario.

In order to explain the parameters' definition of the logical scenarios, the paper takes the scenario that the surrounding vehicle cuts in as the case to analyzing. As shown in Fig. 5, the black vehicle slows down and cuts into the front of the test red vehicle. For keeping a safe distance, the test vehicle needs to decelerate.

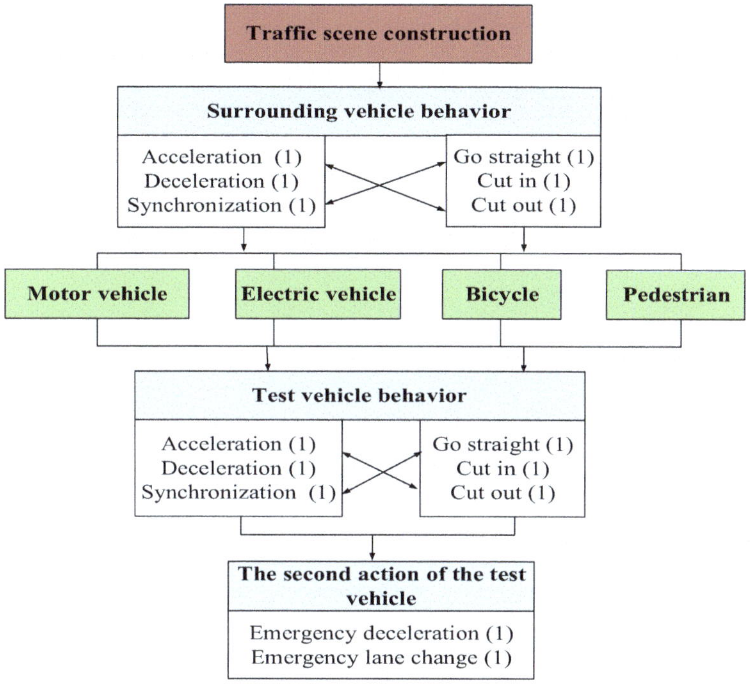

Fig. 3 Function scenario of different traffic conditions

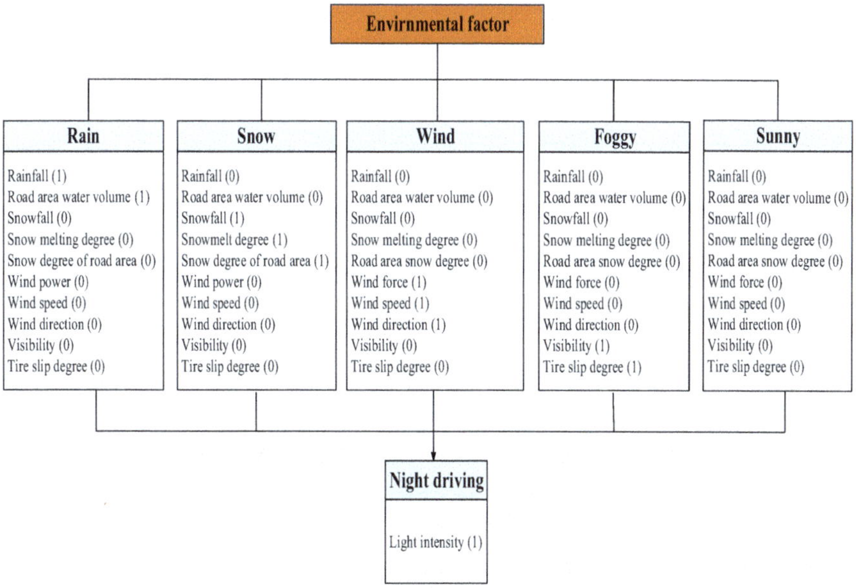

Fig. 4 Function scenario of different environment conditions

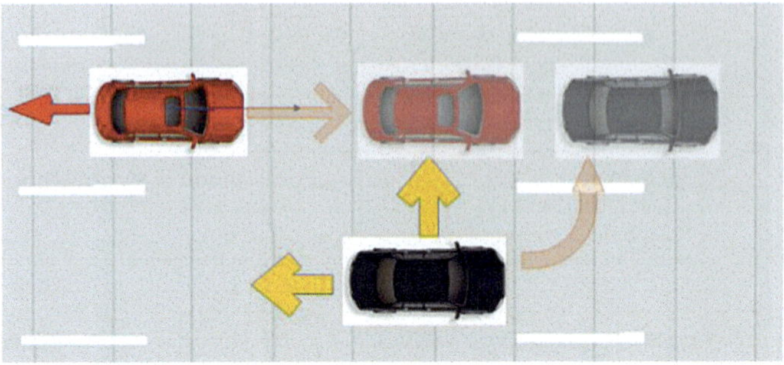

Fig. 5 Location relationship of the vehicles

The specific parameters required for this scenario include: the initial speed of the autonomous vehicle (V0), the initial lateral distance (Dy), the initial longitudinal distance (Dx) between the test vehicle and the surrounding vehicle, lateral velocity (Vy), and relative longitudinal velocity (Ve).

Based on the verification of real traffic data, the parameter distribution is automatically analyzed, and the concrete scenario parameter distribution is extracted. The distribution and range of cut-in scenario parameters are obtained by applying the automatic analysis process to the real traffic data.

For the above parameters, the distribution and value ranges of some of the available parameters are shown in Table 2. Take the proportion of the parameter as more than 5%. The value ranges of the following specific parameters are obtained as in Table 2.

6.3 The Concrete Scenarios Generalization

According to the logical scenario, the specific values of each parameter are analyzed for the concrete scenarios.

(1) Initial Vehicle Speed

The probability density function of the normal distribution can be obtained by the function distribution in Table 2 and can be calculated as: $f(v_0) = \frac{1}{\sqrt{2\pi}\sigma}\exp\left(-\frac{(v_0-\mu)^2}{2\sigma^2}\right)$. The parameters can be obtained such as: $\mu = 37.23, \sigma = 14.96$. Therefore, the probability density function of this parameter is.

$$f(v_0) = \frac{1}{\sqrt{2\pi}14.96}\exp\left(-\frac{(v_0 - 37.23)^2}{2 \times (14.96)^2}\right)$$

The speed values are divided into 7 parts. The range of each part is 10. The probability density function is used to calculate the number of values in each speed part. Assume that the total value is 30, and the specific value in each speed interval is uniformly distributed.

Table 2 Parameter distribution and value range

Parameter	V0	Ve
Function distribution	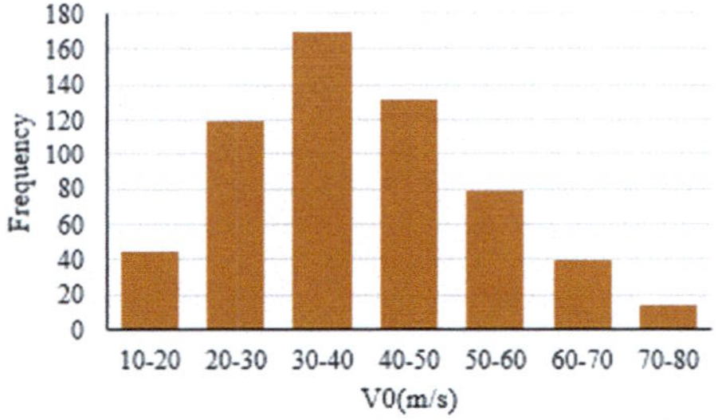	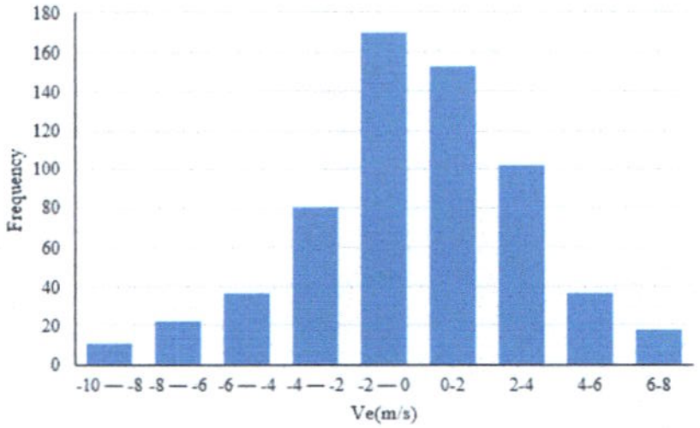
Range	10–80 km/h	−10–−8 m/s
Parameter	Dx	Vy
Function distribution	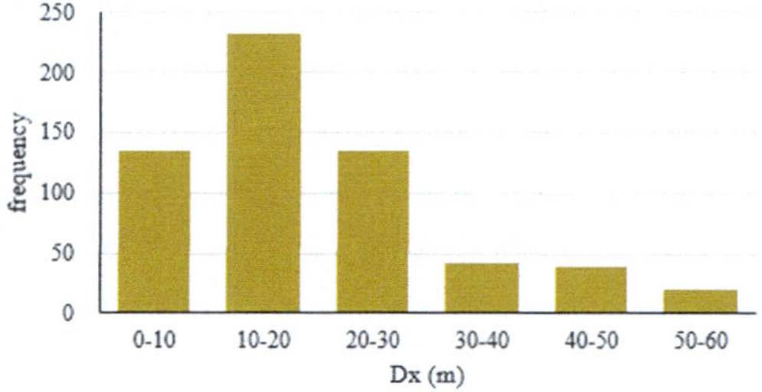	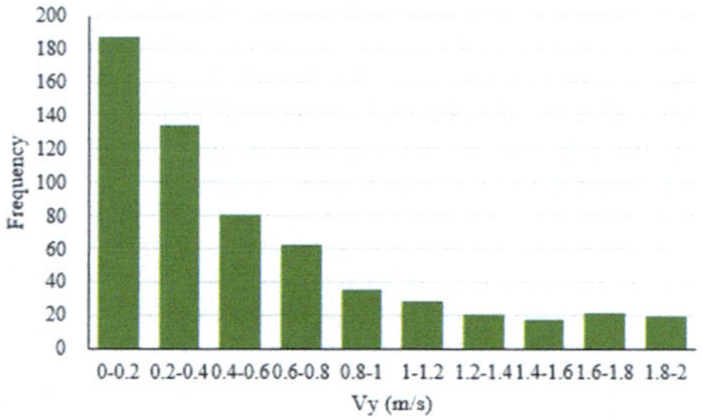
Range	0–60 m	0–7 m/s

(2) Relative Velocity of Vehicles

The probability density function of the normal distribution can be obtained by the function distribution in Table 2. The parameters can be obtained such as: $\mu = 0.094, \sigma = 2.875$. Therefore, the probability density function of this parameter is

$$f(v_0) = \frac{1}{\sqrt{2\pi}2.875} \exp\left(-\frac{(v_0 - 0.094)^2}{2 \times (2.875)^2}\right)$$

The relative speed values are divided into 9 parts. The range of each part is 2. Assume that the total value is 30, and the specific value in each relative speed part is uniformly distributed.

(3) Longitudinal Distance

Assuming that the initial distance of the vehicles obeys a uniform distribution, the distance value is divided into 6 parts, and the range of each part is 10. Then the number of values for each part is calculated as $n_x = N_{sum} * P_{ij}$ by the probability density function. Assume that the total value is 30.

(4) Lateral Velocity

Assuming that the initial distance of the vehicles obeys a uniform distribution, the distance value is divided into 10 parts, and the range of each part is 0.2. Then the number of values for each part is calculated as $n_x = N_{sum} * P_{ij}$ by the probability density function. Assume that the total value is 30.

(5) Scenarios Integration by Parameters' Relationship

According to the above results, the specific value range of each parameter has been obtained. When generalizing concrete scenarios, specific test scenarios can be obtained by sequentially taking values for each parameter. The relationship between the longitudinal distance between the two vehicles and the relative speed of the vehicle can be obtained by calculation as shown in Formula (4).

$$d_x = \frac{1}{2} * |v_0 - v_{e0}| * t \tag{4}$$

When determining the specific value of each parameter, the initial speed of the vehicle can be selected from the specific value set in Table 3 (in this study, it is assumed that the initial speed of each vehicle is the same), and the relative speed of the two vehicles can be selected from the specific value set in Table 4. The longitudinal distance between the two vehicles can be calculated from the selected vehicle relative speed and Formula (4). The calculated longitudinal distance is corrected according to the values listed in the specific value set in Table 5, and the value is ±0.3. (If the difference between the calculated longitudinal distance and a certain value listed in the specific value set in Table 5 is between ± 0.3 (including ± 0.3), the calculated longitudinal distance is set as the acceptable vehicle longitudinal distance.) The value of each parameter in some concrete scenarios is shown in Table 3.

Table 3 Values of various parameters in some concrete scenarios

No.	Initial vehicle velocity (km/h)	Velocity of vehicles (m/s)	Longitudinal distance between two vehicles (m)	Number of concrete scenarios
1	15, 20, 22, 23, 25, 27, 28, 30, 31, 33, 34, 35, 36, 38, 39, 40, 41, 43, 44, 46, 47, 49, 50, 53, 55, 58, 60, 65, 70, 80	±6	3.3, 14.8, 18, 24.2, 27, 29.8, 45	30 * 2 * 7
2		±4	1.7, 8.3, 10, 14, 18, 24.2, 29.8, 40, 50	30 * 2 * 9
3		±3	1.7, 3.3, 10.8, 13.2, 14.8, 16.4, 18, 19.6, 22.8, 24.2, 25.6, 27, 28.4, 29.8, 45	30 * 2 * 15
4		±1.5	1.7, 3.3, 5, 6.7, 8.3, 10, 10.8, 13.2, 14, 14.8, 15.6, 16.4, 17.2, 18, 18.8, 19.6, 22.8, 24.2, 25.6, 27, 28.4, 29.8, 35, 40, 45, 50, 55	30 * 2 * 27
5		±0.5	1.7, 3.3, 5.0, 6.7, 8.3, 10, 10.8, 11.6, 12.4, 13.2, 14, 14.8, 15.6, 16.4, 17.2, 18, 18.8, 19.6, 21.4, 22.8, 24.2, 25.6, 27, 28.4, 29.8, 35, 40, 45, 50, 55	30 * 2 * 30
6		±1.14	1.7, 3.3, 5, 6.7, 8.3, 10, 10.8, 11.6, 12.4, 13.2, 14, 14.8, 15.6, 16.4, 17.2, 18, 18.8, 19.6, 21.4, 22.8, 24.2, 25.6, 27, 28.4, 29.8, 35, 40, 45, 50, 55	30 * 2 * 30
7		±3.2	1.7, 3.3, 5.0, 6.7, 8.3, 24.2, 25.6, 27, 35, 40, 45	30 * 2 * 11
8		±8	8.3, 24.2, 40	30 * 2 * 3

Table 4 Relative speed value

Relative velocity (km/h)	Probability	Number of values	Correction number	Specific value set (km/h)
[−10 ~ −8]	0.017	1	0	{−}
(−8 ~ −6]	0.035	1	1	{6}
(−6 ~ −4]	0.059	2	2	{−5, −4}
(−4 ~ −2]	0.127	4	4	{−3.5, −3, −2.5, −2}
(−2 ~ 0]	0.270	8	8	{−1.75, −1.5, −1.25, −1, −0.75, −0.5, −0.25, 0}
(0 ~ 2]	0.243	7	7	{0.29, 0.57, 0.86, 1.14, 1.43, 1.71, 2}
(2 ~ 4]	0.162	5	5	{2.4, 2.8, 3.2, 3.6, 4}
(4 ~ 6]	0.059	2	2	{5, 6}
(6 ~ 8]	0.029	1	1	{8}

Table 5 Longitudinal distance value

Longitudinal distance (m)	Frequency	Probability	Number of values	Specific value set (m)
[0–10]	135	0.2246	6	{1.7, 3.3, 5.0, 6.7, 8.3, 10}
(10–20]	232	0.3851	12	{10.8, 11.6, 12.4, 13.2, 14, 14.8, 15.6, 16.4, 17.2, 18, 18.8, 19.6}
(20–30]	135	0.2246	7	{21.4, 22.8, 24.2, 25.6, 27, 28.4, 29.8}
(30–40]	42	0.0697	2	{35, 40}
(50–60]	39	0.0642	2	{45, 50}
(50–60]	20	0.0323	1	{55}

7 Conclusion

With the rapid development of autonomous driving technology, the testing and evaluation of autonomous vehicles have become a key issue in technology development. And scenario-based testing methods have become an important problem for autonomous driving technology. This paper studied the generalization method of test virtual scenario for autonomous vehicles. The scenarios were divided into functional scenarios, logical scenarios, and concrete scenarios. The theoretical methods for generating the different types of test scenarios were introduced. This method can make the test for autonomous vehicles easily and quickly by the virtual simulation. And the tester can establish test scenarios and obtain scenario data with a short time and small cost.

Acknowledgements This research is supported by project TC210H02S.

References

1. Pütz A, Zlocki A, Bock J, Eckstein L (2017) System validation of highly automated vehicles with a database of relevant traffic scenarios[C]. In: 12th ITS European congress
2. Staplin L, Mastromatto T, Lococo KH, Kenneth WG (2018) The effects of medical conditions on driving performance[N]. National Highway Traffic Safety Administration, Washington, DC
3. Zhou Q, Zhang S, Luo YQ (2019) Progress in simulation testing and evaluation methods of autonomous vehicles[J]. Auto Digest, pp 48–51
4. Menzel T, Bagschik G, Maurer M (2018) Scenarios for development, test and validation of automated Vehicles[N]. In: 2018 IEEE Intelligent Vehicles Symposium (IV). IEEE
5. Wang F.Y, Li J, Tian B, Hu CY (2018) A scene generation system and method for automatic driving vehicle evaluation[P]. CN107727411A, 2018–02–23
6. Xia Q, Duan JL, Gao F, Hu QX, He YD (2018) Test scenario design for intelligent driving system ensuring coverage and effectiveness[J]. Int J Automot Technol 19(04):751–758

7. Ni JJ, Shen K, Chen YN, Cao WD, Yang SX (2022) An improved deep network-based scene classification method for self-driving cars[J]. IEEE Trans Instrum Meas 71:1–14
8. Zhu B, Zhang PX, Zhao J, Chen H (2019) Research progress of scenario-based virtual testing of autonomous vehicles[J]. China J Highway Transport 32(06):1–19. https://doi.org/10.19721/2019-06-001
9. Shu H, Yuan K, Xiu HL, Xia Q, He S (2019) Research on the construction of basic test scenarios for autonomous vehicles[J]. China J Highway Transp 32(11):245–254. https://doi.org/10.19721/2019-11-025
10. Wei JX, Wu XD, Zhao D (2023) Construction of autonomous driving high-risk scenarios based on association rules mining [J]. Shandong Transp Technol 147–151
11. Zhang MF, Wang XC, Wang X (2023) Construction method of highways road test scenarios for l3 autonomous driving[J]. Automobile Appl Technol 40–44. https://doi.org/10.16638/2023-05-007

MaskMel-Prosody-CycleGAN-VC: High-Quality Cross-Lingual Voice Conversion

Siqi Yan, Senda Chen, Yanyan Xu, and Dengfeng Ke

Abstract Voice conversion aims to change the timber of the source speaker to that of the target speaker without changing the speech content. The cross-lingual voice conversion requires non-parallel training data in two different languages, and the prosody and pronunciation of different languages bring challenges to the cross-lingual voice conversion. Previous voice conversion studies based on CycleGAN only used a single pipeline for spectrum mapping. We train two effective pipelines based on the CycleGAN-VC2 separately, in order to get better speech spectrum mapping and speech prosody mapping and decompose F0 (an important prosodic factor) into different time scales by continuous wavelet transform (CWT), so as to better adapt to the hierarchical characteristics of F0 and retain speaker prosody characteristics. In addition, we use MaskMel as the processing features of the spectrum mapping pipeline to improve the sound quality after speech conversion. We also use the self-trained MelGAN to combine F0 features and mel features to synthesize speech again. In this way, we achieve high similarity and naturalness of cross-lingual voice conversion. The MOS test results show that MaskMel-Prosody-CycleGAN framework, we proposed, is superior to the CycleGAN-VC2 baseline in our experiments.

1 Introduction

Voice conversion (VC) is designed to translate unique characteristics from the source speaker to the target speaker [1]. It serves as a foundational technology for many applications, such as emotional speech conversion, speech synthesis [2], speaking aids [3, 4], and pronunciation [5].

S. Yan · S. Chen · Y. Xu (✉)
School of Information Science and Technology, Beijing Forestry University, Beijing 100000, China
e-mail: xuyanyan@bjfu.edu.cn

D. Ke
School of Information Science, Beijing Language and Culture University, Beijing 100000, China

S. Yadav et al. (eds.), *Proceedings of 3rd International Conference on Artificial Intelligence, Robotics, and Communication*, Lecture Notes in Electrical Engineering 1172, https://doi.org/10.1007/978-981-97-2200-6_2

Effective methods of VC encompass Gaussian mixture models (GMMs) [6], the predecessor of vector quantization (VQ), feedforward neural networks (FNNs) [7–9], convolutional neural networks (CNNs) [10], and recurrent neural networks (RNNs) [11, 12]. However, neither the optimized GMM method nor the neural network method overcomes the two limitations of fixed speaker identity and the necessity of Dynamic Time Warping (DTW) for the frame level alignment of training data. For cross-lingual voice conversion, parallel data acquisition is basically impossible. Hence, this paper mainly deals with non-parallel VC, meaning that our proposed framework does not depend on the parallel discourse, transcription, or the time-aligned procedure.

Generally, non-parallel voice conversion is particularly complicated and may not achieve the same level of quality as parallel VC due to limitations in training conditions. To address these challenges, some studies have taken the approach of using additional modules. For instance, PPG technology [13, 14] incorporates automatic speech recognition (ASR [14]) and text-to-speech synthesis (TTS [15]) modules. While this approach can result in high-quality voice conversion, it introduces additional complexities and costs that restrict its applicability. To mitigate these costs, recent research has introduced probabilistic neural networks, such as variational autoencoders (VAEs [16]). However, VAEs rely on assumptions about data distribution (e.g., normal or Gaussian distribution), which may not accurately reflect the true data distribution, leading to inaccuracies in synthetic acoustic features. To address these limitations, generative adversarial networks (GANs) have been incorporated, allowing for the learning of target distributions without explicit assumptions and avoiding excessive smoothing caused by statistical averaging.

The CycleGAN-VC [17] model, built upon the GAN [18] framework, learns sequence-based mapping functions using CycleGAN with gating CNN and identifying mapping loss. This approach can capture order and hierarchy while preserving speech information. The performance of voice conversion on the same language is equivalent to that of parallel VC methods. Based on CycleGAN-VC, CycleGAN-VC2 [19] improves the one-step adversarial loss to two-step adversarial loss and improves the 1D CNN generator structure to 2–1-2D CNN. Moreover, the FullGAN discriminator structure is optimized to PatchGAN to further improve the quality of mono-lingual voice conversion.

Unfortunately, previous studies [13, 20–24] have not focused on prosody differences in cross-lingual voice conversion, and there is a lack of research on prosody conversion. Therefore, in this paper, we take fundamental frequency, which shows the change of sound pitch in the whole time domain, as the main processing object of prosody conversion [27], and divide F0 into 10 diverse time scales by CWT to represent the hierarchical characteristics of F0 [28, 29]. Then, using mel with the time mask as the processing object of the spectrum conversion, the pipeline enables CycleGAN to learn the time–frequency structure better. Finally, MelGAN [26] is used to resynthesize the converted F0 and mel into speech.

This paper's contributions are mainly as follows: (1) We introduce a cross-lingual speech conversion CycleGAN-based framework for spectrum and prosody conversion, utilizing MelGAN for speech resynthesis; (2) we employ continuous wavelet transform (CWT) [30–32] to analyze F0 at various time resolutions; (3) we use mel with a time mask, instead of sp features, as the input for spectrum mapping, enabling better processing of the time-domain characteristics of speech. As far as we know, this paper represents the first attempt to combine mel features with F0 features, resulting in enhanced cross-lingual speech conversion quality.

The organization of the rest of the paper is as follows: In Sect. 2, we provide the method of obtaining and processing MaskMel features. In Sect. 3, we present the MaskMel-Prosody-CycleGAN framework. In Sect. 4, we give the MOS test results. We conclude the discussion in Sect. 5.

2 MaskMel

We encourage the converter to go through a circular conversion process to complete missing frames relying on the surrounding frames.

Figure 1 shows the conversion mode detail of MaskMel features. We give x to the source mel-spectrogram and make a temporary mask $m \in M$ that matches the size of x. This mask comprises regions with the value 0 (as the blue region shown in Fig. 1) and other regions with the value 1 (as the white region shown in Fig. 1). The determination of masked regions follows a random selection process based on predefined rules. We complete an element-wise product which is between x and m in order to get $\hat{\mathbf{x}}$. By doing this, we artificially create missing frames. Next, the MaskCycleGAN-VC converter synthesizes $\boldsymbol{y}'$ from $\hat{\mathbf{x}}$ and m as follows:

$$y' = G^{mask}_{X\to Y}(\text{concat}(\hat{x}, m)) \quad (1)$$

As m is used as the conditional information, $\mathbf{G_{X->Y}}$ is capable of filling in the frames while being aware of which frames require completion. This process of induction proves valuable for acquiring knowledge about the time–frequency structure within a mel-spectrogram through self-supervised means.

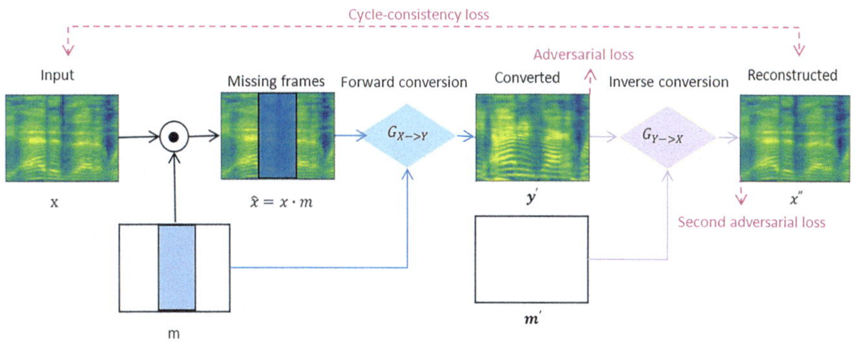

Fig. 1 Conversion mode of MaskMel

3 MaskMel-Prosody-CycleGAN

The proposed framework, MaskMel-Prosody-CycleGAN, introduces an approach based on the CycleGAN-VC2 for cross-lingual VC, which efficaciously captures the mapping between speakers from different languages. The MaskMel-Prosody-CycleGAN-VC framework is shown in Fig. 2.

The framework consists of two pipelines, one for spectrum conversion called MaskMel-CycleGAN and the other for prosodic conversion called Prosody-CycleGAN.

3.1 *Training of MaskMel-Prosody-CycleGAN*

For prosody conversion, we apply the WORLD vocoder [25] to get fundamental frequency (F0) and use the MelGAN to extract mel features from source and target waveforms. For prosody conversion, we extract 10-scale CWT-F0 coefficients and a 10-dimensional prosody frame for the Prosody-CycleGAN. For spectrum conversion, we extract 24-dimensional MaskMel features for the MaskMel-CycleGAN. We train the two CycleGANs separately.

The adversarial loss quantifies the dissimilarity between the data distribution of the transformed data and that of either the source or target data. In the case of forward mapping, it is precisely defined as follows:

$$\begin{aligned} L_{ADV}\left(G_{X_{en}\to Y_{cn}}, D_{Y_{cn}}\right) &= \mathbb{E}_{y_{cn}} P_{Y_{cn}(y_{cn})}\left[\log\left(D_{Y_{cn}}(y_{cn})\right)\right] \\ &+ \mathbb{E}_{x_{en}} P_{X_{en}(x_{en})}\left[\log\left(1 - D_{Y_{cn}}\left(G_{X_{en}\to Y_{cn}}(x_{en})\right)\right]\right. \end{aligned} \tag{2}$$

The adversarial loss solely constrains $G_{X->Y}$ to adhere to the target distribution; thus a periodic consistency loss is proposed to ensure contextual information consistency between input and output. It can be defined as follows:

$$L_{CYC}\left(G_{X_{en}\to Y_{cn}}, G_{Y_{cn}\to X_{en}}\right) = \mathbb{E}_{x_{en}\sim P(x_{en})}\left[G_{Y_{cn}\to X_{en}}\left(G_{X_{en}\to Y_{cn}}(x_{en})\right) - x_{en_1}\right]$$

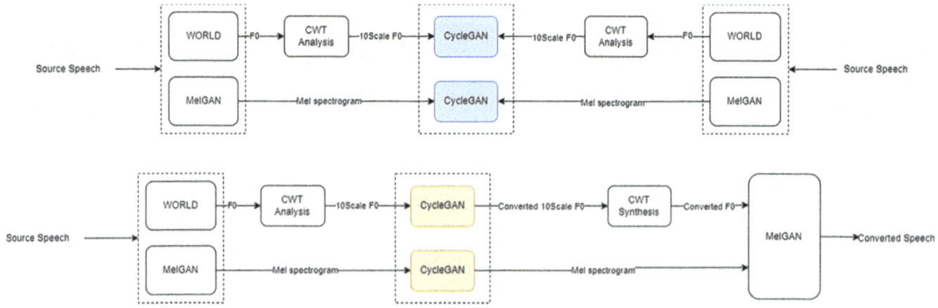

Fig. 2 Proposed MaskMel-Prosody-CycleGAN-VC framework

$$+\mathbb{E}_{y_{cn}\sim P(y_{cn})}\left[G_{X_{en}\to Y_{cn}}\left(G_{Y_{cn}\to X_{en}}(y_{cn})\right)-y_{cn_1}\right] \quad (3)$$

To ensure the preservation of language information without relying on external processes, identity mapping loss is defined as follows:

$$L_{\mathrm{ID}}\left(G_{X_{en}\to Y_{cn}},G_{Y_{cn}\to X_{en}}\right)=\mathbb{E}_{x_{en}\sim P(x_{en})}\left[G_{Y_{cn}\to X_{en}}(\mathrm{X_{en}})-x_{en}\right] \\ +\mathbb{E}_{y_{cn}\sim P(y_{cn})}\left[G_{X_{en}\to Y_{cn}}(y_{cn})-y_{cn}\right] \quad (4)$$

Through the adversarial loss, the periodic consistency loss as well as the identity mapping loss [33], we hope that CycleGAN can effectively learn the bidirectional mapping between the spectral distribution and the prosody distribution of the source language and the target language, improving the similarity and naturalness after conversion without changing the content of the speech.

3.2 Runtime Conversion

The framework used in the runtime conversion stage is almost the same as that used in the training stage. F0 is extracted by WORLD vocoder, processed by CWT, and then inputted into prosody conversion CycleGAN. Mel is processed by Mask and inputted into spectrum conversion CycleGAN. The converted F0 and mel are inputted into the trained MelGAN to synthesize the converted speech at last.

4 Experiments

We experiment to evaluate the performance of our proposed MaskMel-Prosody-CycleGAN-VC framework. VCC 2020 dataset [34] and the Blizzard Challenge 2010 dataset [35], which contains both English and Chinese speech data, are used to train the model. We select four English-speaking women (VCC2SF1, VCC2TF1, VCCTF2, and VCC2TF2) and one Chinese-speaking woman for the experiments. We use cn2en to represent the conversion of speech from Chinese to English and en2cn to represent the conversion of speech from English to Chinese. For en2cn, we conduct two experiments and select VCC2SF1 and VCC2TF2 as the source speakers. For cn2en, we also conduct two experiments, choosing VCC2SF2 and VCC2TF2 as source speakers, respectively. On the basis of the same datasets, we compare the cross-lingual speech conversion performances of MaskMel-Prosody-CycleGAN-VC and CycleGAN-VC2.

In order to evaluate the system under non-parallel wave data, we choose to use 80 non-parallel discourses from the source language and target user for training and 50 discourses for evaluation. We downsampled the speech data to 16 kHz and then extracted F0 features and sp features using the WORLD vocoder and MelGAN,

Table 1 Contrast of MOS test results between the CycleGAN-VC2 and MaskMel-Prosody-CycleGAN-VC for en2cn and cn2en

Method	MOS	
	en2cn	cn2en
CycleGAN-VC2	2.90 ± 0.34	3.28 ± 0.30
MaskMel-Prosody-CycleGAN-VC	2.94 ± 0.32	3.35 ± 0.27

respectively. In the training phase, we set $\lambda_{CYC} = 10$. The initial learning rate for the generator is set to 0.0002, while the initial learning rate for the discriminator is set to 0.0001. The learning rate remains constant for the first 3×105 iterations and linearly decays for the last 1×105 iterations.

We use the mean opinion score (MOS) to assess the voice quality and speaker similarity. We also use en2cn and cn2en to represent two opposite directions of speech conversion. Twenty native bilinguals take part in all the listening tests and listen to a total of 150 converted sentences.

To assess the quality of converted speech, we first run a MOS test between CycleGAN-VC2 and MaskMel-Prosody-CycleGAN-VC. We select 20 sentences randomly from the set for evaluating. In the MOS test, each judge is required to rate each sample on a five-point scale (5: excellent, 4: good, 3: fair, 2: poor, 1: bad). As the results shown in Table 1, MaskMel-Prosody-CycleGAN-VC achieves better results than which CycleGAN-VC2 achieves.

5 Conclusions

In this paper, we proposed a new framework for cross-lingual speech conversion without parallel data. We took mel features and F0 features as processing objects and transformed the spectrum and prosody, respectively, with non-parallel training data based on CycleGAN. In addition, we used MelGAN to process the mel features and F0 features to resynthesize speech to obtain higher quality cross-lingual speech conversion performances. The MOS experimental results show that our proposed framework is very effective in improving speech quality and the similarity between the converted speech and the original speech.

References

1. Sisman B, Yamagishi J, King S et al (2021) An overview of voice conversion and its challenges: from statistical modeling to deep learning[J]. IEEE/ACM Transact Audio Speech Lang Process. https://doi.org/10.1109/TASLP.2020.3038524
2. Liu R et al (2020) WaveTTS: Tacotron-based tts with joint time-frequency domain loss
3. Kain AB, Hosom JP, Niu X et al (2023) Improving the intelligibility of dysarthric speech

4. Nakamura K, Toda T, Saruwatari H et al (2012) Speaking-aid systems using GMM-based voice conversion for electrolaryngeal speech[J].Speech Commun 54(1):134–146.https://doi.org/10.1016/j.specom.2011.07.007
5. Kaneko T, Kameoka H, Hiramatsu K et al (2017) Sequence-to-sequence voice conversion with similarity metric learned using generative adversarial networks[C]. Interspeech. https://doi.org/10.21437/Interspeech.2017-970
6. Benisty H, Malah D (2011) Voice conversion using GMM with enhanced global variance. In: INTERSPEECH 2011, 12th annual conference of the international speech communication association, Florence, Italy, August 27–31, 2011 DBLP
7. Desai S (2010) Spectral mapping using artificial neural networks for intra-lingual and cross-lingual voice conversion [J]
8. Mohammadi Seyed H, Kain A (2015) Voice conversion using deep neural networks with speaker-independent pre-training. In: 2014 IEEE spoken language technology workshop (SLT) IEEE
9. Oyamada K et al (2016) Non-native speech conversion with consistency-aware recursive network and generative adversarial network. In: 2017 Asia-pacific signal and information processing association annual summit and conference (APSIPA ASC)
10. Takuhiro K, Hirokazu K, Kaoru H, Kunio K (2017) Sequence-to-sequence voice conversion with similarity metric learned using generative adversarial networks. In: Procceding Interspeech, pp 1283–1287
11. Nakashika T, Takiguchi T, Ariki Y (2014) High-order sequence modeling using speaker-dependent recurrent temporal restricted Boltzmann machines for voice conversion [J]
12. Sun L et al (2015) Voice conversion using deep bidirectional long short-term memory based recurrent neural networks. In: 2015 IEEE international conference on acoustics, speech and signal processing (ICASSP) IEEE
13. Zhou Y, Tian X, Das RK et al (2020) Many-to-many cross-lingual voice conversion with a jointly trained speaker embedding network. IEEE
14. Ding S, Zhao G, Gutierrez-Osuna R (2020) Improving the speaker identity of non-parallel many-to-many voice conversion with adversarial speaker recognition. In: Proceeding INTERSPEECH 2020, pp 776–780
15. Sun L, Wang H, Kang S et al (2016) Personalized, cross-lingual TTS using phonetic posteriorgrams [J]. https://doi.org/10.21437/Interspeech.2016-1043
16. Hsu CC, Hwang HT, Wu YC et al (2016) Voice conversion from non-parallel corpora using variational auto-encoder[C]// IEEE. IEEE
17. Kaneko T, Kameoka H (2018) CycleGAN-VC: non-parallel voice conversion using cycle-consistent adversarial networks[C]. In: 2018 26th European Signal Processing Conference (EUSIPCO). IEEE
18. Lee S, Ko B G, Lee K et al (2020) Many-to-many voice conversion using conditional cycle-consistent adversarial networks [C]. https://doi.org/10.1109/ICASSP40776.2020.9053726
19. Kaneko T, Kameoka H, Tanaka K et al (2019) CycleGAN-VC2: improved CycleGAN-based non-parallel voice conversion [J]. IEEE
20. Zhou Y, Tian X, Xu H et al (2019) Cross-lingual voice conversion with bilingual phonetic posteriorgram and average modeling[C]. In: International conference on acoustic, speech and signal processing (ICASSP)
21. Zhou Y, Tian X, Yılmaz E, Das RK, Li H (2019) A modularized neural network with language-specific output layers for cross-lingual voice conversion. In: IEEE ASRU, pp 160–167
22. Zhou Y, Tian X, Li H (2020) Multi-Task WaveRNN With an integrated architecture for cross-lingual voice conversion. IEEE Signal Process Lett 27:1310–1314
23. Sisman B, Zhang M, Dong M, Li H (2019) On the study of generative adversarial networks for cross-lingual voice conversion. IEEE Autom Speech Recogn Understand Workshop (ASRU) 2019:144–151. https://doi.org/10.1109/ASRU46091.2019.9003939
24. Ho TV, Akagi M (2021) Cross-lingual voice conversion with controllable speaker individuality using variational autoencoder and star generative adversarial network. IEEE Access 9:47503–47515. https://doi.org/10.1109/ACCESS.2021.3063519

25. Morise M, Yokomori F, Ozawa K (2016) World: a vocoder-based high-quality speech synthesis system for real-time applications. IEICETRANSACTIONS on Inform Syst 99(7):1877–1884
26. Kumar K, Kumar R, De Boissiere T et al (2019) MelGAN: generative adversarial networks for conditional waveform synthesis [J]. https://doi.org/10.48550/arXiv.1910.06711
27. Hel E, Nurminen J, Gabbouj M (2009) 651 Analysis of LSF frame selection in voice conversion [J]
28. Zhou K, Sisman B, Li H (2020) Transforming spectrum and prosody for emotional voice conversion with non-parallel training data [J]. Speech Commun. https://doi.org/10.1016/j.specom.2020.05.004
29. Sisman B, Li H (2018) Wavelet analysis of speaker dependent and independent prosody for voice conversion. INTERSPEECH
30. Vainio M, Suni A, Aalto D (2013) Continuous wavelet transform for analysis of speech prosody [J]. Corros Eng, Sci Technol. https://doi.org/10.1179/1743278215Y.0000000007
31. Tokuda K (2017) Flexible speech synthesis based on hidden markov models[J]. https://doi.org/10.1109/jproc.2013.2251852
32. Suni A, Aalto D, Raitio T et al (2013) Wavelets for intonation modeling in hmm speech synthesis [C]//Speech synthesis workshop (SSW8)
33. Kaneko T, Kameoka H (2017) Parallel-data-free voice conversion using cycle-consistent adversarial networks. arXiv preprint arXiv:1711.11293
34. Zhao Y, Huang W C, Tian X et al (2020) Voice Conversion Challenge 2020: Intra-lingual semi-parallel and cross-lingual voice conversion[J]
35. Kinga S, Karaiskosb V (2009) The blizzard challenge 2010

Design and Research of UAV Fire Emergency and Intelligent Inspection System in the Valve Hall of Converter Station

Zhendong Li, Junjie He, Haiqing An, and Tao Li

Abstract As the core of the entire converter station, the valve hall can only rely on traditional manual methods for daily inspection and lacks intelligent inspection methods. When the converter valve is running, the operation and inspection personnel cannot enter the valve hall in time, and a large number of power electronic components lack effective fire emergency measures. This paper aims to explore the feasibility of the application of UAV in the flexible valve hall, develop a multi-functional UAV system with fire emergency and intelligent inspection, and realize the emergency fire extinguishing and infrared detection and abnormal early warning of the core equipment in the valve hall. Relying on the Zhangbei flexible DC project, this paper has been demonstrated and applied in the Zhangbei converter station and will be gradually promoted and applied to the valve hall of the converter station of different voltage levels, so as to provide effective auxiliary means for on-site operation and inspection personnel and improve the operation safety of flexible straightening equipment.

1 Introduction

With the proposal of the national strategy of "emission peak and carbon neutrality" [1–3], the development of clean energy will inevitably usher in a new climax [4–6]. As the preferred solution for clean energy-friendly grid integration [7, 8], flexible HVDC transmission technology will inevitably become the focus of future construction [9, 10].

As the core of the flexible converter station, the valve hall has high requirements for safety precautions such as fire prevention, heat insulation, shielding electromagnetic interference, and radio interference. As the core equipment of AC/DC conversion, the abnormalities and faults (such as overheating, fire, water leakage, short circuit, etc.) of the converter valve in the valve hall will lead to the forced shutdown of the

Z. Li · J. He · H. An (✉) · T. Li
State Grid Jibei Electric Power Co., Ltd. EHV Power Transmission Company, Beijing 100000, China
e-mail: 245932206@qq.com

S. Yadav et al. (eds.), *Proceedings of 3rd International Conference on Artificial Intelligence, Robotics, and Communication*, Lecture Notes in Electrical Engineering 1172, https://doi.org/10.1007/978-981-97-2200-6_3

DC transmission system and even bring disastrous consequences to the operation of the cross-regional power grid. At present, the inspection of the converter valve hall still adopts the traditional manual first method. There are the following pain points to be solved. First, the fire protection facilities of the valve hall are configured with "ultraviolet + smoke" device, which can only provide fire warning information; once a fire occurs, the operation and inspection personnel cannot enter the valve hall in time; and a large number of power electronic components lack effective fire emergency measures. Second, although a certain number of fixed and orbital cameras are arranged in the valve hall, which can realize visible light and infrared detection functions, the valve tower in the valve hall has a compact structure, and the layout of auxiliary facilities such as HVAC is scattered, and there are many blind areas in the existing detection methods, which cannot achieve accurate monitoring without dead angles.

In this paper, a multi-functional UAV with fire emergency and intelligent inspection is designed to realize special inspection work such as emergency fire extinguishing and infrared detection of core equipment in the valve hall. For the first time, the firefighting equipment is combined with UAV, and the idea of UAV firefighting with flexible DC valve hall is proposed innovatively. Through the UAV to quickly build a three-dimensional monitoring system, instead of the operation inspection personnel the first time to go deep into the central area of the fire to obtain information, the visible light camera can fully present the scene, and the thermal imaging camera can penetrate smoke and part of the building occlusion, obtain temperature distribution, and assist the operation inspection personnel to accurately identify the fire point and high temperature area. The live screen is transmitted back to the rear remote control in real time via the wireless network. Let on-site personnel grasp on-site information and scientifically command and dispatch.

Relying on the Zhangbei flexible DC project, this paper has been demonstrated and applied in Zhangbei converter station and will be gradually promoted and applied to other converter stations, providing effective auxiliary means for operation and inspection personnel and improving the safety of equipment operation.

2 Design and Research of Firefighting UAV System

2.1 UAV Mounting and Remote Control Precision Throwing Technical Scheme

The UAV adopts DJI M300 aircraft and has a flight control development interface and the function of developing mounted equipment. The M300 UAV has a maximum take-off weight of 9 kg, a maximum load weight of 2.7 kg, and dimensions (unfolded, excluding propellers): 810 × 670 × 430 mm (length × width × height), which meets the flight conditions of the valve hall space.

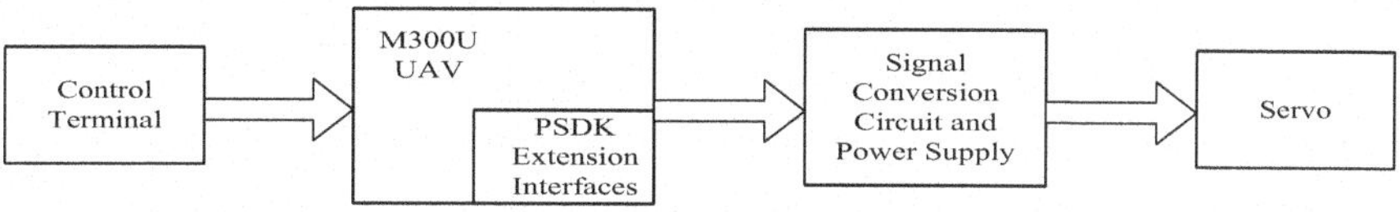

Fig. 1 Mount the control logic diagram

The fire extinguishing device adopts the form of fixed-point fire extinguishing bomb. The fire extinguishing ball is fixed at the bottom of the aircraft. The aircraft flies above the fire source when working. The fire extinguishing ball is thrown through the self-made delivery mechanism. The fire extinguishing ball is blasted to complete the fire extinguishing. During the fire, the drone can maintain a certain safe distance from the fire source, avoiding the damage of the fire to the drone. The automatic delivery device adopts servo-driven crank slider mechanism, which has simple structure, flexible and convenient operation, lightweight, and high stability. The drone realizes the control of the mounted device through an independent wireless link, and the specific implementation scheme is shown in Fig. 1.

- The throwing remote control button is integrated in the centralized control flight control system. The drone arrives above the fire source, and the remote one-key control throwing mechanism works.
- The throwing device is mounted under the drone. The customized development is carried out through the open PSDK development interface to realize the control and power supply of the thrower.
- During firefighting operations, the operator manually controls the drone to reach the air above the fire source, observes the position of the fire source through the drone mounting lens, and adjusts the attitude of the drone. When receiving the throwing control command, the ostrich worked, opened the fire extinguishing bomb bolt at the same time, pulled the fire extinguishing bomb safety, performed the throwing action, and threw the hanging bomb.

2.2 *Efficient Fire Extinguishing Material Configuration Scheme Suitable for Power Electronics in the Valve Hall*

Through the study of high temperature and combustion characteristics, fire extinguishing materials, and properties of power electronic equipment in the valve hall, the selection of efficient fire extinguishing materials in the valve hall is completed.

The fire extinguishing material adopts ultra-fine ABC fire extinguishing dry powder. The dry powder is dry fine solid powder, which is a chemical fire extinguishing agent made of sodium bicarbonate and a small amount of additives (flow agent, moisture-proof agent). Dry powder is characterized by high fire extinguishing efficiency, fast speed, non-conductive, non-corrosive, and low toxicity and is suitable for fighting fires of flammable liquids, combustible gases, and electrical equipment.

Fig. 2 Actual flight of the UAV inspection system is on site

After combustion, there is no harm to human objects and no pollution to air, soil, water sources, etc. The stable pH value of pyrotechnic agent is 6–7, and the pH value of safety performance is between 5 and 9. The pull ring design is safe and reliable, and the outer surface is made of new environmentally friendly material recycled paper.

When working, the drone flies to the fire location, the bomb throwing device on the drone works, the fire extinguishing bomb is thrown to the fire position, the fire extinguishing bomb contacts the fire source to burst, and the instantaneous shock wave will isolate local oxygen, destroy the combustion chain of combustibles, and prevent combustion. At the same time, under the action of the shock wave, the ultra-fine fire extinguishing dry powder quickly covers the local combustibles and prevents re-ignition by isolating the combustibles. Ultra-fine fire extinguishing dry powder has the effect of heat absorption and cooling, and when it drops to a certain temperature, a layer of diaphragm will be formed, and the combustible material will be flame retardant again until the fire is completely extinguished.

3 Design and Research of UAV Intelligent Inspection System

The system consists of intelligent air station, centrally controlled UAV, centralized control flight control system, and indoor positioning system. It realized unmanned and intelligent in the process of automatic inspection. As shown in Fig. 2, the actual flight of the UAV inspection system is on site.

3.1 Intelligent Air Stations

Intelligent air station equipment is the UAV stop, communication station, supply station, and exchange station, providing one-stop support services for UAV, relying on the terminal equipment in the unattended state, to achieve UAV all-day and all-weather application.

The intelligent air station provides a series of functions such as drone storage, release, recovery, and charging.

3.2 Centrally Controlled UAV

As the terminal of the centralized control flight system, the centralized control UAV can be customized and integrated design. The body adopts carbon fiber and composite body design, which has the characteristics of lightweight, high strength, beautiful appearance, large load capacity, simple and flexible operation, and convenient transportation.

The centralized control UAV can carry out remote control and high-definition surveillance video transmission, realize video surveillance and aerial video transmission in designated areas, and can also carry different types of mission loads to meet specific needs according to specific application scenarios and user needs. The UAV terminal can operate unmanned throughout the process, with a high degree of autonomy, and has the ability of automatic vertical take-off and landing, fixed-point hovering, and automatic route flight.

3.3 Centralized Control Flight Control System

The centralized control flight control system adopts modular design and through the software programming concept of high cohesion and low coupling. It can achieve: (1) relative independence. Modules can be designed, manufactured, debugged, modified, and stored separately, which facilitates separate production by different specialized enterprises. (2) Interchangeability. The structure, size, and parameters of the module interface are standardized, and it is easy to realize the interchange between modules, so that the modules can meet the needs of a larger number of different products. (3) Versatility. It is conducive to realize the commonality of modules between horizontal series and vertical series products and realize the commonality of modules between cross-series products.

The centralized flight control system software is divided into five modules: application layer, business layer, communication layer, service layer, and common components. The application layer is responsible for user interaction and processing of input and output. The business layer is responsible for the calculation and distribution of data and organizes the data interaction between the application layer and other modules. The communication layer is responsible for establishing connections with external systems, such as drones, terminals, and other attached products for data interaction. The service layer is used to obtain configuration information, route file information, and communicate with other service systems for data interaction. The common component layer is used to provide the tools that serve the system software.

3.4 Indoor Positioning System

Indoor spatial positioning is very different from outdoor spatial positioning technology; when UAV flying in indoor air, it is also necessary to carry out three-dimensional positioning of UAV and build a three-dimensional coordinate system. Since the location information of the positioning base station is based on the space Cartesian coordinate system, the position information format required by the satellite signal based on pseudo-satellite technology is based on the geodetic coordinate system, the UAV receiving module completes the position solution according to the pseudo-satellite signal, and the position information data of the receiving module is also based on the geodetic coordinate system, and because this paper will use the satellite navigation receiver to measure the real position coordinates of each base station corresponding to the indoor layout, the calculated position information is based on the geodetic coordinate system. Therefore, it is necessary to complete the conversion of the coordinate system of the base station and the coordinate system of the UAV receiving pseudo-satellite signals. In this paper, the UWB positioning module design constructs an indoor three-dimensional coordinate system required for simple UAV indoor autonomous navigation.

3.5 Research on Electromagnetic Interference in UAV Communication Band

A comprehensive spectrum test was carried out for the valve hall of the converter station to investigate whether the indoor electromagnetic situation interfered with the remote control signal (frequency band: 2.4–2.4835; 5.725–5.850 GHz). The survey results are as follows (Table 1).

- Through the spectrum test, a total of 5 main frequency bands were tested, which were 0–300 MHz; 0–600 MHz; 0–1.300 GHz; 1.150–1.650 GHz; and

Table 1 Valve hall electromagnetic environment test results

Start frequency	Center frequency	Cutoff frequency
0	150 MHz	300 MHz
0	300 MHz	600 MHz
0	650 MHz	1.3 GHz
1.15 GHz	1.4 GHz	1.65 GHz
2.15 GHz	2.4 GHz	2.65 GHz

2.150–2.650 GHz. Through the spectrum, you can see that there are interference frequencies in the corresponding frequency bands.

- In the 0–300 MHz frequency band, it can be seen that there are frequency points with strong interference through the spectrum, mainly concentrated in the 200 MHz frequency band, and there are extremely strong interference signals at a single frequency point of 6.5 MHz
- In the 0–600 MHz frequency band, it can be seen that there are frequency points with strong interference through the spectrum, mainly concentrated in 330 and 500 MHz adjacent frequency bands.
- In the 0–1.300 GHz frequency band, it can be seen that there are frequency points with strong interference through the spectrum, mainly concentrated in 700 MHz and 1.100 GHz adjacent frequency bands.
- In the 1.400–1.650 GHz band, it can be seen that there are frequency points with strong interference through the spectrum, mainly concentrated in the 1.440 and 1.550 GHz adjacent frequency bands.
- In the 2.150–2.650 GHz band, it can be seen that there are frequency points with strong interference through the spectrum, mainly concentrated in the 2.440 GHz band, and the signal strength is from −47.00 to 49.25 dBm.

Through on-site radio spectrum acquisition, it can be seen that there are different degrees of interference sources in multiple frequency bands, among which there are strong interference frequency points in the 0–300 MHz frequency band, mainly concentrated in the 200 MHz frequency band, and there are extremely strong interference signals at a single frequency point of 6.5 MHz.

The communication frequency band of the UAV is 2400–2483.5 MHz and 5725–5850 MHz, and the transmission power is 20–33 dBm. The receiving frequency is seen through the drone radio frequency test, and the UAV receives signal strength between −22 and 7.20 dBm. The test results are in the 2.440 GHz band, and the signal strength is from −47.00 to −49.25 dBm.

Comparing the strength of the frequency band that may interfere with UAV communication in the valve hall with the strength of the signal transmitted and received by the UAV itself, it can be seen that the signal in the UAV communication band is much stronger than the power of the interference signal. Combined with the calculation of the field strength distribution of the valve tower, it can be seen that the stable communication between the drone and the remote control can be ensured when maintaining a certain safe distance from the valve tower.

4 Conclusion

The development and application of valve hall fire UAV can make up for the lack of fire emergency measures in the existing valve hall. When a fire occurs, the first time to control the firefighting drone mounted fire extinguishing bomb to fly over the fire source to throw the fire to extinguish, to achieve the purpose of timely control of the fire and even extinguish the fire, greatly improving the safety of valve hall equipment.

The development and application of UAV intelligent inspection system will enrich the indoor inspection technical means of valve hall and supplement the blind area of existing inspection technology. Through the application of UAV airport intelligent power exchange, UAV automatic flight control, UAV dual lens shooting, AI intelligence, and other technologies, the efficiency of operation and maintenance will be greatly improved. For the operation inspection profession, it can improve the level of intelligent operation and maintenance, reduce operation and maintenance costs, and promote the application of emerging operation and maintenance functions, which is a specific practical measure to implement the State Grid Corporation to build a world-class energy internet enterprise with Chinese characteristics.

Acknowledgements This work is supported by the Science and Technology Project of State Grid Jibei Electric Power Co., Ltd., EHV Power Transmission Company (52018H220004).

References

1. International Panel on Climate Change (2018) Global warming of 1.5°C [R]. Cambridge University Press, Cambridge. https://www.ipcc.ch/sr15/
2. Davis SJ, Lewis NS, Shaner M et al (2018) Net-zero emissions energy systems [J]. Science 360(6396):eaas9793
3. Xiaoxin C, Shuyong Lu, Zong X et al (2018) Technology features of the new generation power system in China [J]. Proceed CSEE 38(7):1893–1904(in Chinese)
4. Shaner MR, Davis SJ, Lewis NS et al (2018) Geophysical constraints on the reliability of solar and wind power in the United States[J]. Energy Environ Sci 11(4):914–925
5. Denholm P, Arent DJ, Baldwin SF et al (2021) The challenges of achieving a 100% renewable electricity system in the United States [J]. Joule 5(6):1331–1352
6. Hui L, Dong L, Danyang Y (2021) Analysis and reflection on the development of power system towards the goal of carbon emission peak and carbon neutrality [J]. Proceed CSEE 41(18):6245–6258 (in Chinese)
7. Chongqing K, Liangzhong Y (2017) Key scientific issues and theoretical research framework for power systems with high proportion of renewable energy[J]. Autom Electric Power Syst 41(9):1–11 (in Chinese)
8. Jianbo G (2021) Thinking about challenges and mechanisms faced by new power system [J]. China Power Enterprise Manage 25:8–11
9. Guangfu T (2005) Accelerating the construction of a new power system with "four modernizations" [N]. State Grid News 2021:10–12
10. Xiaorong X, Jingbo H, Hangyin M et al (2021) New issues and classification of power system stability with high shares of renewables and power electronics [J]. Proceed CSEE 41(2):461–475 (in Chinese)

Development Overview and Application Prospects of Advanced UUVs

Zheng Wang, Xinyu Qu, Longmei Li, and Liming Wang

Abstract Unmanned Underwater Vehicles (UUVs) showcase transformative capabilities within various domains, attributable to their inherent characteristics of stealth, intelligent capabilities, and reconfigurability. Currently, a multitude of nations globally engage in a competitive development of various UUV archetypes, extensively employed in intelligence surveillance and hazardous material searching. This paper summarizes the UUV models and their applications being developed or installed by science and technology powers such as the United States, Britain, and Russia. Subsequently, the manuscript scrutinizes the quintessential technologies buttressing the application of UUV and elucidates the interplay between pertinent technologies and capacity to implement mandates. In conclusion, the manuscript proffers a prospectus on the application of UUV in forthcoming complex marine environments, seeking to provide relevant references for the development of advanced UUV in China to accomplish various underwater mission.

1 Introduction

Unmanned Underwater Vehicles (UUVs) can be bifurcated into two predominant categories according to their modality of control: Remotely Operated Vehicles (ROVs) and Autonomous Underwater Vehicles (AUVs) [1]. ROVs are operated by the operator of the mother ship in real time via receiving control signals and power supplies through connecting cables, which can be used for hazardous material searching, undersea equipment layout and withdrawal, underwater manned vehicle assistance, etc. In contrast, AUVs can realize achieve autonomous navigation auto-navigating operation by using their own energy power and carrying corresponding payloads. Since the 1950s, global research endeavors concerning Unmanned Underwater Vehicles (UUVs) have been conducted. Presently, a global zenith of research and application in Unmanned Underwater Vehicles (UUVs) has been inaugurated,

Z. Wang · X. Qu · L. Li · L. Wang (✉)
School of Electrical Engineering, Naval University of Engineering, Wuhan 430033, China
e-mail: 602257804@qq.com

S. Yadav et al. (eds.), *Proceedings of 3rd International Conference on Artificial Intelligence, Robotics, and Communication*, Lecture Notes in Electrical Engineering 1172, https://doi.org/10.1007/978-981-97-2200-6_4

with the United States, as the leader of science and technology power, accentuating its prospective applications in future ocean mission [2]. The research and development of advanced UUVs in China started later than other countries, thus the application fields for mature devices remain relatively limited. By scrutinizing the development and application of UUVs among the world's leading science and technology powers and analyzing the development status of UUV's crucial technologies, this manuscript posits a conception of UUV application in ensuing ocean mission. Such an examination aims to provide a substantive reference for the construction of UUV ability to perform underwater missions within China.

2 Current Development and Application Status of Advanced UUV Abroad

2.1 Overview of the Development of the Advanced UUV in US

The United States is the country with the most comprehensive development, mature technology, and experienced application in the UUV field in the world. The advanced UUV of the United States is developed in the order of small, medium, large, and ultralarge. At present, the development and application of small and medium-sized UUVs are relatively mature, with practical combat capabilities such as searching underwater hazards, intelligence monitoring, and ocean exploration. One typical set of UUVs, named REMUS [3] UUVs include three types: REMUS100, REMUS600, and REMUS6000. The REMUS100 can flexibly configure sensors and related system accessories, which can perform tasks such as searching underwater hazards and hydrological measurement. In 2003, US achieved UUV to search underwater hazards for the first time using the REMUS100 in Umm Qasr Port, Iraq [4]. The REMUS600 is equipped with side-scan sonar, cameras and other equipment, mainly used for tasks such as intelligence surveillance and ocean exploration. In 2015, US successfully launched and recovered the REMUS600 underwater on a Virginia-class nuclear submarine. In 2012, the first REMUS 6000 was delivered to US. The REMUS6000 significant improvement on its diving depth and endurance ability and can perform missions at the depths of 6000 m. The Bluefin-12, one of US autonomous underwater vehicles for battlefield space preparation, mainly performs tasks such as offshore investigations, search and salvage, searching underwater hazards, and etc. [5]. In 2017, the "Knifefish" developed on the basis of the "Bluefin-12" was delivered to US for testing and evaluation. This type of UUV uses low-frequency electromagnetic waves to detect and identify ground hazards and buried hazards in noisy environments. Ships carrying "Knifefish" have effectively expanded their detection range and significantly reduced the risk of hitting hazards. In addition, the United States is actively promoting the development of large and ultralarge UUVs. In 2017, the US large UUV "Proteus" completed adversarial unmanned mission testing in complex marine environments. This type of UUV has a range of up to 600 km [6]. In 2022,

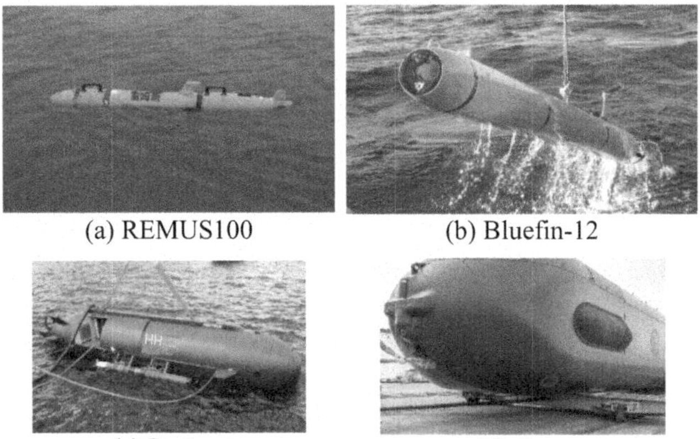

(a) REMUS100 (b) Bluefin-12

(c) Proteus (d) Orca

Fig. 1 US-advanced UUV

the ultralarge UUV "Orca" [7] was delivered for sea trials. The design requires that this type of UUV can be equipped with corresponding loads according to different tasks, with a range of over 2000 nautical miles. The US-advanced UUVs are shown in Fig.1.

At present, advancements in unmanned underwater vehicle (UUV) technology are being explored for a variety of applications. These applications extend beyond traditional reconnaissance and data collection to include underwater infrastructure inspection, marine environment monitoring, and oceanographic research. New operational concepts have emerged that focus on utilizing UUVs for safeguarding underwater communication cables, surveying for scientific purposes, and conducting maintenance on underwater installations, thereby enhancing capabilities in commercial and research activities within relevant maritime regions.

2.2 Overview of the Development of the Advanced UUV in the European Countries

In 2008, the UK successfully developed an anti-mine UUV "Talisman" that could carry four disposable mine suppressors. This aircraft achieved all anti-mine tasks, such as mine detection, identification, and mine suppression. The disposable mine suppressor "Archerfish" is also a small UUV with a diameter of 104 mm and a length of 880 mm. The "talisman" UUV adopts an open structure design and can also be configured with different loads according to the nature of the task [8]. In 2014, the UK launched the "MASTT" ultralarge UUV, which used a titanium alloy shell with a diameter of 2.75 m and a length of 8.85 m. The UUV has a range of over 2500 nautical miles and can carry various types of equipment, and even small frogman carriers [9]. In addition, other European countries have also developed UUVs that are

Fig. 2 Advanced UUV of European countries

(a) Talisman (b) Archerfish

(c) Hugin 1000 (d) AUV62-MR

suitable for their actual needs, such as Norway's "Hugin" series, France's "Alister" series and Sweden's AUV62-MR and "MuMNS" series [10]. The advanced UUVs of European countries are shown in Fig.2.

In response to the UUV's single function, small platform, and other shortcomings, as well as to consider the future needs of new missions and the future marine environment on navigation, image resolution and other higher requirements, European countries began to turn to the development of large, ultra-large UUV.

2.3 Overview of the Development of the Advanced UUV in Russian

Russia has greatly prioritized the research and development of UUV applications of since the era of the former Soviet Union. Since entering the twenty-first century, Russia has developed a series of large-scale UUVs in combination with its own maritime needs [11]. In 2018, the "harpsichore-2" passed testing. This type of UUV reached a depth capacity of 6000 m, had a range of 300 km, and was equipped with an autonomous navigation system that could maintain communication with the mother ship. It was mainly used for ocean exploration and had previously carried out underwater search and rescue missions in Arctic waters. In 2016, Russian Design Bureau of "Ruby" proposed the design concept of a "Surrogate" UUV, which simulated the physical field of various types of underwater vehicle. In 2019, the "Poseidon" nuclear-powered UUV was launched for testing. Using a miniaturized nuclear power device, this type of UUV navigated underwater at a speed of over 100 knots, with a range of up to 10,000 km, and carried various payloads, including nuclear warheads. In addition, it also has excellent acoustic stealth performance. The Russian advanced UUVs are shown in Fig.3.

(a) Harpsichord-2 (b) Poseidon

Fig. 3 Advanced UUVs of Russian

In Russia, the progression of Unmanned Underwater Vehicles (UUVs) encountered a protracted period of sluggish development following the dissolution of the Soviet Union, primarily attributable to economic and political considerations. Nevertheless, concomitant with the gradual ascendancy of unmanned platforms within Russia's vehicle, there has been a substantial augmentation in the quantity of UUVs undergoing development and actively employed in service. Russia's UUVs have, concurrently, delineated a specialized trajectory characterized by the centrality of large and super-large UUV platforms featuring nuclear propulsion systems, complemented by smaller UUVs.

3 The Key Technology of Advanced UUV

Based on the technical parameters and functional characteristics of foreign installed UUV models, the key technologies that constrain and determine the development of UUV applications in related fields should also be considered. The integration and application of UUV technology are essential to develop the ability to perform underwater mission capabilities, and the cluster formed by multiple UUVs has stronger robustness and scalability and can accomplish tasks that are difficult to accomplish by a single UUV. UUV clustering involves the communication and interaction between UUVs in the cluster, task allocation, control, and coordination. The cooperative operation of UUVs in the form of clusters has led a new development direction.

3.1 The Monomer Technology of UUV

Sensor Technology. Both exploratory and operational UUVs need to carry many types of sensors, including the basic temperature and salt depth sensors, altitude sensors, altimeters, and Doppler range meters, as are in line with different mission requirements. These UUVs also need to carry the appropriate detection equipment, such as dual-frequency scanning sonar for mine detection and identification and magnetic detection sensors. High-precision sensing equipment is often a direct guarantee of the high performance of the UUV. The U.S. "Bluefin-21" type UUV with its 455 kHz side-scan sonar can provide a 7.5–10 cm resolution image, and an operation can be found within a few square miles of the underwater hazardous materials. Modularization is also an important direction in the development of sensors. According to the needs of different missions, the UUV configuration can be changed at any time the sensor module shortens the development cycle and reduces costs, which is of great significance.

Energy power technology. The operating radius of UUVs is severely limited by the amount of energy they can carry. Small- and medium-sized UUVs have limited internal space, small battery capacity, low speed, short endurance, and generally poor power performance. Currently, most UUVs are mainly powered by lithium batteries [12] and silver-zinc batteries [13], but there is limited space to improve the energy density of these materials. Therefore, it is also necessary to develop new energy sources with higher power generation efficiency and higher energy density to improve the power performance. At present, the fuel cell technology combined with hydroxide energy, the electric-electric hybrid energy composed of fuel cells and large multiplier primary batteries, and the thermal-electric hybrid energy of diesel engines combined with lithium batteries all represent promising breakthroughs [14]. In addition, the US has conducted demonstration tests of UUV underwater charging stations, and the completion and use of underwater charging stations will also break the spatial and temporal limitations of UUV deployment to a certain extent [15].

Navigation and positioning technology. Accurate navigation and localization is a prerequisite for the safety and successful operation of UUVs [16]. The error of the inertial navigation system (INS) will increase with the accumulation of time, and the accuracy of the hydroacoustic navigation system (SONAR) is heavily dependent on the number and position of the base array and the limited range of action. The combination of the two SINS/SONAR navigation methods reduces the cumulative error of time and enlarges the range of hydroacoustic navigation, but at the same time, it also generates a new problem of difficulty in locating the acoustic markers and the base arrays in unknown sea areas. However, it also creates a new problem of difficulty in locating sound markers and base arrays in unknown sea areas. Through the introduction of a global satellite navigation system (GNSS) for autonomous positioning of buoys at sea, UUV can receive and process the position information sent by buoys with acoustic signals underwater and then combine with the inertial guidance system and depth sounder to determine its own accurate position [17]. At present, most UUVs use SINS/SONAR/GNSS combination navigation. To further reduce the influence of external interference signals, scholars have also studied combinations of geophysical navigation methods, such as seafloor terrain matching navigation [18] and gravity magnetic navigation [19].

Communication technology. Relying on the function of the UUV communication relay, strategic communication technology can reduce the number of times the submarine floats up and thus improve its covertness. The communication methods of UUV mainly include hydroacoustic communication and fiber-optic cable communication, in which the fiber-optic cable communication transmits signals at a fast speed and with high reliability, but it needs to be wired by the fiber-optic cables and cables, which greatly restricts the operation range and flexibility of UUV, and it is generally commonly used in ROVs. Hydroacoustic communication is a subset of general wireless communication techniques. The available distance depends on the carrier frequency and transmission power of the equipment used, and despite its low transmission speed and large noise interference from the seabed, hydroacoustic communication is still the only reliable method of deep-sea communication [20]. At present, many countries have focused on research into hydroacoustic high-speed, long-distance transmission technology and transmission security and other issues. To further improve the communication efficiency to underwater nodes, the United States has carried out research on underwater laser communication and realized the use of blue and green lasers to communicate with submersibles at a depth of approximately 100 m underwater through airborne equipment transmissions, but it has not yet been applied to the UUV platform [21].

Deployment and recovery techniques. Successful deployment and recovery is critical to the successful completion of their missions [22]. At present, UUV deployment and recovery is mainly through surface ships, underwater platforms and other platforms. A small number of portable UUVs used for harbor inspection can be directly deployed and recovered by human power, and a very small number of ultralarge UUVs with long-distance operation capability can also be autonomously driven from the dock to the mission waters for use in operations. After, these UUVs are autonomously returned to the dock to recover. Surface ship deployment includes integrated deployment, A-frame deployment, chute deployment, etc., and most of the recovery is done by boom lifting. To prevent the deployment of the recovery process by the impact, swing protection, wind and wave compensation and other technical means are needed. Underwater platforms deployment not only makes up for the short-range shortcomings of the UUV but also maintains the stealth of the UUV. Underwater platforms deployment includes torpedo/missile tube deployment, docking bay deployment, and backpack carry deployment, and funnel-shaped capture, robotic arm capture, or ROV cable capture can be used for underwater recovery [23].

3.2 The Cluster Technology of UUV

Communication network technology. The most obvious advantage of UUV clustering is the ability to share information detected by multiple UUVs with other nodes in the cluster using a communication network, making the battlefield more transparent. An effective communication network is also the default prerequisite for researching other clustering technologies, such as tasking and cooperative control. Hydroacoustic communication is currently the most mature and reliable method of underwater long-distance communication. For the communication delay and data packet loss problems existing in the hydroacoustic communication network, in the absence of breakthroughs in the existing hardware conditions, it is also possible to improve the efficiency and reliability of underwater communication in UUV clusters by transforming the communication topology, designing more efficient communication protocols, and transforming the communication mode.

Tasking techniques. The tasks performed by UUV clusters are becoming increasingly complex, and reasonable task allocation based on the task complexity, the number of UUVs in the cluster, and their location information is the key to ensuring that UUV clusters can efficiently accomplish their tasks. According to the different decision-making rights, task allocation can be divided into centralized and distributed allocation. Centralized allocation signifies that the master UUV receives the state and environment information of each slave UUV and weighs the overall task situation for allocation. Distributed allocation signifies that each UUV in the cluster autonomously performs local task allocation based on local information. These two allocation methods each have their own advantages and disadvantages; centralized allocation has a high degree of mastery of the overall allocation situation but is more dependent on communication, and distributed allocation is less dependent on communication and has faster parallel computation, but due to the lack of the overall allocation situation, there may be allocation conflicts [24]. In practice, a combination of the two allocation methods is often applied. Currently, task allocation in dynamic uncertain environments under weak communication conditions is an urgent problem for UUV clusters.

Cluster control technology. Cluster control technology is the core technology for building UUV clusters. Effective cluster control technology is the basis for realizing mutual cooperation and synergistic operation of UUVs. Due to the complex underwater environment in the ocean, many control algorithms applicable to land-based UUVs and airborne UAV clusters cannot be directly transplanted to UUV cluster systems because of the weak underwater communication capabilities. On the other hand, the dynamics modeling of UUV vehicles is characterized by strong coupling and nonlinearity, and the structural functions, hardware composition, and intelligence level of each UUV in the heterogeneous clusters are not the same, which causes great difficulties in realizing the coordinated control of UUV clusters. At this stage, improving the existing algorithms to meet the condition constraints of UUV clusters in the marine environment is still the main research direction of cluster control algorithms. At present, due to the lack of a UUV cluster validation platform, most studies are still in the stage of simulation testing.

4 UUV Application Outlook

Toward summarizing and sorting the applications of foreign UUVs and various underlying technologies, we propose and the concept of UUV application in two contexts is proposed based on certain reasonable and somewhat forward-looking performance and functional expectations.

4.1 Exploration of Unknown Marine Environments

Many factors in the marine environment have an extremely important impact on the performance of various types of equipment. It is of great significance to master the seabed topography and hydrographic information of relevant sea areas for occupying information superiority and seizing sea control. In this regard, UUV can carry temperature and salt depth measuring instruments, side-scan sonar and other payloads, cover the search in the mission area according to the preset trajectory, real-time measurement and collection of marine environmental data, and draw the seabed topographic map. However, the UUV can also be used to deploy the sea state information collection device in the sea area that needs to be focused on monitoring and regularly scheduled cruises through hydroacoustic communication to receive the device to collect long-term continuous data to provide the basis for the establishment of a specific sea current model to realize the prediction of sea state forecasts. Both applications give full play to the characteristics of covert operation of UUVs, which can effectively support the construction of unknown marine environments without alerting the enemy.

4.2 Exploration of Unfamiliar Landing Site Environments

Facing unfamiliar nearshore waters, equipment and personnel in the area are constantly at great security risk. Uncertain sea conditions, unknown underwater hazards, and unknown coastal topography pose potential threats. With the development of sonar and 3D imaging technologies, UUVs can detect, identify, and classify underwater hazardous objects. To explore unfamiliar coastal environments, the payload transportation capability of UUVs can be utilized to overcome the range limitations of small drones. UUVs can navigate to nearshore waters and deploy small drones to survey the unknown terrain. It provides strong support for our side to win information superiority and grasp the initiative.

5 Conclusion

Unmanned Underwater Vehicles (UUVs) are capable of executing a myriad of specialized missions by leveraging their inconspicuous and intelligent attributes. The trajectory of UUV development inclines toward standardization, modularization, reconfigurability, and clustering, as indicated by existing research. Advancements in pivotal technologies have catalyzed the augmentation of UUV combat capacities in domains such as reconnaissance and surveillance, detection and identification, and payload delivery, among others. The progression of clustering technology facilitates the optimization of UUV usage methodologies and establishes the prerequisites for the actualization of heterogeneous underwater cluster coordination, cross-domain coordination, and manned-unmanned coordination. The maturation and application of UUVs are poised to significantly expand the operational theatre of surface ships and submarines, subverting conventional naval warfare paradigms. Consequently, UUVs are anticipated to gradually morph into an instrumental means of securing information superiority and executing precision firepower strikes in future maritime conflict, thereby offering a robust mechanism for anti-intervention/area-denial strategies.

References

1. Gao J, Zhang FB (2018) Unmanned underwater vehicle control system [M]. Northwest Polytechnic University Press, Xi 'an, pp 1–9
2. Zhao XF (2017) Research on UUV infrared target recognition method under sea and sky background [D]. Harbin Engineering University
3. Edoardo IS, Manhar RD (2017) A USV-based automated launch and recovery system for AUVs [J]. IEEE J Oceanic Eng 42(1):37–55
4. Miranda MI (2014) Mobile docking of REMUS-100 equipped with USBL-APS to an unmanned surface vehicle: a performance feasibility study [J]. Dissertations & Theses-Gradworks
5. Zhang YY (2014) Concept and technical development of autonomous underwater vehicles--beginning with 'bluefin-21' [J]. China Terminol 000 (0z1):131–133
6. Wang Q (2019) Development and military uses of large unmanned underwater vehicles [J]. Digit Ocean Underwater Warfare 2(4):33–39
7. Li J (2017) Underwater unmanned combat system equipment status and development trend [J]. Ship Sci Technol 1:1–5
8. Ren Y (2014) Current status and development trend of unmanned combat systems [J]. Mine Warfare Ship Self-defence 22(04):58–61
9. Department of the Navy (2015) Large displacement unmanned underwater vehicle innovative naval prototype technology [R]. ONRBAA Announcement
10. Wu SH (2018) Development status and trend analysis of foreign unmanned underwater vehicles [J]. Military Digest 5:18–20
11. Zhong HW, Li GL, Song LH et al (2018) A review of the development of large unmanned underwater vehicles abroad [J]. J Unmann Undersea Syst 26(4):273–282
12. Dai H, Zhang B, Xu YZ (2020) Development status and trend of UUV power battery [J]. Ship Sci Technol 42(12):155–158
13. Jin YZ, Liu ZW (2019) Current situation and development trend of UUV power battery [J]. Chinese J Power Sour 43(6):1073–1076
14. Shi XF, Dang JJ, Liang Y et al (2021) Underwater attack and defense weapon energy power technology development status and trends [J]. J Unmanned Undersea Syst 29(06):634–647
15. Yu SK, Wang F, Li H et al (2021) Research on UUV electric energy supply based on underwater wireless charging [C]//China Ship building Engineering Society, Northwestern Polytechnical University, China Shipbuilding Group Co., Ltd., 705th Institute. ' The 4th underwater unmanned system technology summit forum ' -- human/unmanned cooperative technology proceedings. [Publisher unknown], p 5
16. Xi L (2014) Research on integrated design and navigation control technology of small UUV system [D]. Harbin Engineering University
17. Fan G, Zhang Y, Zhao HM et al (2022) Development status and analysis of underwater robot positioning and navigation technology [J]. J Ordnan Equip Eng 43(03):22–29
18. Gao JX, Du HH, Zhang Y (2022) Efficient analysis method for the suitability of UUV seabed terrain matching navigation [J]. Ship Sci Technol 44(24):114–118

19. Han Y, Chen PY, Shen P (2020) AUV seabed terrain aided positioning method based on improved particle filter [J]. J Intell Syst 15(3):553–559
20. Zhou L Research on underwater acoustic communication simulation method and channel estimation for deep sea environment monitoring [D]. Ocean Univ China
21. He Y, Zhou TH, Chen WB (2016) Research on key technologies of blue-green laser communication between underwater and aerial platforms [J]. Sci Technol Inform
22. Xu TL, Liu F, Xiao YJ et al (2022) Foreign unmanned mine countermeasure equipment and technology development [J]. Acta Armamentarii 43(S2):64–70
23. Yang Z, Jing BR, Zhang P (2021) Dynamics of UUV automatic recovery manipulator based on launch tube [J]. Ship Sci Technol 43(11):163–168
24. Zhang W, Wang NX, Wei SL et al (2020) A review of the development status and key technologies of underwater unmanned underwater vehicle clusters [J]. J Harbin Eng Univ 41(02):289–297

Bisporus Mushroom 3D Reconstruction for Digital Twin of Smart Factory

Rui Jiang, Hongxia Cai, and Tao Yu

Abstract In recent years, digital twin technology has been gradually applied in the field of agriculture. As a key technology for digital twins in agriculture, 3D reconstruction technology has attracted more and more researchers' attention. 3D reconstruction of agriculture is difficult because crops are alive, especially for bisporus mushrooms. Bisporus mushrooms are generally grown intensively in narrow spaces in factories, which poses a huge challenge to the layout of the equipment for 3D reconstruction. But there are still some ways to solve the problem of 3D reconstruction of bisporus mushrooms in the factory, except for a small number of extruded mushrooms, most of the bisporus mushrooms can be replaced by regularly shaped geometry, the mushroom cap can be well adapted to the spatial ellipsoid and the space cylinder can express the characteristics of the mushroom stem well. This paper proposes a new method to reconstruction the bisporus mushrooms based on the YOLO v5s network with CBAM attention module. YOLO v5s network with CBAM attention module is used to generate coordinates of bounding boxes to segment the single bisporus mushroom point cloud from the whole point cloud obtained by depth camera, then the point cloud filter denoising algorithm is involved to remove noise spots and the ellipsoid fitting algorithm is used to reconstruct the model of bisporus mushroom caps. The results show that the proposed method can adapt well to bisporus mushroom plant environment in the factory, achieve high-precision fitting, and can be applied to the digital twin technology of bisporus mushroom smart factory in the future.

R. Jiang · H. Cai
Shanghai Key Laboratory of Intelligent Manufacturing and Robotics, School of Mechatronic Engineering and Automation, Shanghai University, Shanghai 200444, China

T. Yu (✉)
Shanghai University of Engineering Science, Shanghai 201620, China
e-mail: yutao@sues.edu.cn

S. Yadav et al. (eds.), *Proceedings of 3rd International Conference on Artificial Intelligence, Robotics, and Communication*, Lecture Notes in Electrical Engineering 1172, https://doi.org/10.1007/978-981-97-2200-6_5

1 Introduction

The concept of digital twin was first proposed by Professor Glieves of the University of Michigan in 2003, and NASA first established the model of digital twin in 2012 [1]. Digital twin is a simulation process that integrates multi-physical, multi-discipline, and multi-scale simulation by making full use of physical model information, sensor information updates, and operation history information. With the development of digital twin techniques, the modern industry brings a wide range of tasks online, including many aspects of manufacturing & assembly, R&D, and marketing and sales, all of which become intensively connected with each other [2]. This technology implicitly projects real products or systems into virtual space, reflecting the life cycle of physical equipment in real time. In recent years, digital twin technology has slowly been applied to agriculture, and the establishment of agricultural digital twin systems should simultaneously achieve the followings: (1) digital scene mapping driven by physical monitoring data in real time; (2) remote control to physical scenes by digital scenes; (3) decision-making in the planting process by collecting dynamic data in the crop life cycle. However, the application of digital twin technology in agriculture is still in its infancy. In recent years, the global edible fungus industry has developed rapidly, growing at a rate of 7–10% per year. And bisporus mushroom is one of the most popular edible mushrooms in the world. The bisporus mushroom factory is shown in Fig. 1.

Real-time construction of digital twins of bisporus mushrooms is an indispensable part of establishing the digital twin architecture of bisporus mushroom smart factory, and the 3D reconstruction technology of bisporus mushrooms in the smart factory has rarely been studied. Considering the cultivation process of bisporus mushrooms, they are generally growing densely, and the space between the trusses used to grow the bisporus mushroom is narrow. However, the shapes of bisporus mushrooms are

Fig. 1 Bisporus mushroom factory

generally similar, and the ellipsoid and cylinder in the mathematical concept can be used to describe appropriately. In this paper, the depth camera RealSense D435i is used to carry out the 3D reconstruction of the bisporus mushrooms in the factory, the point cloud information of the bisporus mushrooms is obtained by the depth camera. Due to the dense growth of the bisporus mushrooms, most of the traditional point cloud segmentation methods have failed. In this paper, YOLO v5s network with CBAM attention module is used to segment the single bisporus mushroom point cloud from the whole point cloud according to the coordinates of four vertices of the bounding box based on the detection of a single bisporus mushroom, then the 3D reconstruction of the densely grown bisporus mushroom is realized through point cloud outlier removal, point cloud shape completion. The main contributions of this paper are as follows:

1. By analyzing the network structure, optimizing parameters, optimizing loss function, a YOLO v5s network with CBAM attention module is proposed to detect bisporus mushrooms and construct 3D shape of mushroom caps.
2. Due to the dense growth of bisporus mushrooms and the failure of the traditional point cloud segmentation method, this paper uses the bounding box segmentation image in YOLO and image conversion point cloud to segment the point cloud.
3. The point cloud processing module only needs to process the point cloud in the bounding box, avoiding processing the entire point cloud, the speed is fast, and the accuracy is high.
4. A hypothesis is proposed that a bisporus mushroom cap can be described by the ellipsoid model. The ellipsoid fitting method is developed to reconstruct bisporus mushroom point clouds.

2 Related Work

2.1 3D Reconstruction of Crops

In the smart farm digital twin system, the 3D reconstruction of crops is a problem. Compared with industrial equipment, the shape of crops generally has no fixed rules to follow. In addition, crops are essentially living organisms, and their growth state is subject to change over time. Taking the bisporus mushroom as an example, its growth state is mainly divided into four stages: (1) the transition from mushroom loading to medium to mycelium growth to soil surface; (2) mycelium grows to the soil surface to the hyphae kinked; (3) mycelium kinks to grow into small mushrooms; (4) small mushroom to mushroom picking.

The main methods of 3D modeling of crops include two types: one is to simulate the growth process of crops according to certain rules, and at the same time establish models that do not exist in the real world, such as L-System [3], AMAP [4], etc. Although such methods can realistically simulate the 3D morphology of crop organs and individual crops, it is difficult to completely match the rules and entities abstracted by them, mainly because there are many factors influencing crop growth, and its growth process and state are not as controllable as industrial products, it is difficult to grow in the expected state; The other is to imitate real-world plants in real time, and target information is obtained by depth camera or 3D scanner before reconstruction of crop 3D model, the use of 3D scanners is more common, through 3D scanner, we can obtain accurate point cloud information. Although there are many methods to reconstruct crop 3D model by point cloud information, the high cost and spatial layout problems hinder the use of 3D scanner. At the same time, 3D point clouds usually lack spatial relationship information between points, so it is necessary to pay attention to which points are adjacent in physical space, which is the fundamental reason why 3D point cloud processing is more challenging than 2D image analysis based on computer vision. Hu et al. [5] used the camera to take complete plant pictures in different directions, and sparsely reconstructed them based on the SFM algorithm, then the multi-view stereo vision (MVS) algorithm is used to perform dense reconstruction according to the above results. The SFM algorithm and MVS algorithm were combined to obtain a dense point cloud of plants, and the reconstruction quality was evaluated. Wen et al. [6] used 3D scanning technology to study the 3D digital modeling and structure analysis technology of maize plant canopy.

In the industrial field, related companies have also developed some 3D reconstruction equipment of crops. The FASTACK 3D digitizer of Polhemus in the United States can directly measure and obtain 3D data of crops, but there are still some problems such as slow speed and low accuracy [7]. In summary, in the field of 3D imaging of crops, the research started very early, and various methods and theories are mature. From the early rule-based approach to more mature and applicable image-based method with the development of computer vision-related theories. However, effective 3D modeling methods for dense, shaded crops are lacking.

2.2 *Point Cloud Segmentation*

The traditional 3D point cloud segmentation methods include edge-based segmentation algorithm, area-based segmentation algorithm, attribute-based segmentation algorithm, and graph-segmentation-based segmentation algorithm, which can be applied to plant point cloud segmentation, but due to the complexity of plant point cloud morphology, specific improvements are often required to achieve better segmentation effects. At the same time, the morphological differences of different plants of the same variety are also very different, and lack of generalization performance of the above algorithm makes the segmentation effect unstable. In the research of plant 3D point cloud semantic segmentation, many methods have achieved better results in recent years. Wahabzada et al. [8] proposed an unsupervised clustering method to achieve automatic separation of plant organs, obtained a normalized histogram from point cloud data, and mapped it to a Euclidean space for measuring similarity. Spectral clustering is a clustering method based on graph theory, the subgraphs are as similar as possible and the distance between subgraphs are as far as possible to achieve clustering [9]. The steps of spectral clustering are as follows: First, connect the point clouds with edges to form a weighted undirected graph, then divide all points of the graph to make the edge weights between the subgraphs be as low as possible, and the edge weights within the subgraph be as high as possible. Santos et al. [10] used spectral clustering to segment plants and achieved better results.

In recent years, deep learning has developed by leaps and bounds with the growth of algorithm theory and hardware computing power. Deep learning algorithms solve many problems that are difficult to deal with effectively with traditional algorithms, especially with convolutional neural networks and fully convolutional neural networks. Although relevant problems in the field of biology are more complex, deep learning has also achieved good results. For example, Pound et al. [11] used CNNs for structural analysis of root systems. However, there are few relevant research and applications of deep learning in the analysis of 3D models of plants. One problem is that the relevant 3D datasets of plants are difficult to obtain, the complex structure makes the labeling of data labor-intensive, and there is currently a lack of available plant point cloud segmentation datasets, the mushroom point cloud dataset is even rarer. Another problem is that the data form of 3D point cloud as a non-Euclidean data has disorder and rotational invariance, and it is difficult for general convolutional neural networks to be applied to point cloud.

2.3 Point Cloud Completion

The collected point cloud data is usually incomplete with the lack of geometric and semantic information due to factors such as observation angle, occlusion, and measurement instability. Therefore, the completion of 3D point clouds has gradually become a very important topic in the field of computer vision. The following research methods are mainly used for point cloud shape completion: (1) Geometry-based method: This type of method uses the geometric information of the existing input point cloud to complete the completion of the overall shape without any external auxiliary data. Geometric-based methods include surface reconstruction and symmetry. The surface reconstruction method fills in locally mutilated point cloud holes by generating a smooth interpolation algorithm [12]. The symmetry method achieves the purpose of point cloud completion by identifying the axis of symmetry and repeating the rule structure in the input incomplete point cloud so that the point cloud in the non-incomplete area can be copied to the incomplete area. The advantage of such methods is simplicity and ease of implementation with simple algorithms. The disadvantages are also very obvious, first of all, the requirements for the input point cloud data are very high, when the incomplete area of the input point cloud is too large, the geometric shape of the missing area cannot be estimated, so that the missing area cannot be completed; Secondly, the generalization of such methods is poor, when the geometry of different point cloud data is different, it is necessary to build a completion algorithm for each missing geometry, so that the completion of large-scale point cloud data cannot be completed because the cost of the algorithm is very high. (2) Alignment-based method: The basic idea of this type of method is to match the incomplete input point cloud with the model in the point cloud shape database to complete the completion. (3) Learning-based methods: This type of method uses parametric models, such as machine learning models or deep learning models [13, 14], to map the input residual cloud into a complete point cloud. The basic idea is to build an "encoder-decoder" network structure, the encoder is mainly composed of convolutional layer and pooling layer, and its function is to extract feature information from the residual cloud of input; the decoder is mainly composed of a fully connected layer or an additional point grid, and its function is to decode the feature information extracted from the encoder, and finally generate a complete point cloud. The biggest advantage of the learning-based method is that it is much more generalized than the above methods.

Generally, the growth of bisporus mushrooms in the factory is dense, and even if many depth cameras are used for sampling from different angles, there are still dead angles that cannot be photographed, affecting the quality of the 3D reconstruction of the point cloud, so it is necessary to perform point cloud completion processing on the collected point cloud data. Considering that the shapes of bisporus mushrooms are generally similar, the bisporus mushroom cap is similar to ellipsoid, and the bisporus mushroom stem is similar to cylinder, and the shape fitting method is used to reconstruct the 3D model of the bisporus mushroom in this paper.

3 Materials and Methods

Figure 2 shows the overall flowchart of proposed method. First of all, the bisporus mushroom detection dataset is collected for training the YOLO detection model. The YOLO detection model outputs bounding boxes mapped on the point cloud. The image used to generate point cloud is cropped according to the bounding box information, preventing the use of the whole obtained point cloud and accelerating the algorithm. The point cloud in the bounding box then goes through the process of voxel down-sampling, outlier removal and ellipsoid fitting algorithm to obtain the bisporus mushroom cap model.

3.1 Data Acquisition

This paper takes bisporus mushrooms as the research object. The images used in this research are from Lianzhong edible mushrooms, Shanghai Province, China. Figure 3 shows the process that RGBD image obtained by depth camera transmitted by raspberry pi during this study. Deploy realsense2 and opencv2 libraries on Raspberry Pi 4B, and remotely control Raspberry Pi to transfer images to the cloud through

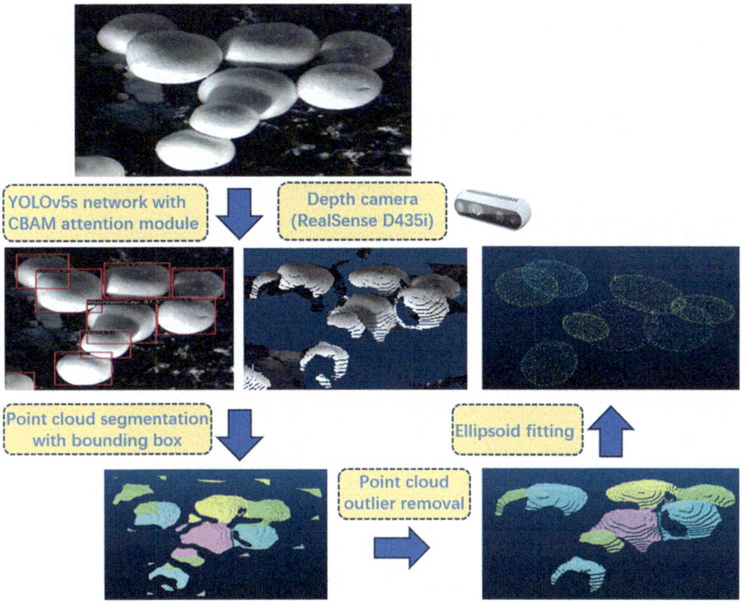

Fig. 2 Overall flowchart of proposed method

Teamviewer software to get RGB images and depth images in real time. There are total two types of datasets used in this study. Dataset 1 was photographed by the normal camera for training the YOLO detection network; Dataset 2 was composed of samples taken with RealSense D435i depth camera, a RGB image and a depth image are included in each sample, used to generate the point cloud of the bisporus mushrooms. The pixels of RGB images and depth images are both 1280 × 720, and the storage format is png.

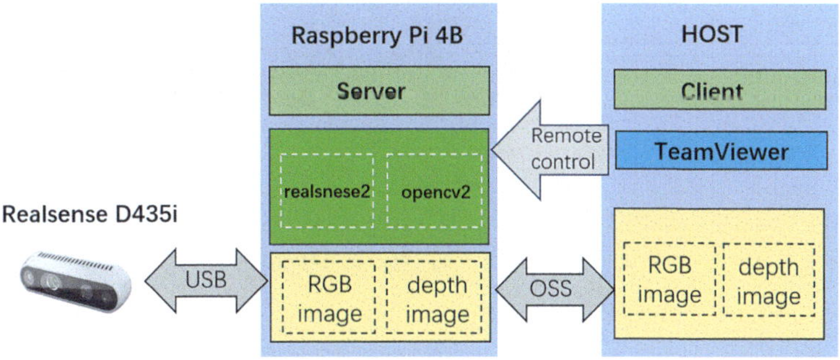

Fig. 3 Raspberry Pi transmits images

3.2 Point Cloud Acquisition

The point cloud is generated by RGB and depth images in this research. The relationship between depth images and point cloud can be calculated using RealSense D435i camera intrinsic parameters. As is shown in Fig. 4, supposing that p is the pixel of the mushroom image, and P is the corresponding 3D point in the real world, the relationship between p and P is shown in Eq. (1). And (x, y) and (u, v) are p's image coordinates and pixel coordinates.

Through Eq. (1), each pixel in the depth image is converted into a 3D point in the camera space. The specific process is as follows: Let $w = \{(u_i, v_i)\}_{i=1}^{N}$ be a set of pixel coordinates of a bounding box, where N is the total number of pixels. Further, w can be converted into a point cloud $X = \{(x_i, y_i, z_i)\}_{i=1}^{N}$:

$$\begin{cases} z_i = I_{\text{depth}}(u_i, v_i) \\ x_i = z_i(u_i - c_x)/f_x \\ y_i = z_i\left(v_i - c_y\right)/f_y \end{cases} \tag{1}$$

where I_{depth} refers to a depth image and $\left(c_x, c_y, f_x, f_y\right)$ are the intrinsic parameters of the depth image. $\left(c_x, c_y\right)$ is the pixel coordinate of the center of the image. $\left(f_x, f_y\right)$ is the focal length of the depth camera.

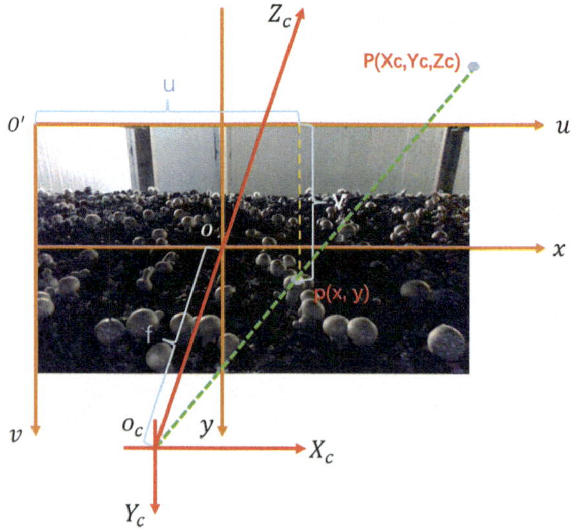

Fig. 4 Schematic diagram of point cloud synthesis

3.3 YOLOv5

A one-stage target detection algorithm YOLO was proposed by Redmon et al. [15] to solve the problem of the slow speed of the two-stage deep learning detection model. Due to its fast detection speed and simple network structure, YOLO has been widely used in the agricultural field since it was proposed. After continuous improvements, the weak detection of small objects and the low detection accuracy in YOLO have been greatly enhanced. The YOLO v5s model architecture is shown in Fig. 5. YOLO v5 is mainly enhanced by adding BottleneckCSP, PANet, Focus and SPP modules based on YOLO v3. YOLO v5 mainly consists of a neck, backbone and three detection heads. The neck is used to feature fusion. The backbone plays the role of extracting image features at different scales. And detection heads obtain three feature maps to predict small, medium, and large objects from large to small. These feature maps are segmented into grids. As for each grid, three prior anchors will be adopted to predict the target bounding box. In the end, every bounding box will output a feature vector, consisting of the predicted bounding box object confidence, coordinate offset and classification probability. Different from other versions, YOLO v5 uses an adaptive anchor strategy [16], the bounding box regression loss uses GIoU Loss, and the backbone part uses the Focus and C3 [17].

3.3.1 Attention Mechanism

The attention mechanism suppresses unnecessary features and improves the representation of important features in the feature map. And adding attention mechanisms can bring performance leaps to the network. This research added convolutional block attention module (CBAM) [18] to the YOLO network to obtain better results. Figure 6 shows the network architecture of CBAM attention mechanism modules. CBAM attention mechanism module contains two sub-modules, SAM (Spatial Attention Module) and CAM (Channel Attention Module), which respectively carry out attention on channels and spaces. This not only saves parameters and computing

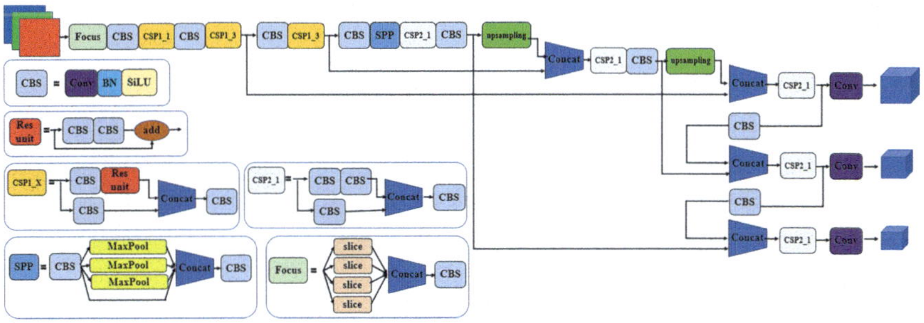

Fig. 5 YOLO v5s model architecture

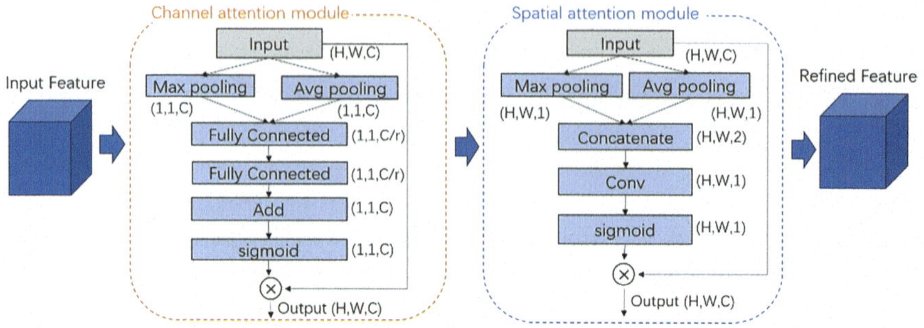

Fig. 6 Network architecture of CBAM attention modules

power, but also ensures that it can be integrated into existing network architectures as plug-and-play modules.

3.3.2 Evaluation Metrics

In order to verify the accuracy of YOLO v5s network with CBAM attention module in mushroom bounding box regression, this paper uses precision (Eq. (2)) and recall (Eq. (3)) as evaluation metrics.

$$\text{Precision} = \frac{\text{TP}}{\text{TP} + \text{FP}} \tag{2}$$

$$\text{Recall} = \frac{\text{TP}}{\text{TP} + \text{FN}} \tag{3}$$

where TP (true positive) is on behalf of the number of bisporus mushrooms which have been correctly detected, FP (false positive) is on behalf of the number of bisporus mushrooms which have been falsely detected, and FN (false negative) is on behalf of the number of missed bisporus mushrooms.

3.4 Bisporus Mushroom Cap Ellipsoid Fitting

The point cloud processing module consists of four parts: voxel downsampling, point cloud segmentation, removing outliers, and ellipsoid fitting. It will obtain the point cloud of the detected mushrooms and approximate 3D position of the mushrooms.

3.4.1 Point Cloud Segmentation

Input a color image to the YOLO v5s network with CBAM attention module to output the target bounding box. Map bounding boxes onto the point cloud. Segment the mushroom target from the point cloud depending on the bounding box.

3.4.2 Voxel Downsampling

The number of point clouds which need to be processed has increased with the improvement of the accuracy of point cloud acquisition instruments. Traditional point cloud algorithms cannot process the point cloud with a large number of point clouds in real time. A 3D voxel grid is created by voxel downsampling based on the original point cloud data. Then in each cube the centroid of all points is calculated to replace all points in the 3D grid, maintain the morphological characteristics of bisporus mushroom point clouds, decrease the number of point clouds, and improve the speed of the algorithm.

3.4.3 Remove Outliers

The algorithm first carries out the statistical analysis of the neighborhood of each point, and some outliers that do not meet the requirements are filtered out based on the distance distribution from the point to all other adjacent points. The algorithm performs totally two iterations: in the first iteration, it calculates the average distance from each point to the k nearest points, and the result conforms to the Gaussian distribution. Next, calculate the mean $\boldsymbol{\mu}$ and standard deviation $\boldsymbol{\sigma}$ of all these distances to determine the distance threshold d and $\boldsymbol{d} = \boldsymbol{\mu} + \boldsymbol{k\sigma}$, and k is a hyperparameter. In the next iteration, if the average neighborhood distance of these points is below or above the threshold, respectively, the points will be classified as interior or outliers.

3.4.4 Bisporus Mushroom Cap Ellipsoid Fitting

The biological morphology of the bisporus mushroom cap is ellipsoids, and in most cases, bisporus mushrooms are occluded over a large area, and the 3D reconstruction of the acquired point cloud data of bisporus mushrooms is not well performed. And in this paper the ellipsoid fitting of the bisporus mushroom cap is raised. First, the convex hull of the point cloud is obtained through the convexHull function in open3d, and then the vertices of the convex hull are used to fit the ellipsoid, which can effectively avoid the interference of abnormal point cloud data points that are difficult to separate out in the point cloud data, abnormal points may be caused by the accuracy of the depth camera itself or the interference of the environment. Then the optimized mushroom cap point cloud is fitted into an ellipsoid by least squares. The general representation equation for ellipsoids is:

$$f(x, y, z) = ax^2 + by^2 + cz^2 + dxy + exz + fyz + gx + hy + iz + j = 0 \quad (4)$$

where a, b, c, d, e, f, g, h, i, j are parameters of the ellipsoidal surface and can be obtained by the least squares method. The initial input data points (x, y, z) of the least squares method are the point cloud data points obtained in the steps above, the error function of the least squares method to measure the distance from the data point to the fitted ellipsoid is defined as follows:

$$(a, b, c, d, e, f, g, h, i, j) = \sum_{k=1}^{N} [f(x_k, y_k, z_k]^2 \quad (5)$$

where N is the numbers of point clouds. (x_k, y_k, z_k) is the coordinate of the kth point cloud. The goal of the least squares method is to find the values of the ellipsoidal parameters *a, b, c, d, e, f, g, h, i, j*, minimize the error function, and it is achieved by solving for the partial derivative of the parameter of the error function and making it zero, then a system of equations consisting of linear equations is derived. By solving this system of linear equations, the optimal solution of the parameters *a, b, c, d, e, f, g, h, i*, and *j* of the ellipsoid surface can be obtained, so as to obtain the fitted ellipsoid model.

4 Results and Discussion

Quantitative experiments were performed to evaluate the performance of the proposed reconstruction algorithm for mushroom cap. Each experiment was implemented on a computer equipped with 32 GB RAM, Intel i7 CPU, Quadro RTX 5000 GPU, and win32 operating system.

4.1 Evaluation of Bisporus Mushroom Cap Detection

The experimental results are shown in Table 1, the YOLO v5s network with CBAM attention module performs better than the original YOLO v5s network, the precision and recall of YOLO v5s network are 81.4% and 85.0%. As for YOLO v5s network with CBAM attention module, the precision and recall are 86.1% and 88.9%. And the Fig. 7 shows the training process of the YOLO v5s network and YOLO v5s network with CBAM attention module.

And Fig. 8 shows the detection result of the bisporus mushrooms, as we can see, almost all the bisporus mushroom caps are correctly detected, and because of the dense growth and occlusion of the bisporus mushrooms, the test results on the bisporus mushroom cap may not be satisfactory.

Table 1 Comparison of proposed network and YOLO v5s

Methods	Precision (%)	Recall (%)
YOLO v5s	81.4	85.0
YOLOv5s with CBAM attention	86.1	88.9

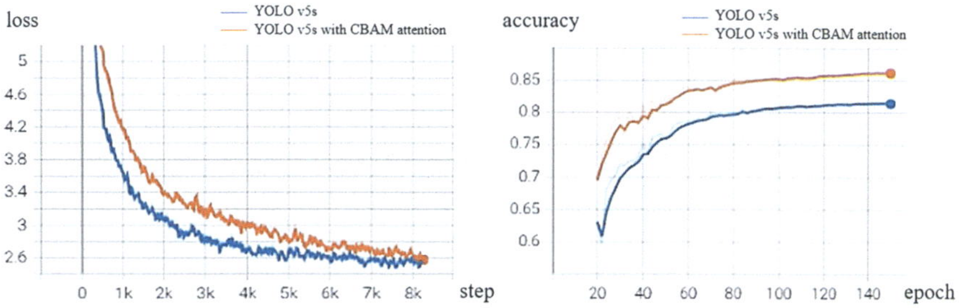

Fig. 7 Training process of bisporus mushroom cap detection

Fig. 8 Bisporus mushroom caps detection result

4.2 *Comparison of Point Cloud Segmentation Methods*

Figure 9 shows the results of traditional point cloud segmentation methods and the proposed method. The traditional point cloud segmentation methods include regional growth segmentation and k-means clustering. Compared with the traditional point cloud segmentation algorithm, the advantage of proposed method is that each point cloud area segmented according to bounding box can ensure that only one bisporus mushroom cap model will be fitted. And the point cloud segmented by the traditional segmentation algorithm is based on the difference in hyperparameter settings, and it is also necessary to determine whether the different categories divided belong to

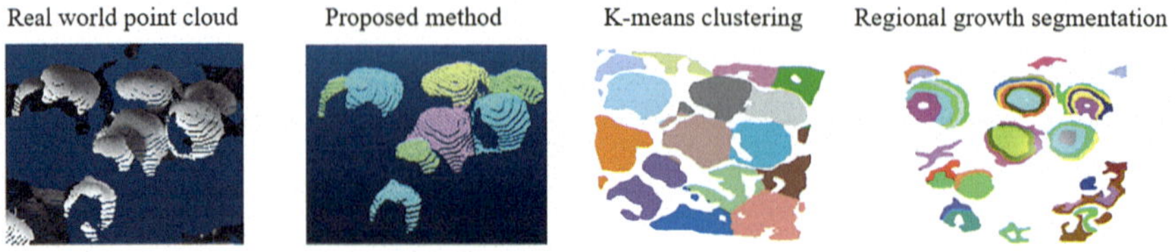

Fig. 9 Visualization of point cloud segmentation comparison with traditional methods

Table 2 Evaluation of bisporus mushroom cap reconstruction

Methods	MAP of IoU	Mean Angle Error
Proposed	90.6	11.7

the same bisporus mushroom, or different bisporus mushrooms, or the background of bisporus mushroom factory.

4.3 Evaluation of Bisporus Mushroom Cap Reconstruction

The bisporus mushrooms on a 1.5 × 1.5 m planting shelf in the bisporus mushroom factory are first modeled to obtain a real-world reference standard, and then the depth camera is used to shoot the bisporus mushrooms in the selected area, and the 3D reconstruction of the bisporus mushroom caps is carried out by the above mentioned method, and finally the reconstructed bisporus mushroom caps are mapped to the real-world coordinate system, and the accuracy of the proposed algorithm is judged by the mean average precision of IoU and the mean angle error. The first metric projected both the reconstructed ellipsoid and the real-world reference standard mushroom cap onto the image and compute the mean IoU between them, and the second metric calculated the mean tilt angle error between the reconstructed ellipsoid and the real-world reference standard mushroom cap. Table 2 shows the evaluation results.

5 Conclusion

In this paper, the 3D reconstruction method of bisporus mushroom caps in the factory is proposed to facilitate the application of digital twin technology in the production plant of bisporus mushrooms. However, the 3D reconstruction algorithm of mushroom caps proposed in this paper also has its own shortcomings, the fitted ellipsoids generally tend to be flattened, which is also a direction that needs to be optimized later.

The construction of the digital twin of bisporus mushrooms can provide the following functions: (1) Obtain the size of the bisporus mushrooms directly from the reconstructed 3D model, avoid the problem of low accuracy in judging the size of the bisporus mushrooms indirectly from the image, and can roughly judge the growth of the bisporus mushrooms, providing a reference for the regulation of the environment parameters such as air temperature and humidity, soil temperature and humidity, Carbon dioxide concentration, etc.; (2) Provide a reference for the trajectory planning of the bisporus mushroom picking robot, under normal circumstances, the bisporus mushroom picking robot only considers the possible volume collision on the plane when moving, and does not consider the distribution of bisporus mushrooms in space, which may cause damage to bisporus mushrooms during the picking process.

Acknowledgements This work was supported by Shanghai Committee of Science and Technology under Grant 22SQBS00800.

References

1. Glaessgen EH, Stargel DS (2012) The digital twin paradigm for future NASA and U.S. air force vehicles. In: Proceedings of the 53rd AIAA/ASME/ASCE/AHS/ASC structures, structural dynamics and materials conference 2012
2. Jiang Y, Yin S, Li K et al (2021) Industrial applications of digital twins. Philos Trans Roy Soc A: Math Phys Eng Sci 379:2207
3. Boudon F, Pradal C, Cokelaer T et al (2012) L-Py: an L-system simulation framework for modeling plant architecture development based on a dynamic language. Front Plant Sci 3
4. de Reffye P, Fourcaud T, Blaise F et al (1997) A functional model of tree growth and tree architecture. Silva Fennica 31(3):297–311
5. Hu P, Guo Y, Li B et al (2015) Three-dimensional reconstruction and its precision evaluation of plant architecture based on multiple view stereo method. Nongye Gongcheng Xuebao Trans Chin Soc Agric Eng 31(11):209–214
6. Paproki A, Fripp J, Salvado O et al (2011) Automated 3D segmentation and analysis of cotton plants. In: Proceedings of the 2011 International conference on digital image computing: techniques and applications (DICTA 2011)
7. Fang H, Hu L, He R et al (2012) Research on plant three-dimensional information acquisition method. Nongye Gongcheng Xuebao Trans Chin Soc Agric Eng 28(3):142–147
8. Wahabzada M, Paulus S, Kersting K et al (2015) Automated interpretation of 3D laser scanned point clouds for plant organ segmentation. BMC Bioinform 16
9. Ng AY, Jordan MI, Weiss Y (2002) On spectral clustering: analysis and an algorithm. In: Proceedings of the 15th annual conference on neural information processing systems (NIPS). Vancouver, Canada, Dec 03–08 2001.
10. Santos TT, Koenigkan LV, Barbedo JGA et al (2015) 3D Plant modeling: localization, mapping and segmentation for plant phenotyping using a single hand-held camera. In: Proceedings of the 13th European conference on computer vision (ECCV). Zurich, Switzerland, Sep 06–12, 2014
11. Pound MP, Atkinson JA, Townsend AJ et al (2017) Deep machine learning provides state-of-the-art performance in image-based plant phenotyping. Gigascience 6(10)
12. Retsinas G, Efthymiou N, Anagnostopoulou D et al (2023) Mushroom detection and three dimensional pose estimation from multi-view point clouds. Sensors 23(7)
13. Wen X, Xiang P, Han ZZ et al (2023) PMP-Net plus plus: point cloud completion by transformer-enhanced multi-step point moving paths. IEEE Trans Pattern Anal Mach Intell 45(1):852–867

14. Yuan W, Khot T, Held D et al (2018) PCN: point completion network. In: Proceedings of the 6th International conference on 3D vision (3DV). Verona, Italy, Sep 05–08, 2018
15. Redmon J, Divvala S, Girshick R et al (2016) You only look once: unified, real-time object detection. In: Proceedings of the 2016 IEEE conference on computer vision and pattern recognition (CVPR). Seattle, WA, Jun 27–30, 2016
16. Gao MY, Du YJ, Yang YX et al (2019) Adaptive anchor box mechanism to improve the accuracy in the object detection system. Multimedia Tools Appl 78(19):27383–27402
17. Wang CY, Liao HYM, Wu YH et al (2020) CSPNet: a new backbone that can enhance learning capability of CNN. In: Proceedings of the IEEE/CVF conference on computer vision and pattern recognition (CVPR). Electrical network, Jun 14–19, 2020
18. Woo SH, Park J, Lee JY et al (2018) CBAM: convolutional block attention module. In: Proceedings of the 15th European conference on computer vision (ECCV). Munich, Germany, Sep 08–14, 2018

Research on Automatic Music Generation Based on LSTM

Bei Xie and Yuan Zheng

Abstract The traditional way of music creation has a long cycle and requires the creator to have professional music knowledge. With the rapid development of artificial intelligence, the use of artificial intelligence algorithms to automatically generate music melodies can be used to assist the creation of musicians, providing musicians with new creative ideas, the use of computer understanding of music has become one of the hot spots of research. MIDI music files are converted music sequence information, with small size, easy to edit and other characteristics, very suitable for training using machine learning methods. Based on this feature, this paper is oriented to Musical Instrument Digital Interface (MIDI) music, combined with word vector technology and Long Short-Term Memory (LSTM) algorithm, constructed a deep learning model, trained the model, and realized the music automatic generation system, and the experimental results show that the music automatically generated by the system can basically satisfy the initial creation needs of the creators, and it has certain reference value.

1 Introduction

At the end of the twentieth century, the development of computers gave rise to a new era of machine music. Computers could be programmed to generate music, for example, using algorithms, mathematical models, and artificial intelligence techniques. Early experiments with computer music focused on music synthesis and sound processing, but over time, machines became involved in the process of composing music. Generative music is a form of music composition based on computer algorithms and artificial intelligence. These algorithms analyze and mimic the styles of human musicians to generate new musical compositions. Some early generative music systems include David Cope's Experiments in Musical Intelligence (EMI) [1] and Iamus [2].

B. Xie (✉) · Y. Zheng
Guangdong University of Science and Technology, Dongguan 523000, China
e-mail: 491053801@qq.com

S. Yadav et al. (eds.), *Proceedings of 3rd International Conference on Artificial Intelligence, Robotics, and Communication*, Lecture Notes in Electrical Engineering 1172, https://doi.org/10.1007/978-981-97-2200-6_6

In recent years, developments in deep learning and neural networks have driven new advances in machine music composition. Neural networks can learn large amounts of music data and generate musical compositions with a high degree of complexity and creativity. Some projects such as Google's Magenta and OpenAI's MuseNet utilize deep learning to create music. How to let computers learn and understand the "emotion" in music, and create information containing "emotion" through deep learning has also become a hot research topic in the field of artificial intelligence. Sun et al. [3] proposed a new music generation method for the symbolic music generation model mostly using recurrent neural network structure recursive generation, generation over time sequential processing, there is a lack of spatial correlation, poor parallelism, slow training speed, etc. The method effectively learns the music data, and generates music in line with the naturalness of the distribution of the notes of the human compositions, and the model's convergence speed is better than other models. Zhou [4] explores the effect of explicit structural encoding of musical structures by solving the constrained problem of melody generation based on two models, LSTM and WaveNet. Using a bunch of extended convolutional layers to encode the structure more explicitly can significantly improve the performance, and better results can be obtained by globally encoding the underlying chord levels during the generation process. Jun et al. [5] propose a method that combines the Transformer deep learning model with generative adversarial networks (GANs) to explore more competitive music generation algorithms. Compared with the problem that LSTM cannot handle long sequences of music, the model based on Transformer and GANs can well extract the relationships in the notes of long sequences of music samples and learn the laws of music composition. Meanwhile, the optimized transformer and GANs model has obvious advantages in terms of system complexity and accuracy of generated notes. Gullapalli et al. [6] utilized Bi-directional Long Short-Term Memory (Bi-LSTM) neural network with attentional mechanism to generate old-time music with recapitulated melodic structure, and the experiments showed that the Bi-LSTM with attentional was able to maintain the richness and technical subtlety of the music played. Cong et al. [7] propose a new generation network based on Transformer. The decoding block of the Transformer is used to learn the internal information of the music in a single track and the cross-track Transformer is used to learn the information between the tracks of different instruments. A music theory based reward network is proposed to optimize the global and local loss objective functions while training and discriminating the network so that the reward network can provide a reliable tuning method for network generation. A combination of reward network and cross-entropy loss is used to guide the training of the generator to produce high-quality music works.

The use of artificial intelligence techniques for music generation is mainly manifested by allowing computers to deeply learn and capture the characteristics of real music, and automatically generate new scores based on these characteristics. However, music is not just a pile of notes, notes not only need to follow the time gradually unfold, after the combination of each other need to produce melody to let the audience produce auditory enjoyment, but also bear the role of conveying the composer's emotions. MIDI is the most widely used standard format for music arrangement, which can be called "computer-understandable sheet music". In this paper, we construct note feature vectors according to the MIDI track extraction method, choose the long short-term memory network and note feature vectors to construct a music model, and design a music generation system according to the model.

2 Related Theory

2.1 *MIDI Music Files*

MIDI file is a standard digital music format, which is not a digital audio waveform, but a digital control signal or music code that encodes the elements of music, i.e., pitch, tone, timbre, and intensity, and expresses them in a coded form, which numerically represents the notes to record the music [8]. MIDI production music can produce virtually any kind of music at this stage of the game. MIDI technology allows computers to communicate with electronic musical instruments so that the music can be turned into instructions that the computer can understand, and in the communication are transmitted control commands that are needed for the computer to understand the music and for the computer to generate the music.

MIDI files, which are converted musical sequence messages, are characterized by their small size and ease of editing, making them ideal for training using machine learning methods.

2.2 *Neural Network*

RNN Network Architecture. The structure of RNN is shown in Fig. 1 [9].

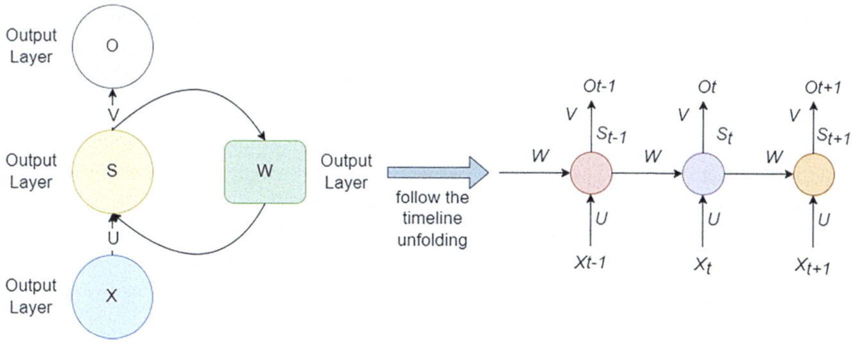

Fig. 1 Structure of RNN

The value of hidden layer is S_t and the output value is O_t after this network receives the input x_t at the moment t. And the value of S_t is related to x_t and S_{t-1} so the related formulas of calculating O_t and S_t are shown in (1) and (2):

$$O_t = g(V \cdot S_t) \tag{1}$$

$$S_t = f(U \cdot X_t + W \cdot S_{t-1}) \tag{2}$$

The fully connected layer of the network structure is the output layer, each node is connected to each other to form the hidden layer, f and g denote the activation function, and the cyclic layer is distinguished from the fully connected layer by the weight matrix W. The value of O_t is related to the x of the input at each time period t. This is the reason why the RNN is able to solve the time series problem.

LSTM Network Architecture. In application, RNN solves the problem of time series, but it does not perform well for long sequences, such as the problem of gradient disappearance and gradient explosion when dealing with long languages or texts. LSTM network was proposed by Hochreiter in 1997, and later improved and popularized by Alex Graves [10]. LSTM network changed the hidden layer based on RNN [11]. LSTM adds three stages to the recurrent layer of the RNN, namely the forgetting stage, the selection memory stage, and the input stage, which allows the RNN to solve the problem of gradient vanishing and gradient explosion during the training of long sequences.

The LSTM network preserves the chained form of the RNN which is spliced by the same modules, but the difference between the two network structures is that the LSTM changes the same modules by adding three important structures, namely, the forgetting gate, the input gate and the output gate, to the modules. The oblivion gate determines the information that is discarded by the module instead of keeping all the information, which saves storage space and is not affected by previously useless data at this stage. The input gate is updated based on the union of vectors generated by the activation function. The input gate combines the information that passes through the forgotten gate with the new information. The last important output gate. The value output from the output gate will ultimately determine the result. The output gate will root the sigmoid function to determine the output portion of the cell state, which is processed through the tanh layer, and the product of the two is the final information.

In summary, the internal structure of LSTM is to determine the information through three gates, which can remove the unwanted information and memorize the needed information. This is especially useful for many tasks that require "long-term memory".

3 Note Feature Vector Construction

3.1 MIDI Music Track Extraction

In the construction process of the note feature vector, the music generation data is the note information in the main melody track, so the first step needs to extract all the MIDI tracks, the MIDI file note start and end time and other information combined to form the track information matrix; and then extracted on the basis of the track information matrix, the process is mainly divided into three steps:

(1) Get all the tracks in the MIDI music;
(2) Extract the desired audio tracks;
(3) Generate the audio track file.

Eventually, the main melody track of the MIDI music is extracted. And now the music file is the file of the main melody track and no longer the same as the MIDI music file, the extracted main melody track file is used to study the music representation problem, the use of word vectors to represent the music features for subsequent music generation to lay a good foundation.

3.2 Note Feature Vector Generation Method

This note vector feature generation method is specifically used to jump word algorithm, its network structure as shown in Fig. 2, the feed-forward neural network structure has a four-layer structure, respectively, the input layer, the projection layer, the softmax layer and the output layer, the input layer input is a data set of notes, which is represented by an array of $\boldsymbol{x} = [x_1, \ldots, x_i, \ldots, x_n], \boldsymbol{y} = [y_1, \ldots, y_i, \ldots, y_n]$, is the predicted output value of the output layer output, the value is obtained by the input layer input, projection layer judgment and the softmax layer operation and obtained. y_n is the predicted output value output from the output layer, which is obtained from the input layer input, projection layer judgment and softmax layer operation. Using the dimension 129 parameter to the input layer, the projection layer will output a K*129 dimensional matrix, the K represents the number of different notes in the input, the matrix is processed by the softmax function layer to turn the multidimensional notes into the corresponding note representation array, which is based on the most probable dimensions in the multidimensional vectors to be combined, and also facilitates the subsequent automatic generation of the modeling algorithms.

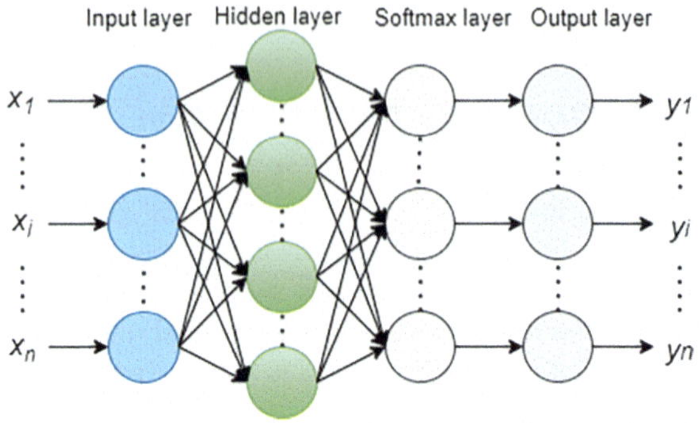

Fig. 2 Network of note vector feature generation

4 Model Design

4.1 Model Structure

The music generation model can be divided into three layers on the large module, namely, as input layer, hidden layer and output layer. The input and output note sequences of the model must be greater than 1 and of equal length, e.g., if the input is a note sequence of length s ($s > 1$), then the length of the output note sequence must also be s. The input data required in the input layer are the main melody note sequences represented as note feature vectors.

The input data required for the input layer is the main melody note sequence is represented as a note feature vector. The input data is accomplished by first taking the acquisition of the MIDI main melody note sequences and then generating the note feature vectors from the main melody note sequences based on the word vector model. The final input data is a kind of main melody note sequence with contextual information and represented as note feature vectors, which enhances the expression of note features in music generation and improves the accuracy and effectiveness of the music generation model.

The hidden layer is the implementation of the LSTM network, the model in this paper adds dropout operation in LSTM network, LSTM network stores the important dependency information in the input note sequence, calculates the influence of the note on the current moment state from the note feature vector, and then gets the corresponding weight probability distribution of the LSTM network, and finally obtains the vector of the note sequence and inputs the vector to the output layer through the fully connected layer.

The output layer uses a softmax function, which maps the output of multiple neurons, into the (0, 1) interval, which can be viewed as a probability to understand. Thus the output of the hidden layer is mapped to the note numbers to get the probability distribution of the note numbers, and the final predicted note is the note with the highest probability among them.

4.2 *Training Process*

The training process of the whole music generation model is shown in Fig. 3.

Music generation model training first to carry out the acquisition of data, the acquisition of data from the Internet to find down more than 500 MIDI music. The algorithm is used to extract all the tracks, and then the comparison method is used to obtain the track with the highest number of notes as the MIDI main melody sequence. The final sequence of note feature vectors input to the model is the sequence calculated by the model after inputting the MIDI main melody sequence into the word vector model, which represents the note characteristics of a piece of music.

After obtaining the dataset, the data will be randomly divided into two parts, one is 70% as the training set, and the other is 30% as the test set, and then the network will learn the features of the 70% training set, and in the process of feature learning of the neural network, each iteration will use the 30% test set as a validation, to test the accuracy of the model as well as the effect of the model, that is, the accuracy of model prediction of notes from the current test set. The model accuracy and effect, i.e., the accuracy of the model prediction for the current test set of notes, are tested to improve the accuracy of the model and to make the music generated by the model more realistic and pleasant.

The rules of the training process are as follows: the LSTM network receives a vector representation of the processed note sequences described above, and outputs the next most probable occurrence of a musical phrase as judged by the network. After each iteration, 30% of the test set data as the reference of the output music segment, compare their note sequences, calculated by the cross-entropy function, to get the error between them, the neural network and then through the back-propagation algorithm to update the weight parameters in the network structure, repeat the above feature learning process until the model convergence, and ultimately get the music generative model that can predict the notes.

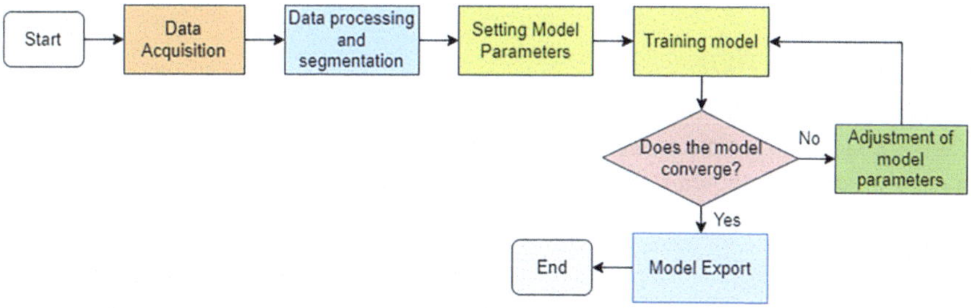

Fig. 3 Model training flowchart

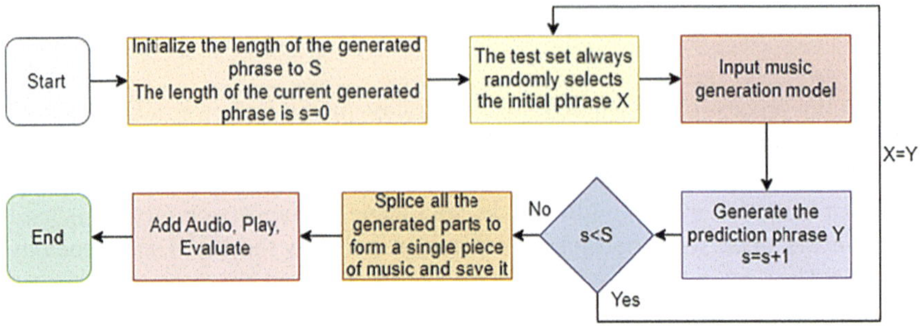

Fig. 4 Flowchart of music production

4.3 Music Generation Process

The model in the system process, you can input note sequences, can also be randomly selected in the data set, and then by the model to determine the probability of outputting a section of the input sequence and the input sequence of note sequences of equal length, if the input note sequence length of *S*, the length of the output section of the music is also *S*, and then the output of the note sequence again as an input to predict the next section of note sequences, the cycle of the process until the note sequence length reaches a threshold, and finally, the output section length reaches a threshold, and then the output section length reaches a threshold. The above process is repeated until the length of the note sequence reaches the threshold value, and finally the system outputs the whole music sequence, and the generation process of a piece of music is finished. The whole music production process is shown in Fig. 4.

5 System Design and Result Analysis

5.1 System Design

In accordance with the different users of the music base, this music generation system process structure is shown in Fig. 5.

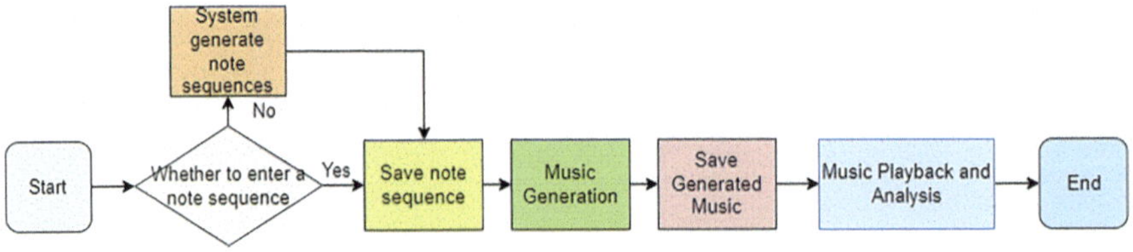

Fig. 5 System design flow structure

5.2 Result Analysis

In the control module by the page input note sequence, input C, D, E, F notes. When the system gets the input notes, it will turn the input notes into an array note = ['C', 'D', 'E', 'F']. The array back to the display will be in the middle of the note plus the arrow symbol, when the music generation, the data will be converted to the corresponding vector array, the system music generation, the system generates music with a certain degree of randomness in the input of the same array of the result of the music may also be different, after the model generates corresponding to the sheet music as shown in Fig. 6.

If you don't enter notes on the Input Notes page, you will default to letting the system automatically select notes to compose, and when automatically generating music, the system will randomly generate music that doesn't output exactly the same sequence of notes every time you don't enter notes. Without input from the system, the result of the music score is shown in Fig. 7.

Fig. 6 Partial score map for custom generated music

Fig. 7 System automatically generates a partial score map of the music

In the music generation transaction, the note sequence is given to the music generation transaction for note feature vector transformation, the note sequence is transformed into the corresponding note sequence vector, and the note sequence vector is input into the music generation model for the corresponding music. After the music is completed save operation is performed on the music file data. Through the analysis of the generated music, the music has a certain melody and comfort and can provide a reference for music creation.

6 Conclusion

The development of deep learning and neural networks makes machine participation in composing a reality. The thesis based on the LSTM model has the characteristics of long-term memory, for MIDI music generation system design, so that people who do not know how to create music can customize the music works, who write music can get inspiration from it, and have a certain reference value for the computer intelligent music creation.

References

1. Cope D (1989) Experiments in musical intelligence (EMI): non-linear linguistic-based composition. J New Music Res 18(1–2):117–139
2. Vico FJ, Alabau V (2013) Iamus: Autonomous composition. Artif Intell Eng Des Anal Manuf 27(3):189–202
3. Sun FX, Sun RC (2022) A symbolic music generation algorithm based on WGAN and dilated convolution. J Fudan (Nat Sci Ed) 61(05):536–545
4. Zhou L (2022) Automatic music generation based on deep neural networks. Inf Technol 05:53–61
5. Jun M, Zhao QL, Lei W et al (2022) Music generation system for adversarial training based on deep learning. Processes 10(12)
6. Gullapalli K, Vaishnavi AN, Prerana M et al (2022) Attentional networks for music generation. Multimedia Tools Appl 81(4)
7. Cong J, Tao W, Xiao BL et al (2021) A transformer generative adversarial network for multi-track music generation. CAAI Trans Intell Technol 7(3)
8. Gu Y (2018) Research on MIDI sound source and its application. Popular Lit Art 03:139–140
9. Wang S Y, Zhou J G. (2019) A char RNN-based polyphonic music generation method. Comput Eng 45(05):249–255+260
10. Xin MZ (2019) Automatic music generation based on LSTM. Electron Prod (10):41–42+62
11. Ma PC, Lu SQ, Wang XB, Li XD, Song HY (2020) Music creation based on recursive neural network artificial intelligence technology. Electron Technol Softw Eng 03:176–180

Specific Emitter Identification Method Based on Aggregation Model with Metric Learning in Complex Field

Siyang Xu, Ruokun Wang, Qinghao Zhu, and Xiaodong Wu

Abstract In the field of Specific Emitter Identification (SEI), due to the time-varying characteristics of the ionosphere, the radio short-wave communication channel is a complex variable parameter channel, and there are multi-path fading, polarization rotation, Doppler shift and other phenomena, resulting in the short-wave radiation source individual fingerprint features are covered by the channel features, and the recognition algorithm based on a single machine learning model or deep learning model is often difficult to achieve good recognition results. To solve this problem, based on the idea of metric learning and ensemble learning, this paper proposes an aggregation model of metric complex value CNN (AM-MCVCNN) for SEI and tests it on 2G-ALE signal. We generate multiple metric learning models, which can be used for SEI, and use a sort cumulative aggregation method (SCAM) to generate aggregation model. The experimental results show that the recognition accuracy of AM-MCVCNN can reach 89%, compared with the single model, the recognition accuracy is improved by 10–15%. Testing AM-MCVCNN on the data after one week, AM-MCVCNN still has a high accuracy (89.56%). For the data after two weeks, the accuracy of AM-MCVCNN decreased slightly (81.23%). For data collected after four months, some individuals still have a high identification accuracy.

1 Introduction

Specific emitter identification (SEI) is a technique to identify emitter by detecting fingerprint features in the signal. The fingerprint features of the emitter are generated by their own hardware defects, which is manifested in that the transmitted radio signal must have a certain degree of distortion. The distortion of the signal transmitted

S. Xu (✉) · R. Wang · Q. Zhu · X. Wu
China Shipbuilding Industry Corporation No. 722 Institute, Wuhan 430200, China
e-mail: 13662138985@139.com

R. Wang
e-mail: wrk1908@163.com

S. Yadav et al. (eds.), *Proceedings of 3rd International Conference on Artificial Intelligence, Robotics, and Communication*, Lecture Notes in Electrical Engineering 1172, https://doi.org/10.1007/978-981-97-2200-6_7

by different emitters is not the same. Therefore, each signal contains the unique fingerprint features of the emitter.

Since the specific emitter identification was proposed by the Naval Research Laboratory (NRL) in 1977, many researchers have carried out extensive and in-depth research on it. Amani et al. [1] found that wireless channels significantly affect classification accuracy, that is, the classification accuracy of wireless channels in experimental data sets and DARPA data sets is 85–9% and 30–17% respectively. Tong et al. [2] used a one-dimensional ResNet50 network to classify IQ signals of Wi-Fi and ADS-B, and expanded the number of devices to 1000. Trevor et al. [3] used an improved multiple discriminant analysis (MDA) fusion classifier to achieve device classification, and the results showed that for ZigBee signals, the method based on multiple discriminant analysis fusion had better robustness compared with similar methods. Kevin et al. [4] developed a framework for training convolutional neural networks using time-domain complex baseband error signals, using Cognitive Communication Networks (CCN), and demonstrated recognition accuracy of over 90% in seven 2.4 GHz commercial ZigBee devices. Ma [5] proposed an optimal combination algorithm based on local mean decomposition (LMD), two schemes for radio frequency fingerprint extraction based on bispectrum and two methods based on deep canonical correlation analysis (DCCA), and the recognition accuracy of a single walkie-talkie reached more than 90%. Wang [6] proposed an RF fingerprint recognition method using bispectrum-based method using energy entropy and color moments (BEECM) as RF fingerprint features and support vector machine as a classifier. Simulation proved that when the signal-to-noise ratio was 0, the accuracy of RF fingerprint recognition was close to 80%. With the increase of the signal-to-noise ratio, the classification accuracy would be improved. Qing et al. [7] designed a specific convolutional neural network (CNN) with a short-time Fourier transform (STFT) preprocessing module that can efficiently identify differences in feature graphs learned from signals from various RF devices. Yu et al. [8] focus on the low-lens SEI (FS-SEI) for aircraft identification based on automatic Dependent Surveillance Broadcast (ADS-B) signals, and propose a FS-SEI method based on depth metric ensemble learning (DMEL). Based on the metric learning of complex valued convolutional neural networks (CVCNN), the discriminant features with compact intra-class distance and separable inter-class distance are extracted.

At present, individual recognition of radiation sources is more applied to ADS-B, Bluetooth signals or intercom signals, mainly for network security protection. However, there are still few practical applications for SEI of radio short wave signals transmitted by sky wave, which plays an important role in the field of electronic countermeasures.

2 2G-ALE Signal and Data Processing

Second Generation Automatic Link Establishment (2G-ALE) is a shortwave automatic link establishment technology based on MIL-STD-188-141A. 2G-ALE is one of the most widely used radio communication protocols in the world.

Due to the advantages of wide use, easy data acquisition, and accurately identify emitter information, we select 2G-ALE signal as the training data.

The 2G-ALE data used in this paper was collected by our research group (without using any public dataset) from February 27, 2023 to March 24, 2023, with a sampling rate of 9600 Hz. The emitter to which each sample belonged to was determined by call sign. Considering the practical application requirements, we do not make clear requirements on SNR, and all the signals that can be demodulated and decoded will be used as training data to improve the universality of the model. Signals are processed to ensure as far as possible only signals are retained:

1. Using double sliding window algorithm to obtain the start point and end point of the signal.
2. Frequency offset correction, use peak detection algorithm to determine frequency offset, and move the center frequency to 0 Hz.
3. Filter to remove out-of-band noise, taking into account the error of frequency offset estimation, filter bandwidth is set to 2200 Hz.

Due to the wide variety of emitter in the collected data and the huge difference in the number of samples from different emitters, it is necessary to carry out data cleaning. The cleaning method is to delete the data of emitters whose sample size is less than 500, and use the remaining data to construct the dataset. After data cleaning, the dataset contains a total of 16,209 samples, including 21 emitters.

The duration of 2G-ALE signals varies greatly, and the neural network requires the same length of input data, so it is impossible to use the complete 2G-ALE signal as the input. Considering that the shortest length of samples in dataset is about 1.5 s, fragments with a duration of 1.5 s from the start point detected by double-slide window algorithm are selected for interception (see Fig. 1).

At the same time, this fragment contains transient characteristics, steady-state characteristics and the process of gradual transition from transient characteristics to steady-state characteristics of the individual radiation source. The intercepted fragment covers the fingerprint features of the emitter to the greatest extent.

The number of samples of these 21 types of emitters varies greatly among individuals, and there is an obvious imbalance in the number of samples.

The unbalanced number of samples will obviously affect the tendency of the model, that is, the model pays more attention to the categories with a large number of samples. There are many methods to eliminate sample imbalance, such as Synthetic Minority Over-sampling Technique (SMOTE) [9], changing data weights, and combinatorial ensemble learning. The model proposed in this paper combines the two methods of deep metric learning and aggregation model and needs to train multiple submodels, each of which has little demand for the number of samples.

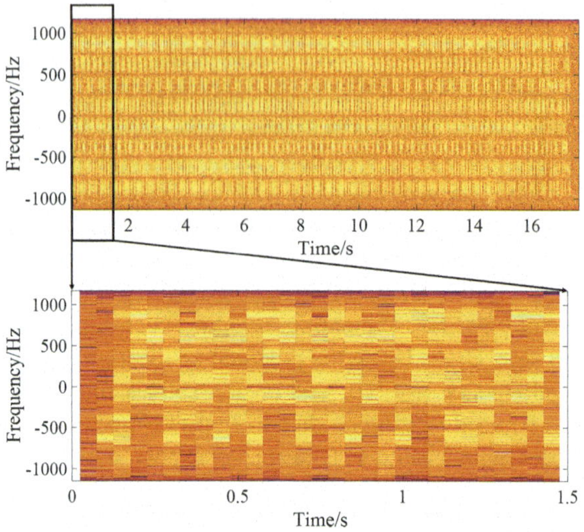

Fig. 1 An example of a signal spectrum and its corresponding intercepted signal fragment in 2G-ALE dataset after double-slide window algorithm, frequency offset correction and filtering. The 2G-ALE data used in this paper was collected by our research group

Therefore, SMOTE is chosen to avoid the negative impact of sample imbalance. The specific processing method of SMOTE is: randomly select a certain number of samples for each category as the training set, and use the remaining samples as the verification set to train the model. Repeat this process several times to train multiple models. The obtained models are combined, and the polymerization method is generally used to obtain an aggregation model that eliminates the influence of unbalanced sample number.

To sum up, the generation steps of 2G-ALE data train dataset, valid dataset and test dataset are as follows:

1. 1100 samples of each type are randomly selected as the test dataset (named D_{test}), and this part of data will not participate in the subsequent training.
2. Before each submodel training, 400 samples were randomly selected from the remaining samples and divided into train dataset (named D_{train}) and valid dataset (named D_{valid}) according to the ratio of 3:1.
3. Repeat step 2 until all submodels are trained.

3 Model Construction

Aggregation model is a large model (also called strong classifier) formed by integrating multiple submodels (also called weak classifier) based on small sample measurement learning methods. The block diagram of aggregation is mainly divided into two steps (See Fig. 2):

First, build multiple submodels based on deep metric learning; second, aggregate multiple submodels to form an aggregation model of metric CVCNN (AM-MCVCNN).

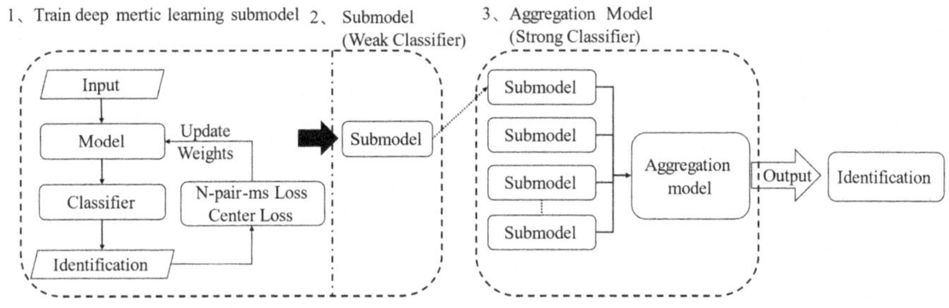

Fig. 2 Flowchart for generating aggregation model

3.1 Submodel

The submodel is generated by deep metric learning. The goal of measurement learning is to make the features of the same kind of targets gather as much as possible and the features of different types of targets disperse as much as possible, that is, to highlight the differentiation of feature vectors in vector space, and to achieve a generalized and effective classification model with only a small number of samples.

For the task of SEI, the fingerprint features of emitters are hidden in IQ data which is very subtle, and will be interfered by channel features in the propagation process [12].

In this paper, the feature extraction process of depth metric learning is completed by complex convolutional neural network (CVCNN). Complex arithmetic is more complex than real arithmetic, and neural networks based on real numbers cannot process IQ data well and is hard to extract fingerprint features [10]. The commonly used functions to measure learning loss are Contrastive loss, Triple loss, n-pin-MS loss, Center loss, etc. This paper introduces two loss functions, center loss and N-pair-ms loss, to realize deep metric learning.

The N-pair-ms loss [11] defines an anchor point so that the distance between the positive sample and the anchor point is minimized. The distance between all negative samples and anchor points should be enlarged as far as possible, and the distance should be larger than the preset parameter margin, so as to achieve the purpose of expanding the feature distance between classes. The Center loss [12] sets a class center for each class. By minimizing the distance between the feature in the class and the center point, the distance between the feature in the class is reduced.

The training process was performed with the N-way-K-shot task. Different combinations of hyperparameters are selected for multiple tests, and the model complexity is smaller and the performance is better. The submodel hyperparameters are set as follows: the number of convolutional layers is 8, the number of convolutional cores is 16, and the size of convolutional cores is 4. Finally, a total of 6 submodels are generated.

All submodels adopt same model structure. Due to the different training data, the fingerprint features extracted by different submodels are also different (see Fig. 3).

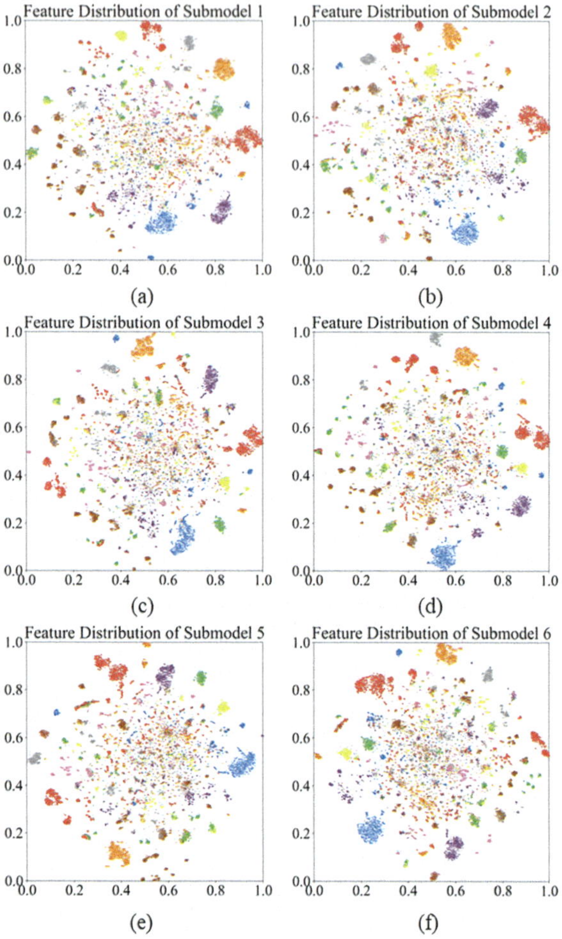

Fig. 3 Distribution of features extracted by six submodels, **a** for submodel 1, **b** for submodel 2, **c** for submodel 3, **d** for submodel 4, **e** for submodel 5, **f** for submodel 6

T-SNE method was used to reduce the dimensionality of the extracted features. The features extracted from the six submodels are obviously different, but the dimensionality reduction projections of category features are locally similar. This result is in line with expectations, that is, all six models can extract fingerprint features, but there are differences in the expression of fingerprint features extracted by different models, and these differences directly affect the distribution of feature vectors in hyperspace. It is by integrating these different features that the aggregation model improves the comprehensiveness of fingerprint features.

The accuracy of the six submodels on D_{train}, D_{valid} and D_{test} is listed in Table 1, sorted from high to low accuracy of D_{test}, and named as submodel 1–6, respectively:

Table 1 Accuracy of six submodels on D_{train}, D_{valid} and D_{test}

Model name	Accuracy of D_{train}	Accuracy of D_{valid}	Accuracy of D_{test}
Submodel 1	0.8339	0.8262	0.8085
Submodel 2	0.8065	0.7930	0.7751
Submodel 3	0.8216	0.8306	0.7699
Submodel 4	0.7979	0.8027	0.7623
Submodel 5	0.8012	0.7922	0.7566
Submodel 6	0.8082	0.8022	0.7441

3.2 Aggregation Model

The classification probability results of several submodels are mixed in the sum average way to obtain the final aggregation model results. The calculation formula is as Eq. (1):

$$R = \arg\max\left(\frac{1}{N}\sum_{i=1}^{N} P_i\right) \tag{1}$$

where P_i representing the classification probability of submodel *i*, *N* representing the number of submodels, argmax() returns the most likely category, *R* is the output result, that is, the category with the greatest probability.

In order to obtain the optimal aggregation model efficiently, a sort cumulative aggregation method (SCAM) is proposed in this paper. Before model aggregation, the trained submodels need to be sorted according to accuracy of D_{test} from high to low. During model aggregation, submodels with better performance on D_{test} are preferentially selected. By increasing the number of submodels aggregated, the accuracy of the aggregate model on D_{test} changes with the increase of the number of models, and the optimal number of submodels is determined according to the change curve (see Fig. 4).

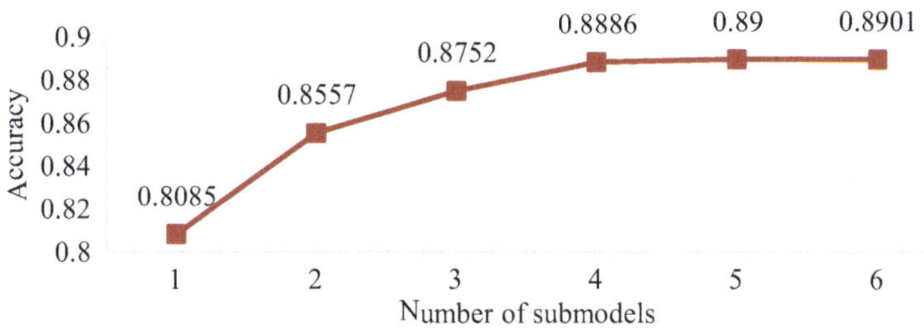

Fig. 4 Accuracy of AM-MCVCNN on D_{test} varies with the number of submodels

When the number of submodels is 5, the accuracy of the aggregation model is gradually stable, and the performance improvement is not obvious when the number of submodels is greater than 5. Setting the number of submodels to 5 is the most appropriate.

4 Experimental Results and Analysis

2G-ALE data, which collected from March 25, 2023 to March 31, 2023 (named D_{t1}), from April 1, 2023 to April 7, 2023 (named D_{t2}) and from August 8, 2023 to August 11, 2023 (named D_{t3}), was used to test the trained AM-MCVCNN model. The statistics are shown in Fig. 5.

Figures 6 and 7 shows the confusion matrix of AM-MCVCNN on D_{t1} and D_{t2}.The accuracy of the model on D_{t1} and D_{t2} was 89.56% and 81.23%, respectively. The model still maintains a high recognition accuracy on dataset D_{t1}, but for dataset D_{t2}, the recognition accuracy is significantly reduced.

For D_{t3}, only 9 types of emitters have relatively good recognition effect, and the overall recognition accuracy of these 9 types of individuals is 64.6% (see Table 2). Compared with the test samples from February to March, the accuracy rate is generally decreased by about 20–30%, while the remaining 7 types of individuals are completely unrecognizable.

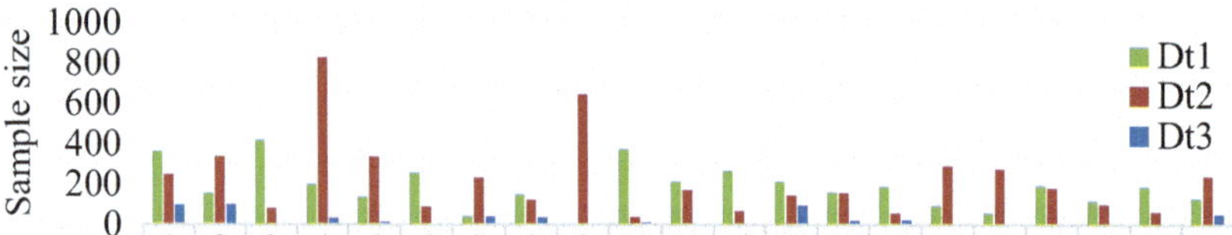

	1	2	3	4	5	6	7	8	9	10	11	12	13	14	15	16	17	18	19	20	21
Dt1	362	153	419	203	140	259	45	151	4	376	215	269	214	160	190	96	61	196	120	191	132
Dt2	249	335	81	827	335	89	231	123	643	37	170	68	144	157	58	290	276	181	101	65	239
Dt3	96	100	0	32	14	0	39	35	7	11	2	0	96	20	23	1	2	2	0	0	54

Sample size of 21 emitters on $Dt1$, $Dt2$ and $Dt3$

Fig. 5 Number of samples from each emitter in D_{t1}, D_{t2} and D_{t3}. Dt1 represents D_{t1}, Dt2 represents D_{t2}, Dt3 represents D_{t3}. Below the horizontal axis are listed the sample size of each emitter in D_{t1}, D_{t2} and D_{t3}, respectively

Fig. 6 Confusion matrix of AM-MCVCNN on D_{t1}

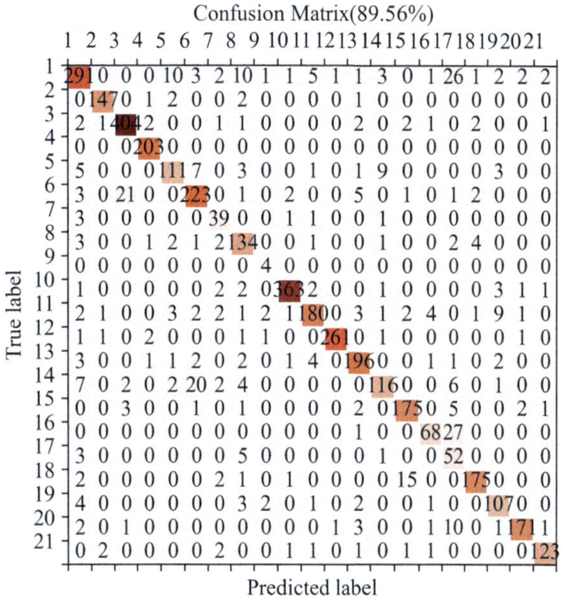

Fig. 7 Confusion matrix of AM-MCVCNN on D_{t2}

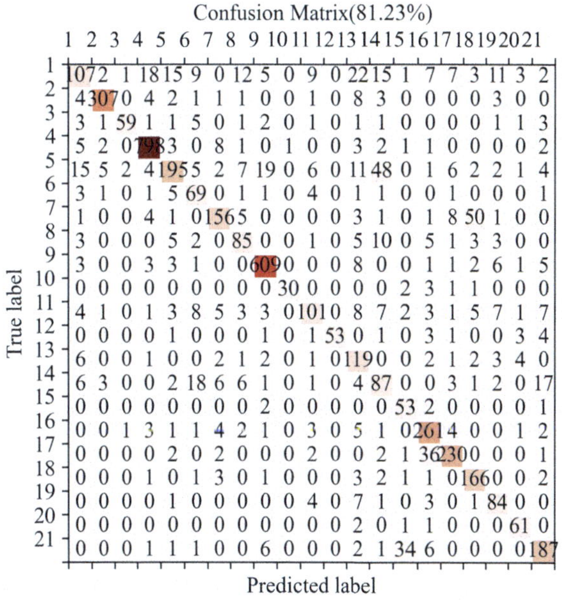

Table 2 Accuracy of 9 categories in D_{test} and D_{t3}

Category	D_{t3}	D_{test}
1	0.760	0.970
2	0.500	0.736
5	0.714	0.942
8	0.686	0.937
11	0.500	0.983
13	0.677	0.893
14	0.600	0.898
16	1.00	0.911
21	0.630	0.917

5 Conclusion and Deficiency

The training process of the aggregation model proposed in this paper is easier, the required sample size is relatively small, and the final recognition accuracy is obviously better than that of a single complex model. It is a reliable solution for individual identification of radio short-wave radiation sources, and has a good application prospect.

However, in practical applications, it is difficult to obtain a large number of effective data, which is an important reason restricting the application of deep learning in the identification of individual radiation sources. In this paper, although a large amount of 2G-ALE data is actually collected, considering the requirements of deep learning for sample size and sample balance, the actual available data is still limited, and the existing data volume does not support the recognition accuracy of the test model under different signal-to-noise ratio and different number of categories. The characteristics of the model performance with the change of SNR and the number of classes still need to be further explored.

In addition, although the number of samples used in this paper has been reduced, in some special scenarios, it is very difficult to obtain hundreds of samples of a certain radiation source. How to reduce the demand for sample size of the model is still one of the key problems for specific emitter identification.

References

1. Al-Shawabka A, Restuccia F, D'Oro S, Jian T, Melodia T (2020) Exposing the fingerprint: dissecting the impact of the wireless channel on radio fingerprinting. In: IEEE INFOCOM 2020—IEEE conference on computer communications. Toronto, pp 646–655. https://doi.org/10.1109/INFOCOM41043.2020.9155259
2. Jian T, Rendon BC, Ojuba E, Soltani N, Ioannidis S (2020) Deep learning for RF fingerprinting: a massive experimental study. IEEE Internet Things Mag 3(1):50–57. https://doi.org/10.1109/IOTM.0001.1900065

3. Bihl TJ, Bauer KW, Temple MA (2016) Feature selection for RF fingerprinting with multiple discriminant analysis and using ZigBee device emissions. IEEE Trans Inf Forensics Secur 11(8):1862–1874. https://doi.org/10.1109/TIFS.2016.2561902
4. Merchant K, Revay S, Stantchev G, Nousain B (2018) Deep learning for RF device fingerprinting in cognitive communication networks. IEEE J Sel Top Signal Process 12(1):160–167. https://doi.org/10.1109/JSTSP.2018.2796446
5. Ma S (2020) Fingerprint feature extraction and recognition of wireless communication equipment based on multi-feature fusion. https://kns.cnki.net/kcms2/article/abstract?v=7P_nOixU6lXCKo8QfuTqJ_5fTsKdM06aYbYwr2AHepjb23IO9FrK7mrtDOr0hGXvlpubHJMw9vpdYhfZgztayQOC7gTpeGrrFBtBoUPPkY5Ucojly2ZvQZbDwmL3203vHzvG40T8HyQ=&uniplatform=NZKPT&language=CHS
6. Wang X (2018) Radio frequency fingerprinting identification method research based on steady state signals. https://kns.cnki.net/kcms2/article/abstract?v=7P_nOixU6lUARGEsXov4YwgdkU2QSigi1AmWnumS0YS6eUEN5CqVBhKwtSYSU0UtqmOoPeeuf17guxyyOF9a7EE9ow3xszEWNeR2oYWVuhQs-5J2zTad-SKKb_lCCUAUEU83D_iO5wI=&uniplatform=NZKPT&language=CHS
7. Wang Q, Liu Q, Zhang Z, Fang H, Zheng X (2021) Deep open set identification for RF devices. https://doi.org/10.48550/arXiv.2112.02536
8. Wang Y, Gui G, Lin Y, Wu HC, Yuen C, Adachi F (2022) Few-shot specific emitter identification via deep metric ensemble learning. IEEE Internet Things J 9(24):24980–24994. https://doi.org/10.1109/JIOT.2022.3194967
9. Chawla NV, Bowyer KW, Hall LO, Kegelmeyer WP (2002) SMOTE: synthetic minority over-sampling technique. https://doi.org/10.1613/jair.953.
10. Nils M, Manandhar S (2018) Evaluation of complex-valued neural networks on real-valued classification tasks. https://doi.org/10.48550/arXiv.1811.12351
11. Sohn K (2016) Improved deep metric learning with multi-class N-pair loss objective. In: Neural information processing systems. Changsha, pp 1857–1865. https://dl.acm.org/doi/10.5555/3157096.3157304
12. Wen Y, Zhang K, Li Z, Qiao Y (2016) A discriminative feature learning approach for deep face recognition. In Leibe B, Matas J, Sebe N, Welling M (eds) Computer vision—ECCV 2016. ECCV 2016, Amsterdam, pp 499–515. https://doi.org/10.1007/978-3-319-46478-7_31

Fake Comment Detection Based on Generative Adversarial Networks

Xiaofei Deng, Tianya Cao, Jian Chen, and Jingwei Li

Abstract Fake comments are a widespread problem in various online activities. This issue not only impacts the consumer experience and service quality but also makes it difficult to distinguish genuine information from the vast pool of comments. Therefore, automated detection of fake comments has become crucial, and this challenging task has garnered considerable attention. Machine learning and deep learning have offered several effective approaches for automatically detecting fake comments, with deep learning, in particular, significantly advancing the field of fake comment detection. However, deep learning often requires a substantial amount of labeled data as support, and the lack of high-quality labeled data has long been a major obstacle in fake comment detection. To address the issue of insufficient labeled data, generative adversarial networks (GANs) have been introduced to compensate for the shortage by generating high-quality fake samples. This method utilizes a limited number of labeled samples and generated fake samples for adversarial learning, thereby enhancing the performance of fake comment detection methods. In this approach, a specific discriminator network structure is designed, combined with long short-term memory networks (LSTM), to optimize feature propagation between network layers. This approach not only maximizes the learning of contextual features but also helps alleviate the potential performance degradation that deep networks may encounter. Ultimately, the proposed method is evaluated using two publicly available high-quality fake comment datasets. The results show that this method achieves a high validation accuracy of 97.87% and an average validation accuracy of 91.04%. Furthermore, metrics such as F1-score, recall, and precision also outperform other comparative methods, with average values of 91.26%, 92.4%, and 95.65%, respectively. A comprehensive analysis of various evaluation metrics suggests that this method outperforms other approaches.

X. Deng (✉) · T. Cao · J. Chen · J. Li
College of Computer Science and Engineering, Northwest Normal University, Lanzhou, Gansu 730070, China
e-mail: 2021212127@nwnu.edu.cn

S. Yadav et al. (eds.), *Proceedings of 3rd International Conference on Artificial Intelligence, Robotics, and Communication*, Lecture Notes in Electrical Engineering 1172, https://doi.org/10.1007/978-981-97-2200-6_8

1 Introduction

Online shopping has become an important mode of consumption. However, compared to physical shopping, online shopping lacks a real-world experience. Online comments, as a crucial means of information exchange, contain a wealth of behavioral information from past users. An increasing number of users are using comment information as a reference when making purchase decisions [1, 2].

Online comments provide crucial reference points for consumers when making purchase decisions. However, some businesses deliberately create comments containing deceptive information to influence users' perceptions of products by exaggerating or damaging a user's reputation [3] to gain greater profits. Over time, this can have a significant impact on consumer experiences, the credibility of businesses, and even the reputation of the platform itself. Therefore, detecting fake comments to protect the genuine interests of consumers, businesses, and the platform is an important task.

In recent years, the detection of online comment spam has been a hot topic of research [4]. Both the academic and industrial sectors have made significant contributions to the task of detecting fake comments. Some service platforms have already taken necessary measures to identify these fraudulent fake comments [5]. Numerous scholars have also made significant contributions to the task of fake comment detection and have proposed many effective detection methods [6–8].

This article views fake comment detection as a text classification problem, that is, identifying whether a comment is fake or genuine using intelligent methods. In recent years, deep learning techniques based on natural language processing have proven successful in text classification tasks. However, the lack of a large volume of genuine and effective datasets, which is required for most supervised machine learning techniques, has been a major challenge in fake comment detection. This problem becomes more prominent for methods based on neural networks and deep learning. To address this issue, generating data has provided an effective avenue for exploration. Some researchers have turned their attention to the significant contributions of generative adversarial networks (GANs) [9] in the data generation field and have introduced them to the task of fake comment detection. Generative adversarial networks are a deep learning framework that allows for the integration of suitable deep learning models based on task requirements, thus obtaining intelligent task models for the desired objectives.

2 Related Research

Fake comment detection is an important research area in the field of natural language processing. Over the past decade, research on fake comments has garnered widespread attention from numerous researchers. In this section, we primarily focus on the related research of Generative Adversarial Networks (GANs) in fake comment detection.

Since Goodfellow and his colleagues introduced GANs, they have been widely applied in research areas such as visual tasks [10], and speech and language processing [11]. They have also gained significant attention from many scholars in tasks related to fake comments and spam detection.

Aghakhani et al. were among the first to apply Generative Adversarial Networks to the task of fake comment detection [12]. They used a network framework with two discriminators and one generative model. This method not only avoids mode collapse issues but also provides good stability. Danilo Croce and others introduced GAN-BERT for fake comment detection. This method is based on a semi-supervised learning architecture with the Transformer [13]. Gao and colleagues developed an unsupervised fake comment detection model with an attention mechanism. This model extracts statistical features from comments and identifies the subjective opinions expressed by users in movie comments. It introduces an attention mechanism into comment embeddings and uses conditional generative adversarial networks to learn users' comment styles for different types of movies [14].

Qu and others addressed the issue of posting fake comments on the Internet to deceive users for commercial gain. They proposed a Sybil detection model based on Generative Adversarial Networks. The model consists of a feature extractor, a domain classifier, and a Sybil detector. It uses a TextCNN neural network model to build the feature extractor, extracting text features from comments. It introduces a domain classifier to learn common features in different domain comments and ultimately detects fake comments using the Sybil detector [15]. Most fake comment research is based on text features, but some studies have found that behavioral features can also effectively identify fake comments. Tang and others investigated the cold start problem in fake comment detection based on behavioral features. They used generative adversarial networks to generate synthetic behavioral features (SBF) from easily accessible features for new users (EAF) [1].

3 Methodology Introduction

This section introduces the problem statement of the proposed model and its corresponding module framework. The overall architecture of the model presented in this paper is depicted in Fig. 1. First, pre-trained models are utilized to obtain text-embedding representations of true and fake comments. These embedding representations are then used as inputs for the discriminator, which is trained with pre-labeled labels to discern the authenticity of comments. Random noise is used as input for the generator, producing generated comment samples that are convincing enough to deceive the discriminator. The loss is calculated based on the discriminator's output of sample class probabilities, which in turn adjusts the parameters of both the discriminator and the generator. Adversarial learning between the generator and the

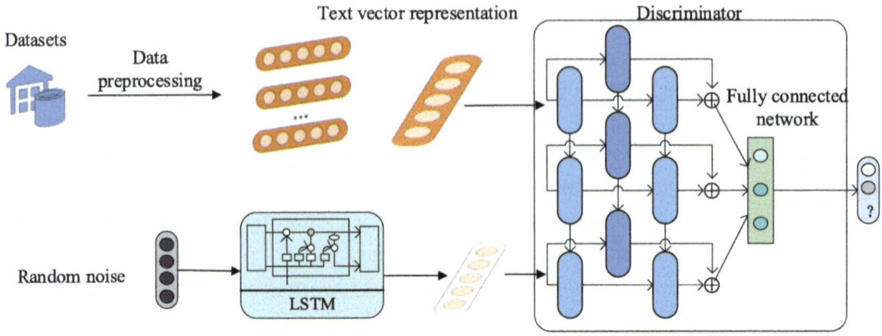

Fig. 1 Methodology framework

discriminator is employed to achieve the predefined training objectives for obtaining a high-performance fake comment recognition model.

3.1 Problem Statement

Fake comment detection aims to train a supervised model that can detect the authenticity of comments in complex comment sequence datasets, which include mostly genuine comments and a small number of fake comments.

This paper, $R = \{R_F, R_T\}$ represents all the data used for model training and validation. The data samples consist of fake comments R_F and true comments R_T. Each comment, $R = \{r_1, r_2, \ldots, r_n\}$, is composed of a sequence of n words. r_i represents the labels for each word in the comments. The data labels, $Y = \{y_1, y_2\}$, are divided into two categories based on the task. The ultimate goal of the model is to accurately predict the probability $p = \{Y_j|R, j = 1, 2\}$ of the label to which the target comment belongs in category by learning features from parts R_F and R_T of the samples R.

3.2 Generator Network

The generator is a deep network composed of multiple layers of long short-term memory networks. Long short-term memory network is a variant of recurrent neural network, and long short-term memory network can retain information in historical periods more persistently than recurrent neural network. Through a more complex structure than recurrent neural networks, the gradient vanishing problem in network learning is mitigated.

The goal of the generative network is to obtain the representation H of the generated sample that is close to the representation of the real sample. The introduction of long short-term memory networks can lead to richer contextual representations:

$$H = G(N, Y, \theta_g) \tag{1}$$

where $G(N, Y, \theta_g)$ is a generator network built with a long short-term memory network. N represents the random noise input to the generator, Y represents the target class of the generated sample, and θ_g is the parameter of the generated network.

In this paper, the cross-entropy loss function is used to participate in the optimization of network parameters, because the generator network only needs to generate enough samples to deceive the discriminator, and cannot change the state of the discriminator. In the joint distribution of comment sentence vectors for a given real sample R (including R_F and R_T), the goal of the generator is to learn the parameter distribution $P(r, y_i)$, which is close to the distribution of real samples. N represents the random noise vector and $P_N(N, y_i)$ represents the target distribution of the random vector. So the generator's loss function can be defined as:

$$L_G = -\mathbb{E}_{N \sim P_x(N, y_i)} \log[D(H, y_i, \theta_D)] \tag{2}$$

Under the goal of the generator network, $D(H, y_i, \theta_D)$ should be as close to 1 as possible, i.e., L_G is maximized. θ_D is the network parameter of the discriminator.

3.3 Discriminator Network

The goal of the discriminator is to distinguish between real and fake samples, and this paper uses a long short-term memory network to form a discriminator. The long short-term memory network can retain the memory of historical moments for a long time and learn richer contextual representations. To further avoid network degradation as the network depth increases. In this paper, a new network structure is designed. Based on the long short-term memory network, the output of the double-layer long short-term memory network and the output of the single-layer long short-term memory network are integrated through the residual method, which constitutes the input of the next-layer network, and the structure of the network is shown in Fig. 2.

L_i represents the input of layer i of the network, which can be represented in the following form depending on the network structure of the discriminator:

$$L_i = \delta(L_{i-1}) + \sigma(L_{i-2}) \tag{3}$$

where $\delta(L_{i-1})$ is the output of the previous layer of L_i double-layer network, and $\sigma(L_{i-1})$ represents the output of the previous layer of L_i's single-layer network.$\oplus$ indicates that the current output of L_i is superimposed as output and the sum of the residuals $\sigma(r_{i-1})$ calculated by the previous layer of the network.

The discriminator learns the distribution of features represented by real samples from the ratings of a given real sample. Enter the generated sample and the real

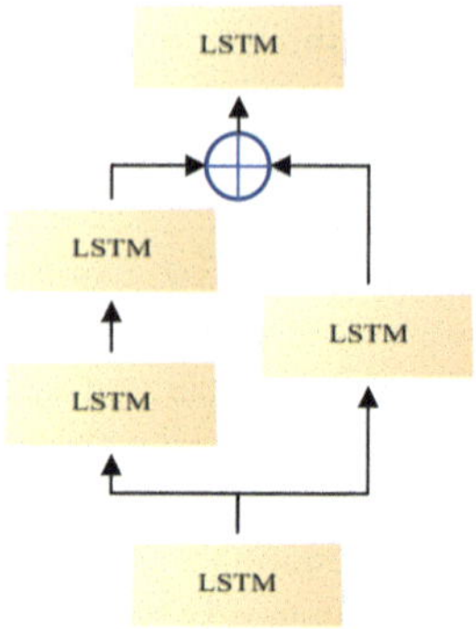

Fig. 2 The network structure of the discriminator

sample into the probability $D(r)$ that the input sample belongs to Y, respectively:

$$\mathrm{D}(r) = \sigma(\omega \cdot L_i + b_d) \quad (4)$$

The discriminator needs to learn not only the characteristics of the real sample, but also the ability to identify fake samples. Therefore, the loss function for the discriminator can be defined as:

$$L_D = -\mathbb{E}_{R\sim P(R,Y)} \log[D(R, \theta_D)] - \mathbb{E}_{N\sim P_N(N,Y)} \log[1 - D(H, \theta_D)] \quad (5)$$

where $D(H, \theta_D)$ represents the probability that the sample generated under additional conditions will be identified by the discriminator of the sample type of interest.

4 Results and Analysis

In this section, we first introduce the evaluation datasets and evaluation metrics. Then, we analyze the performance of the method proposed in this paper and machine learning methods through experimental results on two datasets. Finally, we analyze the performance of the method proposed in this paper and existing methods of the same type by examining the average validation metric values.

4.1 Datasets

Dataset 1 (D1): Mukherjee et al. used GPT-2 [16] to create a dataset of fake comments for a classification task. This dataset consists of 20,000 forged (artificially generated) comments and 20,000 genuine (original) comments from the Amazon comment dataset. The genuine comments were randomly selected, while the fake comments were randomly generated using the original dataset and a stratified sampling method [17].

Dataset 2 (D2): The 'gold standard' dataset developed by Ott et al. This dataset comprises 1600 comment data, with an equal distribution of positive and negative sentiment comments, each containing 400 positive (negative) sentiment true (fake) comments [18].

4.2 Evaluating Indicator

This passage evaluates the proposed method through commonly used metrics in classification models. These metrics include accuracy, precision, recall, and the F-score index. Typically, these metrics are defined as the ratio of mathematical statistics values for predicted labels and true labels by the model.

Accuracy is the most basic and commonly used evaluation metric. It provides a comprehensive assessment of the ratio of correctly predicted instances to the total number of samples in the dataset:

$$\text{Accuracy} = \frac{TP + TN}{TP + TN + FP + FN} \tag{6}$$

where TP represents the number of reviews where the model correctly predicted true positive (true labeled as Positive comments), FP represents the number of reviews where false positive (false labeled as Positive comments) occurred, FN represents the number of reviews where false negative (false labeled as Negative comments) occurred. TN represents the number of reviews where true negative (true labeled as Negative comments) was predicted among reviews labeled as fake comments.

Precision reflects the probability of accurately predicting reviews labeled as true (fake) among those predicted as true (fake):

$$\text{Precision} = \frac{TP}{TP + FP} \tag{7}$$

Recall, in contrast to precision, reflects the probability of accurately predicting samples in the true labels:

$$\text{Recall} = \frac{TP}{TP + FN} \tag{8}$$

The F-score is the weighted harmonic mean of recall and precision. It balances the ratio of correctly predicted samples in both the model's predicted labels and the true labels, providing a better reflection of the model's performance.

$$F - score = \frac{(\alpha^2 + 1) * Precision * Recall}{\alpha^2 * (Precision + Recall)} \tag{9}$$

when $\alpha = 1$, the F-score is commonly known as the F1-score. The F1-score is positively correlated with the model's performance, meaning that a higher F1-score indicates better performance of the model.

4.3 Result and Analysis

The lack of a sufficient amount of high-quality labeled data has been a significant challenge in the task of detecting fake reviews. This paper introduces the use of GANs to leverage generated samples to enhance the performance of the discriminator. Two publicly available datasets, D1 and D2, were employed in this study. Firstly, we validated the impact of data volume on model performance, thereby demonstrating the necessity of employing technical methods to augment the quantity of learning samples for the model.

Figure 3 clearly demonstrates the performance trends of the proposed method under the conditions of datasets D1 and D2 across four evaluation metrics. Combining the information from Table 1, it is evident that the proposed method performs better on dataset D1, with all metrics surpassing those of the model trained on dataset D2. Notably, the model trained on dataset D2 exhibits a significantly higher recall rate than other metrics. As the task here is to identify fake reviews, the proposed method shows robust performance regardless of dataset size, making it applicable even in scenarios with limited sample data.

The network architecture introduced in this paper facilitates the continuous transmission of textual features within the network. For a small number of fake reviews, their features are consistently retained by the network, preventing premature discarding of low-value features. In other words, the features of fake reviews obtained by the proposed network structure from a limited dataset end up being richer than those obtained by a conventional network.

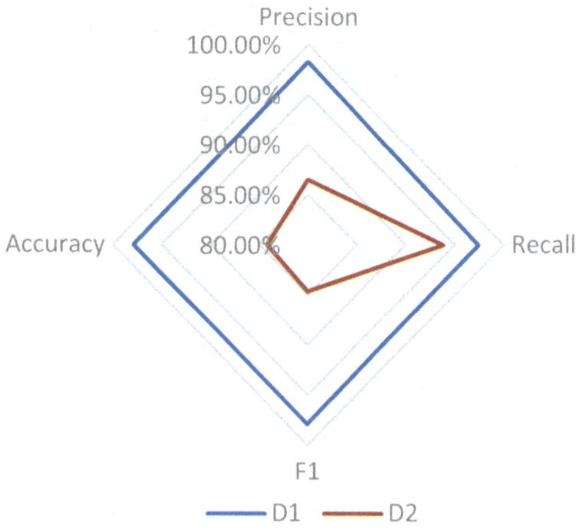

Fig. 3 The impact of sample size on model performance

Table 1 The experimental results of our method's evaluation metrics

	Precision (%)	Recall (%)	F1 (%)	Accuracy (%)
D1	98.29	97.45	97.87	97.87
D2	86.51	93.85	84.62	84.20

In situations where the model is tasked with classifying both fake and genuine reviews, superior overall performance is required. This necessitates having a larger set of learning samples. The generator trained in this study can produce additional samples, serving as supplementary data to enhance the quantity of training samples. This process contributes to improving the model's performance and generalization capabilities.

Figure 4 illustrates the average values of precision, F1-score, recall, and accuracy for decision trees, random forests, as well as some existing methods compared to the model proposed in this paper. In the task of identifying fake comments, the objective is to recognize as many fake comments as possible, requiring the model to have a higher recall. As evident in Fig. 4, the model proposed in this paper outperforms the comparison models significantly in terms of recall, demonstrating its superior performance in identifying fake comments.

For tasks that involve further actions on genuine comments, the requirement is a lower prediction error rate for genuine comments, demanding a higher precision for the fake comments model. The model proposed in this paper also exhibits significantly higher precision compared to the comparison methods, as shown in the precision analysis graph in Fig. 4.

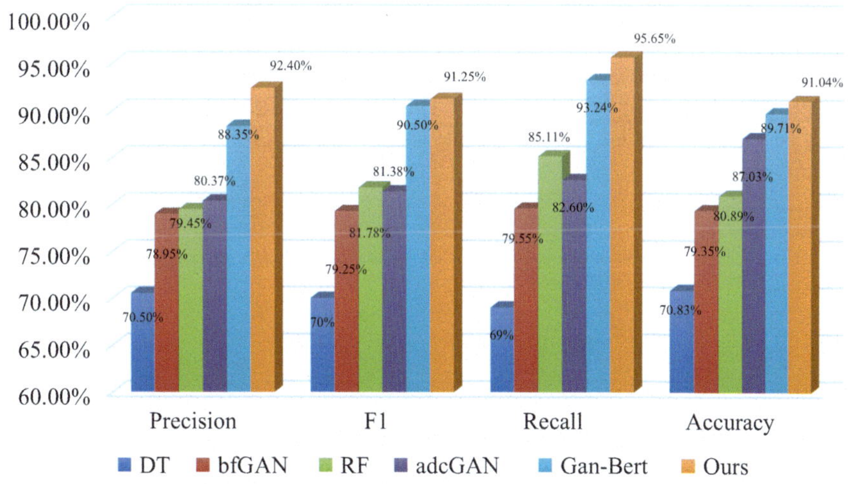

Fig. 4 Comparison of our methods with existing methods under different evaluation indicators

F1-score combines precision and recall, providing a more comprehensive evaluation of the overall model performance and being positively correlated with the model's performance. In the F1-score analysis in Fig. 4, the proposed method surpasses the comparison methods, indicating its superior performance.

Accuracy reflects the ratio of correctly predicted fake and genuine comment samples to the total comment samples. In Fig. 4, the proposed method excels in accuracy compared to the comparison methods. Through the comprehensive analysis of precision, recall, and F1-score, the model proposed in this paper demonstrates better performance in the task of identifying fake comments and exhibits superior stability.

5 Conclusion

Fake comments not only impact consumers' buying experiences but also have a significant effect on the reputation of businesses and platforms. This paper addresses the issue of sample scarcity by generating fake comment samples, thus complementing traditional and deep learning methods. The paper presents a Generative Adversarial Network framework for fake comment detection. This model takes into account the challenges of learning from limited samples, as well as the context features of text and potential network degradation in deep learning models. The advantages of this approach are validated through two publicly available fake comment datasets. The fake comment identification model proposed in this paper outperforms existing models in multiple evaluation metrics, displaying better performance in most criteria.

Despite the superior performance of the proposed method in most metrics, there are still some areas for improvement. Model performance when trained on different dataset sizes is not worse than the comparative methods but still exhibits noticeable differences compared to the model itself. Therefore, ensuring the overall stability of model performance under varying sample sizes will be a critical aspect to consider in future work.

Acknowledgements This work was supported by the National Natural Science Foundation of China (No. 62241207); This work was supported by the Natural Science Foundation of Gansu Province (21JR1RA229, 23JRRA686).

References

1. Tang X, Qian T, You Z (2020) Generating behavior features for cold-start spam review detection with adversarial learning. Inf Sci 526:274–288
2. Zhao G, Lei X, Qian X, Tao M (2019) Exploring users' internal influence from reviews for social recommendation. IEEE Trans Multimedia 21(3):771–781

3. Hu N, Liu L, Sambamurthy V (2011) Fraud detection in online consumer reviews. Decis Support Syst 50(3):614–626
4. Paul H, Nikolaev A (2021) Fake review detection on online E-commerce platforms: a systematic literature review. Data Min Knowl Disc 35(5):1830–1881
5. Maurya SK, Singh D, Maurya AK (2023) Deceptive opinion spam detection approaches: a literature survey. Appl Intell 53(2):2189–2234
6. Hajek P, Sahut JM (2022) Mining behavioural and sentiment-dependent linguistic patterns from restaurant reviews for fake review detection. Technol Forecast Soc Chang 177:121532
7. Alsharif N (2022) Fake opinion detection in an e-commerce business based on a long-short memory algorithm. Soft Comput 26(16):7847–7854
8. Deshai N, Rao BB (2023) Transparency in healthcare and e-commerce: detecting online fake reviews using a dense neural network model with relevance mapping. Soft Comput 27(14):9861–9875
9. Goodfellow I, Abadie JP, Mirza M, Xu B, Farley DW, Ozair S, Courville A, Bengioet Y (2014) Generative adversarial nets. Adv Neural Inf Proc Syst 27
10. Salimans T, Goodfellow I, Zaremba W, Cheung V, Radford A, Chen X, Chen X (2016) Improved techniques for training GANs. Adv Neural Inf Proc Syst 29
11. Yu L, Zhang W, Wang J, Yu Y (2017) SeqGAN: sequence generative adversarial nets with policy gradient. In: Proceedings of the AAAI conference on artificial intelligence 31
12. Aghakhani H, Machiry A, Nilizadeh S, Kruegel C, Vigna G (2018) Detecting deceptive reviews using generative adversarial networks. In: 2018 IEEE security and privacy workshops (SPW), San Francisco, CA, USA, pp 89–95
13. Croce D, Castellucci G, Basili R (2020) GAN-BERT: generative adversarial learning for robust text classification with a bunch of labeled examples. ACL
14. Gao Y, Gong M, Xie Y, Qin AK (2021) An attention-based unsupervised adversarial model for movie review spam detection. IEEE Trans Multimedia 23:784–796
15. Qu Z, Jia Q, Lyu C, Liu J, Liu X, Zheng K (2022) Detecting fake reviews with generative adversarial networks for mobile social networks. Secur Commun Netw
16. Radford A, Wu J, Child R, Luan D, Amodel A, Sutskever I (2019) Language models are unsupervised multitask learners. OpenAI blog 1(8):9
17. MukherjeeA, Venkataraman V (2014) Opinion spam detection: an unsupervised approach using generative models. Techincal Rep, UH
18. Ott M, Chio Y, Cardie C, Hancock JT (2011) Finding deceptive opinion spam by any stretch of the imagination. arXiv 1107.4557

Shipping Risk Assessment Ship Inspection Method Based on Improved YOLOv5

Ziqi Wang, Haibin Li, Mingzhu Sun, and Louyu Zhu

Abstract Since its inception, the maritime transport industry has occupied a prominent position among the five modes of transport. The study of shipping risk in sea and waterway area, especially the decision-making analysis of ship collision risk, is of great significance for shipping safety engineering, reducing Marine accidents, pollution prevention, and emergency planning. Existing risk assessment models require experts to interpret the model results, which is time-consuming and laborious. In order to solve this problem, a new method of ship encounter scene recognition is proposed according to the collision risk. The complex Marine environment has high requirements for the target detection algorithm, which requires not only a high accuracy rate, but also a high real-time performance, which also requires the algorithm to have good robustness. Aiming at the shortcomings of the traditional target detection algorithm with low detection accuracy and poor robustness, this paper proposes a detection method based on the improved YOLOv5 algorithm model. This method uses convolutional neural network in deep learning to process the images of two or more ships, and makes use of the powerful fitting ability of convolutional neural network. More accurate use of image features to identify boats on the water.

1 Introduction

The strategic risk decision-making method involves the risk control and emergency measures of the ship, which requires high real-time and accuracy. With the rapid development of big data and information technology, this kind of risk decision-making model is showing a trend of data driving. With the rapid development of the field of artificial intelligence and big data in recent years, the methods used for strategic risk decision-making are increasingly favored by evolving statistical algorithms and machine learning algorithms. Machine learning facilitates the study of

Z. Wang · H. Li (✉) · M. Sun · L. Zhu
School of Information Engineering, Zhejiang Ocean University, Zhoushan, China
e-mail: lihaibin@anlantech.cn

S. Yadav et al. (eds.), *Proceedings of 3rd International Conference on Artificial Intelligence, Robotics, and Communication*, Lecture Notes in Electrical Engineering 1172, https://doi.org/10.1007/978-981-97-2200-6_9

strategic risk decision-making methods and enhances the learning and understanding of voyage data.

With the increase of computer computing power, scene object detection in images [1–3] has become a research hotspot in the field of computer vision. The traditional object detection method adopts the region selection method based on the sliding window [4], which lacks pertinence and has low computational efficiency. Meanwhile, the design and selection of features rely on manual labor, which restricts the objectivity and robustness of the algorithm. The emergence of AlexNet network in 2012 opened the research boom of deep learning, and in 2014 R-CNN [5] introduced deep learning into the field of object detection. In 2015, Redmon et al. [6] proposed the YOLO algorithm, which has a much higher rate than other object detection algorithms. The proposed algorithm represents the birth of a series of algorithms for one-stage target detection, and it is also the first time to use an algorithm based on regression [7].

In recent years, the YOLOv5 algorithm has also been continuously improved: Cheng [8] proposed the optimization of the YOLOv5 detection terminal. For small targets such as ships, the detection head dedicated to large targets cannot play a full role, and even interferes with detection. Therefore, the detection head responsible for large targets is removed, and the detection speed is increased on the basis of the original network. Zhang [9] replaced the GIOU LOSS function used in YOLOv5s with CI-OU LOSS, so that the GIOU LOSS function could not judge the position state of the prediction frame when the prediction frame was in the real frame, thus improving the detection accuracy.

The main content of this paper is based on ship detection, research, and improvement based on YOLOv5s, and starts with improving the accuracy and speed of YOLOv5s algorithm for ship detection. By adding SE attention mechanism to the backbone network of YOLOv5s algorithm to improve important features, BIFPN module was introduced to optimize the two parts of spatial pyramid pooling to improve the detection accuracy. Convolutional neural network is used to process multi-ship images, complete ship detection work, and conduct preliminary image processing for subsequent shipping risk assessment work. Apply machine learning algorithms to improve water traffic conditions and enhance the real-time reporting of maritime safety and hazardous situations.

2 Research on Target Detection Algorithm

2.1 YOLOv5 Object Detection Algorithm

YOLOv5 algorithm is currently the algorithm that most researchers are focusing on. Compared with YOLOv1–YOLOv4, the core detection idea of YOLOv5 algorithm has not changed. The input image is divided into grids, and the grids are responsible for detecting objects falling on the center point of the grid.

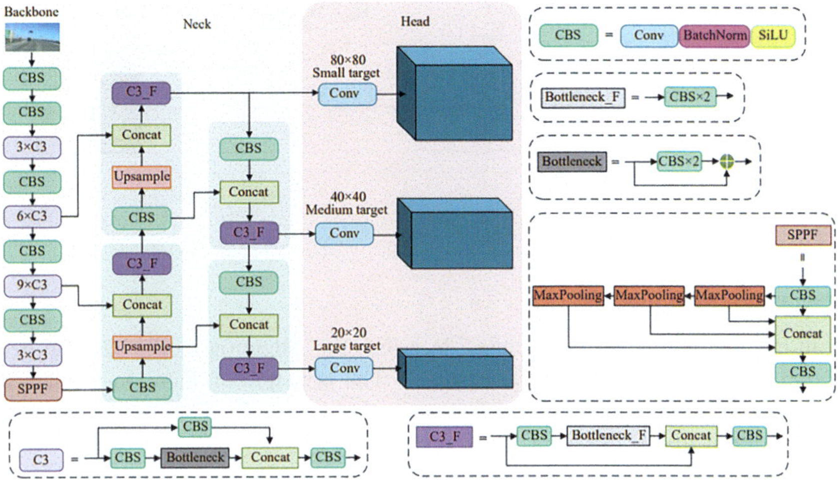

Fig. 1 YOLOv5 network structure

The structure diagram of YOLOv5 is shown in Fig. 1, which is composed of Input, Backbone, Neck, and Head. The input image is scaled to an appropriate size through Mosaic enhancement and input into the convolutional neural network. The location, size, and category of the detected object are obtained through Backbone, Neck, and Head processing. The gap between them is calculated using the loss function, and reduced through gradient descent iteration. Finally, the weight matrix and bias of the minimum value of the loss function in the total number of iterations are taken to obtain the prediction information of the object to be detected.

2.2 Improved YOLOv5 Object Detection Algorithm

Although YOLOv5 has been widely studied and applied in the field of target detection due to its small size structure and fast processing speed, its detection accuracy rate is still insufficient and the detection effect is not satisfactory in the maritime ship detection scenario, so it is necessary to improve YOLOv5 appropriately to meet the accuracy detection requirements of ships.

In this experiment, the neck network integrated with BIFPN thought was used to carry out feature fusion, and the low and medium dimensional feature map containing small target feature information was efficiently used to improve the detection accuracy of the model for ship targets. At the same time, the SE net attention mechanism is added before the SPPF of the backbone of the network to realize the automatic acquisition of the importance of the location of each feature channel by self-learning, and the obtained importance degree is used to promote the features and suppress the features that are not important to the current task.

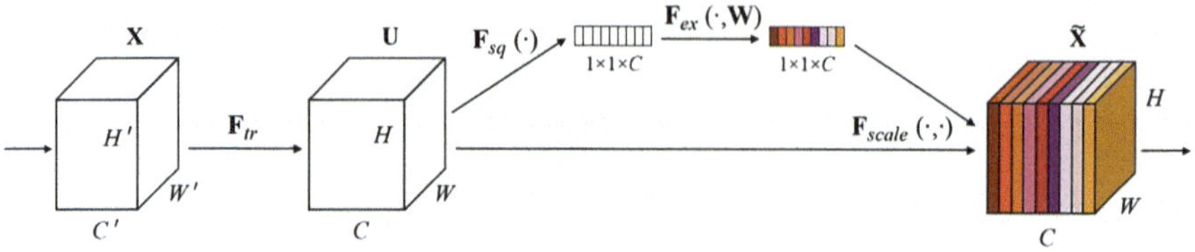

Fig. 2 SE module

Fusion attention module

With the continuous development of deep learning, attention model has also been introduced into deep learning for research. The principle of attention mechanism is mainly to amplify important information features and suppress non-important information features. In recent years, it has been widely used in the field of target detection.

In this experiment, the SE Net module mentioned in CVPR2017's Squeeze and Excitation Networks [10] is introduced. Figure 2 shows the working principle of the SE module.

As shown in the figure, the SE module mainly includes Squeeze and Excitation parts. Firstly, the squeeze operation is used to map the feature information spatially and generate a global distribution embedded in the feature response at the channel level, thus allowing the information from the global receptive field of the network to be used by all its layers.

The aggregation is followed by the excitation operation, which is embedded as input and produces a set of corresponding weights for each channel. These weights are applied to U to generate the output of the SE Block, which is then sent to subsequent networks.

In this experiment, SEnet attention mechanism is integrated in Backbone's SPP module to improve the detection accuracy of the algorithm.

The improvement of space pyramid pool

In this paper, Bidirectional feature pyramid network we adopt is improved on the basis of PANet. The PANet in the original YOLOv5s is replaced, and the feature representation ability is enhanced by simple residual operation.

Based on the advantages of this structure, the PAN structure in YOLOv5 feature pyramid network is changed to BIFPN [11] structure to enhance feature fusion and improve detection accuracy. The feature pyramid of YOLOv5 network fused with BIFPN reduces the loss of small target information during feature fusion.

3 Experiment and Result

3.1 Experimental Dataset

This experiment is a self-built dataset. The dataset is divided into 6 categories according to the target category, then the 6 types of data are sorted and divided according to the sequence of 8:1:1, and then put into the training set, verification set and test set, respectively. The default training cycle of the experiment is 300 Epochs.

3.2 Experimental Platform and Training Parameter Setting

This experiment is a cloud platform, using the cloud platform server as the basic environment for training. Hardware configuration: GPU RTX 3080(10 GB) * 1; AMD EPYC 7543 32 core processor; Memory 40 GB System disk: 25GB in the system and configuration of the cloud server, a unified Linux system is adopted, and Jupyterlab is installed as a code interaction tool. The deep learning software required for the experiment includes Python 3.8, Pytorch 1.10, etc.

3.3 Evaluation Index

The experimental results of this paper evaluate the performance of the improved algorithm from three aspects: average accuracy mean (mAP), recall, and precision. The average calculation formula of recall rate, accuracy rate, and average accuracy rate is as follows:

$$\text{Recall} = TP/(TP + FN) \tag{1}$$

$$\text{Precision} = TP/(TP + FP) \tag{2}$$

$$mAP = \frac{1}{N}\sum_{i=1}^{n} AP_i \tag{3}$$

where TP represents the number of ship targets in the forecast pair, and the IOU ratio between the forecast box and the label box is greater than 0.5. FP represents the number of ship targets predicted incorrectly, indicating the number of detection boxes that meet IOU $\leq$ 0.5. FN represents the number of missed ship targets. AP represents the area of the precision rate-recall curve calculated by interpolation. N means there are N classes.

3.4 Analysis of Experimental Results

Ablation experiment

This study evaluated the influence of different modules on the detection performance of the final target through ablation experiments, and the ablation experiment results are shown in Table 1.

It can be seen from the ablation experiment that each improvement point can have a good impact on the YOLOv5 algorithm and improve the detection performance. Therefore, the validity of each improved module proposed in this paper is verified, and the improved method in this chapter can play a role in optimizing the detection algorithm.

Check test

In order to further verify the performance of the YOLOv5s improved algorithm model proposed in this paper, compared with other current mainstream convolutional neural network models, YOLOv3, YOLOv4 models were used in the experiment to further verify the generalization performance of the proposed algorithm under the same parameter setting conditions above. The comparative experimental data are shown in Table 2.

It can be seen from the data in the table that the algorithm in this paper has been improved in all aspects and is superior to other algorithms in the table with high confidence. This indicates that the improved YOLOv5s model can extract key information more intently, make full use of low-level information, and fuse richer and more comprehensive feature maps. At the same time, it can reduce the interference of useless information on the recognition effect, thus improving the recognition accuracy and speed.

Table 1 Ablation experiment

Model	Precision	Recall	mAP@0.5	mAP@0.5: 0.95
Yolov5	78.12	75.64	75.92	58.72
Yolov5 + SE	80.73	76.39	79.24	63.45
Yolov5 + BiFPN	80.48	78.1	78.97	60.73
Yolov5 + SE + BiFPN	82.26	78.79	79.68	64.04

Table 2 Comparison of different algorithms

Model	Precision	Recall	mAP@0.5	mAP@0.5:0.95
Yolov3	77.71	72.77	74.1	58.37
Yolov4	71.6	68.45	68.05	50.45
Yolov5	78.12	75.64	75.92	58.72
Textual algorithm	82.26	78.79	79.68	64.04

Comparative analysis before and after improvement

Performance comparison

In order to further verify the detection performance of the improved YOLOv5 algorithm, the original YOLOv5 algorithm was compared with the improved YOLOv5 algorithm, and the performance results of each index were shown in Fig. 3.

As the figure shows, the values of P, R, mAP@0.5, mAP@0.5:0.95 of the improved YOLOv5 model have all been improved, which makes the model convergence speed faster and the index has also been improved to a certain extent.

Comparison of test results

The comparison of ship test results before and after improvement is shown in Fig. 4.

As can be seen from Fig. 4, the improved algorithm can better detect various types of ships, improve the recognition accuracy of the basic YOLOv5 model, and achieve a more balanced recognition effect for various types of ships. To a certain extent, the high speed and high precision ship detection is effectively realized. It is proved that the algorithm can be applied to the actual detection environment.

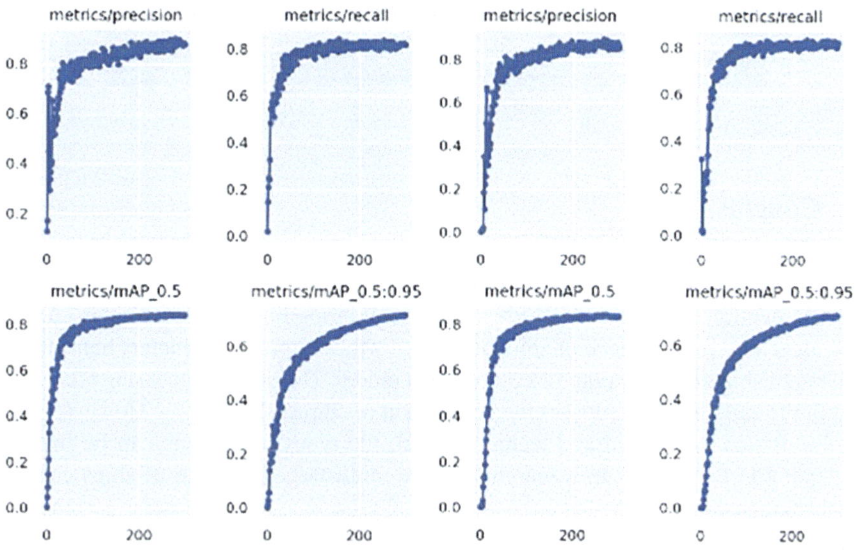

Fig. 3 Performance comparison before and after improvement

Fig. 4 Comparison of test results

4 Conclusion

Based on YOLOv5s, a lightweight ship target detection model is proposed. By using SEnet attention module and BIFPN space pyramid module, the detection accuracy and algorithm performance of ship target are further improved, which is helpful for the establishment of shipping risk assessment model. The image processing is carried out in the early stage to support the acquisition of ship collision data. However, due to the limited theoretical and technical level, the types of datasets can be further expanded in the follow-up research, and the multi-target tracking of ships can be further carried out.

References

1. Girshick R, Donahue J, Darrell T et al (2016) Region-based convolutional networks for accurate object detection and segmentation. IEEE Trans Pattern Anal Mach Intell 38(1):142–158
2. Girshick R (2015) Fast R-CNN. In: Proceeding of the IEEE international conference on computer vision, pp 1440–1448

3. Ren SQ, He KM, Girshick R et al (2017) Faster R-CNN: towards real-time object detection with region proposal net-works. IEEE Trans Pattern Anal Mach Intell 39(6):1137–1149
4. Dalal N, Triggs B (2005) Histograms of oriented gradients for human detection. In: Proceeding of 2005 IEEE computer society conference on computer vision and pattern recognition, pp 886–893
5. Girshick R, Donahue J, Darrell T et al (2014) Rich feature hierarchies for accurate object detection and semantic segmentation. In: Proceeding of IEEE conference on computer vision and pattern recognition, pp 580–587
6. Redmon J, Divvala S, Girshick R et al (2015) You only look once: unified, real-time object detection. In: Proceeding of 2015 IEEE conference on computer vision and pattern recognition, pp 779–788
7. Redmon J, Divvala S, Girshick R et al (2016) You only look once: unified, real-time object detection. In: The IEEE conference on computer vision and pattern recognition. Las-Vegas: IEEE, pp 779–788
8. Cheng Q, Li J, Du J (2023) Ship target detection algorithm in optical remote sensing images based on YOLOV5. Syst Eng Electron 45(5):1270–1276
9. Zhang HQ, Ban YM, Guo LL et al (2021) Ship detection method in remote sensing images based on YOLOV5. Electron Measure Technol 44(8):8792
10. Lin TY, Dollar P, Girshick R et al (2017) Feature pyramid networks for object detection. In: 2017 IEEE conference on computer vision and pattern recognition (CVPR). IEEE Computer Society
11. Ruiz-Ponce P, Ortiz-Perez D, Garcia-Rodriguez J et al (2023) Poseidon: a data augmentation tool for small object detection datasets in maritime environments. Sensors 23(7):3691

Spatiotemporal Prediction of Harmful Algal Blooms Based on Spatiotemporal Attention and Transformer

Peirui Wang, Xiaoqing Luo, and Zhancheng Zhang

Abstract Accurate HAB prediction is crucial for water pollution management, and the application of spatiotemporal prediction methods in HAB has not been fully explored. To further enhance prediction performance, considering the complex HAB motion patterns, a spatiotemporal prediction method named STA-MIMO based on spatiotemporal attention and transformer is proposed. The proposed spatiotemporal attention module (STA) comprises Temporal Attention Block (TA) and 4D Attention Block (4DA). TA is designed to select critical spatial features along the temporal dimension by assigning temporal attention weights. 4DA is designed to finely select critical spatiotemporal feature points by assigning 4D attention weights. The STA can adaptively highlight key local spatiotemporal features, benefiting the prediction of complex motion patterns. Experiment results on the Taihu HAB 2019 dataset demonstrate that the proposed method significantly outperforms the comparative methods in all metrics, validating its superior HAB prediction performance.

1 Introduction

Expansions on industrial activities and climate change have led to frequent HAB events recently, causing economic losses, health threats, and fish killing [1]. Therefore, how to predict HAB accurately has become a hot research topic in water pollution management. Currently, mainstream HAB prediction methods primarily utilize time series data which overlooking the spatial dynamic and spatiotemporal correlations of HAB [2]. In recent years, many spatiotemporal prediction methods have

P. Wang · X. Luo (✉)
School of Artificial Intelligence and Computer Science, Jiangnan University, Wuxi, Jiangsu 214122, China
e-mail: xqluo@jiangnan.edu.cn

Institute of Advanced Technology, Jiangnan University, Wuxi, Jiangsu 214122, China

Z. Zhang
School of Electronic and Information Engineering, Suzhou University of Science and Technology, Suzhou 215000, Jiangsu, China

S. Yadav et al. (eds.), *Proceedings of 3rd International Conference on Artificial Intelligence, Robotics, and Communication*, Lecture Notes in Electrical Engineering 1172, https://doi.org/10.1007/978-981-97-2200-6_10

been proposed and successfully applied in various domains, especially in precipitation nowcasting [3]. However, the spatiotemporal prediction of HAB has not been fully explored. The existing spatiotemporal sequence prediction models are mainly based on LSTM, which belong to single–input–single–output (SISO) models. Shi et al. [4] proposed ConvLSTM by introducing convolutional operators into the LSTM structure, which enabled the LSTM structure to learn spatiotemporal data. Based on ConvLSTM, researchers proposed many novel models, such as PredRNN [5], MIM [6], and MotionRNN [7]. SISO models take the current frame as input to predict the next frame in a recursive manner, which leads to severe error accumulation. To overcome the error accumulation problems, Ning et al. [8] proposed MIMO-VP, a novel multi–input–multi–output (MIMO) model based on extending the pure Transformer. However, HAB is influenced by multi nature factors, making its prediction challenging. The design of the local spatiotemporal block (LSB) of MIMO-VP is too simple to capture the complex HAB motion pattern.

To improve HAB prediction performance, this paper further considers the unbalanced influence of spatiotemporal features on prediction according to the prone variability of HAB motion, propose a HAB spatiotemporal prediction method STA-MIMO which based on spatiotemporal attention and transformer structure. A temporal attention block (TA) is designed to select spatial features adaptively that have a greater impact on HAB prediction along the time dimension. A 4D attention block (4DA) is further designed to select critical spatiotemporal feature point by considering global spatiotemporal features. By incorporating TA and 4DA, a spatiotemporal attention module (STA) is constructed to assist the model in finely select critical spatiotemporal features along both spatial and temporal dimensions, thus enhancing the prediction of prone variability HAB motion trend.

2 Methods

2.1 Spatiotemporal Attention Multi–input–multi–output Prediction Network

The HAB spatiotemporal prediction aims to generate the most likely length-N HAB sequence in the future given the previous length-M sequence. The structure of the proposed prediction network STA-MIMO is shown in Fig. 1. Its encoder and decoder are all composed of 2D convolution multi-head attention (2D MHA), layer normalization and add, and STA module. In each sample learning process, the model takes m frames as input and output n frames prediction.

Firstly, the model generates high-dimensional spatiotemporal features by performing patch and downsample operation on an input image sequence. Then, the positional encoding is generated along the temporal dimension and embedded into the features to maintain the temporal order of the sequence, which is vital for MIMO prediction models. The encoder first utilizes 2DMHA to learn global

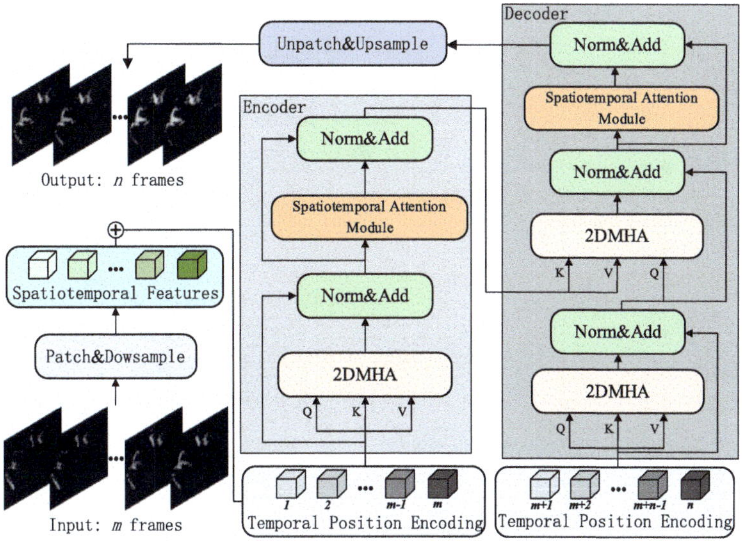

Fig. 1 The structure of the proposed spatiotemporal prediction network STA-MIMO

spatiotemporal features while retaining fine spatial features, then learns critical local spatiotemporal features through the proposed STA module, capturing the right motion trend. The decoder interacts with the encoder output features through 2DMHA and takes the temporal positional encoding for the future n frames as input, learns the correlation between historical and future data, generate prediction spatiotemporal features. Finally, prediction frames are generated by performing unpatch and upsample operation on decoder output features.

2.2 *Spatiotemporal Attention Module*

Due to the HAB motion pattern is intricate, it is vital for the model possess critical spatiotemporal features extraction ability to better capture the right short-term motion trend. Thus, we proposed STA module, which employs attention mechanisms to make the model adaptively focus on key local spatiotemporal features that contribute to prediction from both spatial and temporal dimensions. The structure of STA is shown in Fig. 2. It consists of 3D convolution, group normalization, SiLU, TA, and 4DA blocks.

Temporal Attention Block The spatial features at each timestamp contribute differently to the HAB prediction. Thus, the TA block is proposed to make the model learn temporal features that have a high correlation with future frames, its structure is shown in Fig. 2. Firstly, 3D average pooling is employed to compress and aggregate information from the channel dimension (C), spatial dimensions (H and W)

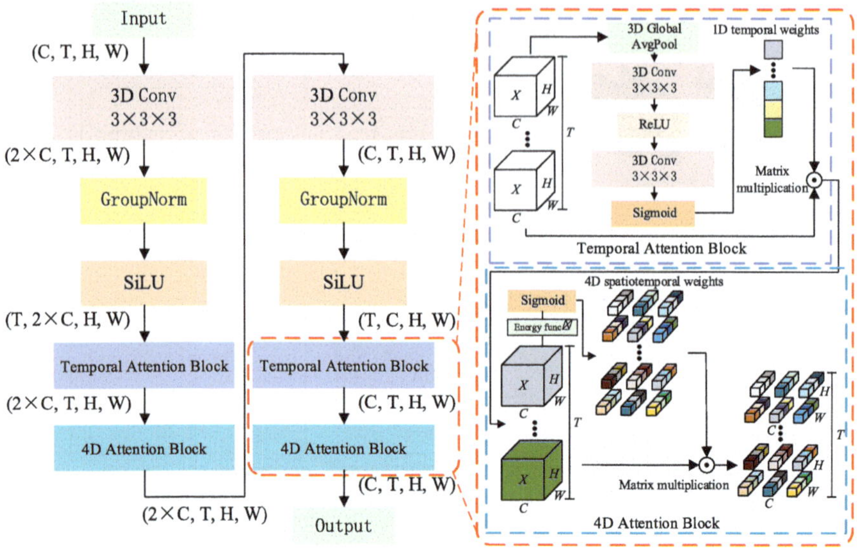

Fig. 2 The structure of the proposed spatiotemporal attention module

of the input. Subsequently, two cascaded $1 \times 1 \times 1$ convolutional and a ReLU are utilized for feature extraction. Finally, Sigmoid is utilized to obtain 1D temporal weights, and through matrix multiplication along the temporal dimension (T), thus the spatiotemporal features are endowed with corresponding weights, enhancing key temporal features.

4D Attention Block Traditional attention mechanisms typically generate coarse weights in the channel and spatial dimensions separately and require parameters [9]. Considering each spatiotemporal feature point contributes differently to the HAB prediction, inspired by SimAM [10], this paper proposed a parameter-free block 4DA to generate fine-grade 4D spatiotemporal attention weights by optimizing the energy function, assigns distinct weights to each spatiotemporal feature point within high-dimensional spatiotemporal feature space, guiding the model to learn critical spatiotemporal feature points, thus enhancing the prediction performance.

The structure of 4DA is shown in Fig. 2. The 4DA block takes the output features of the TA block as input, then generates fine-grade 4D weights through energy function ψ and Sigmoid that have the same size $T \times C \times H \times W$ with the input. Finally, matrix multiplication is applied to assign corresponding weights to each spatiotemporal feature point. As a result, the redundant feature points are suppressed, while the potential contributable feature points are enhanced. The equation of ψ is expressed as Eq. 1.

$$\psi(x) = \frac{(x-\mu)^2}{4 \times \left(\frac{\varphi((x-\mu)^2)}{T \times H \times W - 1} + \lambda\right) + 0.5} \tag{1}$$

where x is the input spatiotemporal features. T, H, and W are the temporal dimension, height and width of spatial dimension of x, respectively. μ is the average of x along the channel dimension. $\varphi(\cdot)$ is the calculation of the sum of the features along the channel dimension. λ is a hyper parameter.

3 Experiment

To validate the effectiveness of the proposed method, Taihu HAB 2019 dataset is created from the HAB normalized difference vegetation index (NDVI) data which collected by satellite in every 10 min of Taihu lake in 2019 that provided by Suzhou Meteorological Bureau. Each image contains three types of HAB intensities: mild, moderate, and severe that represented by pixel values 100, 150, and 200. The dataset comprises HAB data from May to September in 2019, totaling 10,141 frames, 80% are used as train set, while the remaining 20% are used as the test set. We use the first 6 frames as input to predict the next 6 frames.

The comparison results are shown as Table 1, α is intensity threshold. MSE, SSIM [11], the Critical Success Index (CSI), and Heidk Skill Score (HSI) [12] are used as metrics. The best one of each metric is highlighted in bold. For the four SISO models, MotionRNN outperforms other models due to its complex motion modeling ability by MotionGRU units. MIMO-VP has significant improvements in all metrics compared with MotionRNN, due to the parallel structure of transformer can effectively eliminate accumulation errors. Our proposed STA-MIMO achieved the best performance, compared to MIMO-VP, the MSE, CSI-200, and SSIM of STA-MIMO improved by 15.23%, 15.69%, and 0.65%, respectively. Indicating the proposed model can finely select key spatiotemporal features by the STA module, significantly improve HAB prediction accuracy, proved its effectiveness and superiority.

To intuitively show the performance of the proposed model, a prediction result is visualized, shown as Fig. 3. Due to the error accumulation, result of MotionRNN exhibit clear noise where MIMO-VP and STA-MIMO not, mild HAB are viewed in water areas where they should not appear. From overall, the prediction motion

Table 1 The comparison HAB spatiotemporal prediction results of different models

Methods	CSI			HSS			MSE	SSIM
	$\alpha = 100$	$\alpha = 150$	$\alpha = 200$	$\alpha = 100$	$\alpha = 150$	$\alpha = 200$		
ConvLSTM	0.359	0.372	0.118	0.503	0.519	0.196	158.2	0.856
PredRNN	0.360	0.385	0.119	0.504	0.532	0.199	153.8	0.864
MIM	0.367	0.396	0.138	0.512	0.544	0.227	151.1	0.862
MotionRNN	0.387	0.415	0.147	0.533	0.565	0.241	134.9	0.866
MIMO-VP	0.485	0.451	0.204	0.635	0.604	0.323	105.9	0.927
STA-MIMO	**0.524**	**0.480**	**0.236**	**0.671**	**0.633**	**0.366**	**91.9**	**0.933**

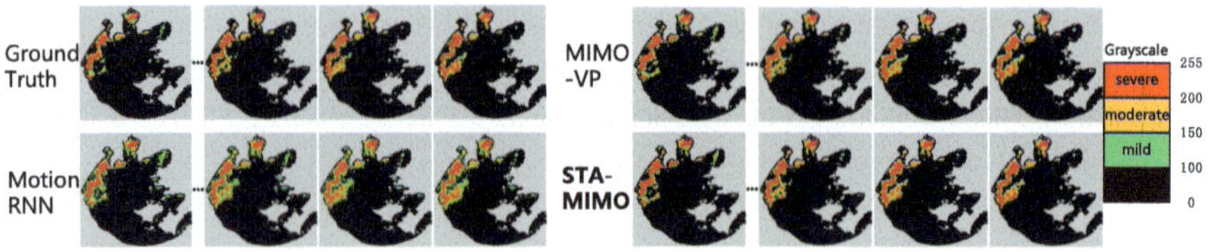

Fig. 3 Visualized HAB spatiotemporal prediction result

Table 2 The ablation study results of different spatiotemporal learning modules

Modules	CSI			HSS			MSE	SSIM
	$\alpha = 100$	$\alpha = 150$	$\alpha = 200$	$\alpha = 100$	$\alpha = 150$	$\alpha = 200$		
LSB	0.485	0.451	0.204	0.635	0.604	0.323	105.9	0.927
TA	0.509	0.470	0.232	0.657	0.622	0.361	96.4	0.930
4DA	0.515	0.465	0.211	0.663	0.619	0.334	96.9	0.931
STA	**0.524**	**0.480**	**0.236**	**0.671**	**0.633**	**0.366**	**91.9**	**0.933**

trend by the proposed method aligns most closely with the actual evolution, especially for moderate and severe HAB occurrences, further intuitively demonstrates its superiority.

To verify the effectiveness of the proposed TA, 4DA, and STA, the comparison results of different spatiotemporal learning modules are shown as Table 2. Substituting TA for LSB results in significant improvement in CSI-200, exhibit that TA can significantly enhance prediction performance for severe HAB by capturing critical temporal features. Substituting 4DA for LSB results in significant improvement in CSI-100, exhibit that 4DA can significantly enhance prediction performance for mild HAB by capturing critical spatiotemporal feature points. STA can achieve the best improvement in all metrics, indicates that select features from both the spatial and temporal adaptively can more comprehensively and significantly enhance the prediction performance, and further validates the effectiveness and rational design of STA.

4 Conclusion

In response to the intricate HAB motion patterns, this paper proposed a HAB spatiotemporal prediction method STA-MIMO based on spatiotemporal attention and transformer. 2DMHA and the proposed STA module are core spatiotemporal learning components of STA-MIMO. TA can adaptively select key spatial features along the temporal dimension. 4DA can adaptively select key spatiotemporal feature points by generating refined 4D spatiotemporal attention weights through an energy function. Thus, STA enables the model to effectively learn the spatial and temporal correlations in HAB dynamics, thereby optimize its spatiotemporal feature representation, enhancing prediction performance. The experiment on Taihu HAB 2019 shows that the proposed method significantly improved HAB prediction in both accuracy and image quality, holds promising application prospects in the field of water pollution control. In the future, external influencing factors will be considered for better HAB prediction.

Acknowledgements National Natural Science Foundation of China under Grant 61772237.

References

1. Zepernick BN et al (2023) Climate change and the aquatic continuum: a cyanobacterial comeback story. Environ Microbiol Rep 15(1):3–12
2. Nguyen XC, Bui VKH, Cho KH, Hur J (2023) Practical application of machine learning for organic matter and harmful algal blooms in freshwater systems: a review. Crit Rev Environ Sci Technol 1–23
3. Wikle CK, Zammit-Mangion A (2023) Statistical deep learning for spatial and spatiotemporal data. Annu Rev Stat Appl 10:247–270
4. Shi X, Chen Z, Wang H et al (2015) Convolutional LSTM network: a machine learning approach for precipitation nowcasting. Adv Neural Inf Process syst 28
5. Wang Y, Long M, Wang J et al (2017) PredRNN: Recurrent neural networks for predictive learning using spatiotemporal LSTMs. Adv Neural Inf Proc Syst 30
6. Wang Y, Zhang J, Zhu H et al (2019) Memory in memory: a predictive neural network for learning higher-order non-stationarity from spatiotemporal dynamics. In: Proceedings of the IEEE/CVF conference on computer vision and pattern recognition, pp 9154–9162
7. Wu H, Yao Z, Wang J, Long M (2021) MotionRNN: a flexible model for video prediction with spacetime-varying motions. In: Proceedings of the IEEE/CVF conference on computer vision and pattern recognition, pp 15435–15444
8. Ning S, Lan M, Li Y, Chen C et al (2023) MIMO is all you need: a strong multi-in-multi-out baseline for video prediction. In: Proceedings of the AAAI conference on artificial intelligence, pp 1975–1983
9. Guo MH, Xu TX, Liu JJ et al (2022) Attention mechanisms in computer vision: a survey. Comput Vis Media 8(3):331–368
10. Yang L, Zhang RY et al (2021) Simam: a simple, parameter-free attention module for convolutional neural networks. In: International conference on machine learning, pp 11863–11874
11. Wang Z, Bovik AC et al (2004) Image quality assessment: from error visibility to structural similarity. IEEE Trans Image Process 13(4):600–612
12. Hogan RJ, Ferro CA, Jolliffe IT, Stephenson DB (2010) Equitability revisited: why the "equitable threat score" is not equitable. Weather Forecast 25(2):710–726

Large-Scale Information Dissemination Algorithm Based on Random Mapping

Yuntian Bai, Xingchen Liu, and Danyu Diao

Abstract In light of the inefficiency in managing and distributing large-scale information, this paper proposes a large-scale information publish-subscribe workflow model based on the publish-subscribe pattern. Furthermore, it introduces an evaluation indicator system for distribution efficiency, leveraging the Analytic Hierarchy Process to guide decision-making in large-scale information distribution processes. Additionally, a text matching algorithm based on random mapping is devised. This algorithm capitalizes on the effectiveness of hash tables and state information to enable efficient matching between documents and subscription information. Experimental results validate a significant improvement in matching efficiency by this algorithm, highlighting its practical value.

1 Introduction

Content-based publish/subscribe systems have received significant attention in both research and practical applications. These systems facilitate matching of message publication and subscription based on their content. The existing body of research primarily focuses on two key aspects. Firstly, researchers aim to enhance system performance and scalability by leveraging distributed algorithms, network communication, and related technologies to reduce transmission load and improve data processing speed [1, 2]. Secondly, researchers strive to achieve precise, fast, and personalized message matching by employing techniques such as predicate matching and fuzzy matching. These techniques enhance the accuracy and efficiency of message recommendation and relevance calculation [3, 4]. Additionally, a number of studies explore security and privacy protection mechanisms to ensure data confidentiality and integrity during transmission and storage processes [5, 6]. However,

Y. Bai (✉) · X. Liu
Army Engineering University of PLA, Nanjing 210001, Jiangsu, China
e-mail: byt825990802@foxmail.com

D. Diao
Air Force Engineering University, Xi'an 710077, Shaanxi, China

S. Yadav et al. (eds.), *Proceedings of 3rd International Conference on Artificial Intelligence, Robotics, and Communication*, Lecture Notes in Electrical Engineering 1172, https://doi.org/10.1007/978-981-97-2200-6_11

currently there is relatively little research on matching algorithms for large-scale information.

In this paper, we propose an information publish-subscribe model that aligns with the context of the big data era. We conduct an analysis using the Analytic Hierarchy Process to assess the impact of each indicator on information distribution. Subsequently, we design an algorithm suitable for large-scale information dissemination. This algorithm leverages random mapping to efficiently process substantial amounts of data in memory, and its efficiency is validated through rigorous experimentation.

2 A Large-Scale Information Publish-Subscribe Workflow Model Directed Toward Distribution Efficiency

2.1 *Publish-Subscribe Pattern*

The Publish-Subscribe Pattern serves as a mechanism for sharing information and pushing messages between various components within a distributed system. In this pattern, publishers disseminate messages to one or more designated topics, whereas subscribers can selectively subscribe to topics and retrieve messages through middleware acting as an intermediary responsible for accepting and delivering information [7]. Upon publishing, all subscribers to the topic receive the message, as illustrated in Fig. 1. Thus, the Publish-Subscribe Pattern embodies attributes such as asynchronous communication, real-time delivery, low coupling, and high scalability [8].

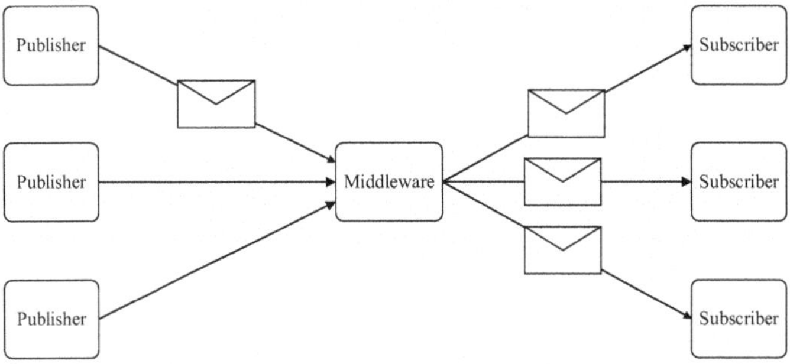

Fig. 1 Information subscription and publishing model

2.2 *Large-Scale Information Subscription and Publication Process*

The large-scale information subscribe-publish flow is a mechanism employed to handle substantial data transmission and real-time data distribution. It comprises four main components, namely, data sources, subscription services, message middleware, and subscribed data processors, as illustrated in Fig. 2.

To begin with, a wide array of systems, devices, or applications serve as the primary data sources, generating a considerable volume of data that is subsequently dispatched to a message queue. The message queue, functioning as the core infrastructure for message delivery, assumes responsibility for data caching. The subscription service retrieves the data from the message queue and converts it into a format amenable to further processing. Furthermore, the subscription service accommodates user subscription requests and retrieves the corresponding data from the message queue in response to these requests. Lastly, the subscribed data processor executes operations such as classification, filtering, and aggregation on the transmitted data, generating results tailored to meet the users' requirements. These results are then transmitted to the users via push notifications or active queries.

Fig. 2 Large-scale information subscription and publication process

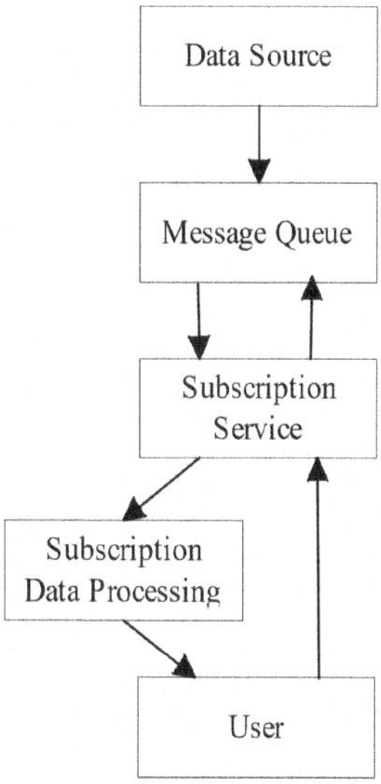

2.3 Large-Scale Information Subscription and Publication Model

In conjunction with the design principles of large-scale information subscription and publishing processes, we present the ensuing model for information subscription and publishing, as depicted in Fig. 3.

The model encompasses the following modules:

(1) Subscription Demand Management Module: Empowering users to independently modify their subscription information, which is stored as a collection of keywords.
(2) User Status Management Module: Employing compulsory offline handling for users who remain unresponsive after multiple calls and extended periods of inactivity, thereby mitigating excessive user occupancy and enhancing information distribution efficiency.
(3) Information Queue Management Module: Given that numerous users consistently publish information at each information dissemination node simultaneously, the effective matching of published information necessitates the use of a designated waiting queue.
(4) User Pre-screening Module: Eliminating users who have not participated in the current subscription round to reduce memory consumption and improve subsequent matching efficiency.
(5) Information-User Matching Module: Swiftly matching information and users based on the execution outcomes of the matching algorithm and the information priority derived from the information queue management module.
(6) Distribution Directory Generation Module: Generating an executable matching projection table that expedites information transmission across the network.

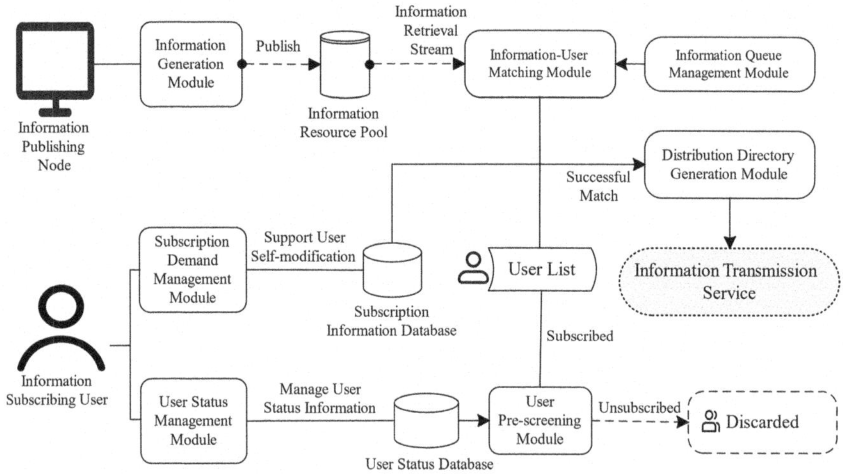

Fig. 3 Large-scale information subscription and publication model

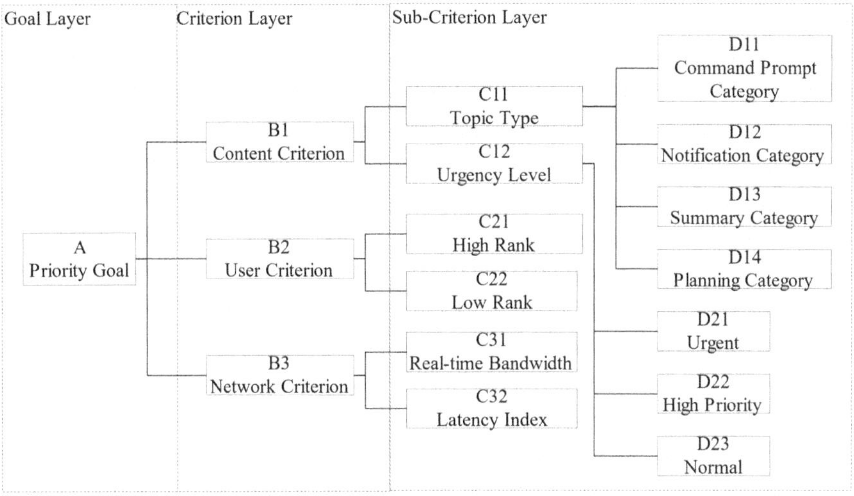

Fig. 4 Hierarchical structure diagram

3 The Evaluation Index System for Distribution Efficiency

3.1 *Construction of the Evaluation Index System for Distribution Efficiency*

The hierarchical structure of the system consists of a goal layer and a criterion layer, as shown in Fig. 4. The criterion layer is divided into three sub-layers: content criteria, user criteria, and network criteria. Content criteria consist of two sub-criteria: topic type and urgency. Topic types include command prompts, notifications, summaries, and plans, while urgency is divided into several levels: extremely urgent, urgent, and normal. User criteria are divided into high-ranking and low-ranking types, each of which is assigned different additional attributes and distribution requirements. Network criteria include real-time bandwidth and latency index.

3.2 *Establishment of Evaluation Index System*

Based on expert opinions, a 1–9 scale method was used to construct the judgment matrices. The judgment matrix *P* for the first level is presented in Table 1, while the judgment matrices *P* for the second level are shown in Tables 2, 3, and 4. Similarly, the judgment matrices *P* for the third level can be found in Tables 5 and 6.

To determine the weights of each criterion for the priority objective, the geometric mean method was applied. The weight vector for B1, B2, B3 with regards to the priority objective was calculated as $\omega_1 = [0.661, 0.208, 0.131]^T$. Similarly, the

Table 1 Priority goal judgment matrix (*A*)

A	B1	B2	B3
B1	1	4	4
B2	1/4	1	2
B3	1/4	1/2	1

Table 2 Content criterion judgment matrix (B1)

B1	C11	C12
C11	1	1/4
C12	4	1

Table 3 User criterion judgment matrix (B2)

B2	C21	C22
C21	1	8
C22	1/8	1

Table 4 Network criterion judgment matrix (B3)

B3	C31	C32
C31	1	5
C32	1/5	1

weight vectors $\omega_2 = [0.2, 0.8]^T$, $\omega_3 = [0.889, 0.111]^T$, $\omega_4 = [0.833, 0.167]^T$, $\omega_5 = [0.66, 0.177, 0.078, 0.085]^T$, and $\omega_6 = [0.661, 0.272, 0.067]^T$ were obtained for the content criterion (C11, C12), user criterion (C21, C22), network criterion (C31, C32), theme type (D11, D12, D13, D14), and urgency level (D21, D22, D23), respectively.

Table 5 Topic type judgment matrix (C11)

C11	D11	D12	D13	D14
D11	1	7(9)	7	5
D12	1/7(1/9)	1	3(5)	3
D13	1/7	1/3(1/5)	1	1(3)
D14	1/5	1/3	1(1/3)	1

Table 6 Urgency level judgment matrix (C12)

C12	D21	D22	D23
D21	1	3(7)	8
D22	1/3(1/7)	1	5
D23	1/8	1/5	1

3.3 Consistency Assessment

Consistency check of judgment matrices usually uses consistency index (CI) and consistency ratio (CR) to evaluate the consistency and reliability of the matrix. The calculation steps are as follows:

(1) Calculate the maximum characteristic root $\lambda_{\max}$ of each judgment matrix, where A is the judgment matrix, n is the order of the matrix, and the specific formula is shown in Eq. (1):

$$\lambda_{\max} = \sum_{i=1}^{n} \frac{(Aw)_i}{nw_i} \tag{1}$$

(2) Calculate the consistency index CI, where $\lambda_{\max}$ is the maximum characteristic root of the judgment matrix, and n is the order of the matrix, and the specific formula is shown in Eq. (2):

$$\mathrm{CI} = \frac{\lambda_{\max} - n}{n - 1} \tag{2}$$

(3) Calculate the consistency ratio CR, where CI is the consistency index, and RI is the average random consistency index. When the order of the judgment matrix is 0–10, the corresponding RI values are 0, 0, 0.52, 0.89, 1.12, 1.26, 1.36, 1.41, 1.46, 1.49, and the specific formula is shown in Eq. (3):

$$\mathrm{CR} = \frac{\mathrm{CI}}{\mathrm{RI}} \tag{3}$$

The consistency check results of each level of criterion judgment matrix are calculated and shown in Table 7.

Both the CI and CR are less than 0.1, indicating good consistency among the weights of elements in the judgment matrix for each level criterion and ensuring the reliability of the decision outcome.

The calculation of the total ranking weight of the sub-criterion layer to the objective layer is shown in Fig. 5.

Table 7 Consistency check results of criteria judgment matrices at all levels

Judgment matrix	Consistency index	Consistency ratio
B1–B3	0.027	0.052
C11–C12	0	–
C21–C22	0	0
C31–C32	0	0
D11–D14	0.072	0.081
D21–D23	0.022	0.042

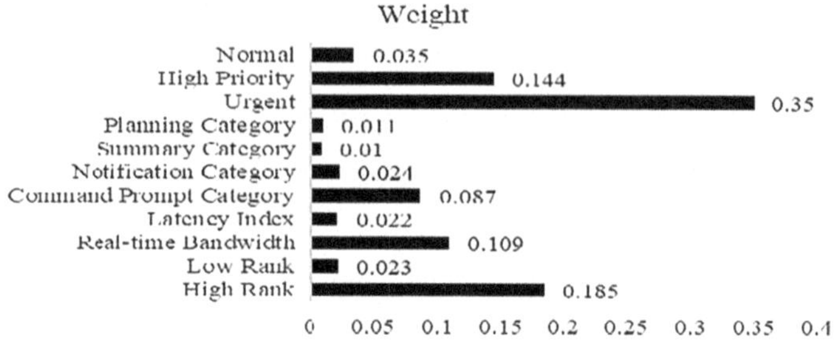

Fig. 5 Hierarchical total ordering

It can be observed that the factor of "urgency" has the greatest impact on the decision outcome, while the factor of "meeting minutes type" has the least influence. Additionally, factors such as high rank, urgency, and real-time bandwidth also carry significant weights, indicating their crucial role in the decision-making process.

4 Large-Scale Information Dissemination Algorithm Based on Random Mapping

Efficient processing and matching of massive data is crucial for accurate and rapid information distribution on a large scale. String-based algorithms like KMP and AC algorithms are used for small-scale text matching, while machine learning-based algorithms handle semantic-level matching [9, 10]. However, there is a current lack of efficient text matching algorithms for scenarios involving massive data. To address this issue, we propose a novel algorithm for large-scale information distribution.

4.1 Principles of Algorithm Design

To efficiently distribute tagged documents to subscribed users, intelligent algorithms utilizing hash tables and state information are essential. Our proposed algorithm utilizes hash tables and state information to achieve efficient distribution. The matching phase utilizes a hash table comprising the original and process components. Matching words transfer subscription information to the matched table, while unmatched data is stored in the process hash table and rehashed based on state information. This ensures swift and accurate information matching and distribution, as depicted in Fig. 6.

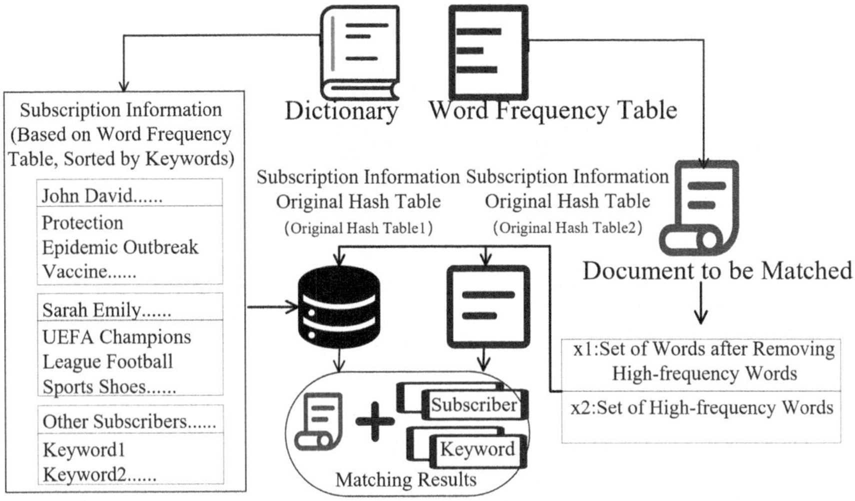

Fig. 6 Schematic diagram of information dissemination algorithm

4.2 Algorithm Workflow

During the preprocessing phase, the first step involves identifying the top n words that appear with high frequency, which typically comprise common vocabulary in the network. Once these n words are determined, subscription information, and documents undergo preprocessing. Subscription information is processed by sorting each word in every subscription information set according to decreasing frequency, followed by adding a state information element N to each set. The state information N denotes how many elements of the subscription information set have been matched during a specific document matching process. $N = 0$ signifies that none of the elements in the subscription information set have yet been matched with the document.

Documents undergo a similar sorting process as subscription information sets. However, unlike subscription information sets where each word appears only once, a document can contain duplicate words. Therefore, duplicates in the document are removed before processing. If the document includes multiple high-frequency words, they are sorted in ascending order of frequency, forming a sequence called x_1. After removing the high-frequency words, the remaining words are sorted lexicographically, resulting in a sequence called x_2. The document's vocabulary is initialized as $< x_2, x_1 >$.

Two hash tables are constructed: hash Table 1 stores subscription information sets with $N = 0$, and hash Table 2 stores subscription information sets with $N > 0$. Hash Table 1 initially stores all subscription information sets that have not yet been matched with a document, while hash Table 2 stores the sets of subscription information already partially matched. In hash Table 2, the value of N for these sets must be at least 1 but cannot exceed the number of words in the sets themselves.

For each subscription information set w in the subscription information, the following operations are performed:

Let $h = \text{hash}(w)$ and find the bucket with index h in hash Table 1. The bucket contains a subscription information set $X = \{x_1, x_2, \ldots, x_n\}$. For each subscription information x_i in set X with only one word, add it to the table of matched subscription information and remove it from the bucket in hash Table 1 with index h. For each subscription information x_i in set X with more than one word, set the value of N for x_i to 1. Use the new value of N to calculate the new hash value $h' = \text{hash}(w_2)$, where w_2 is the first word in x_i. Remove x_i from the bucket in hash Table 1 with index h and add it to the bucket in hash Table 2 with index h'.

Let $h = \text{hash}(w)$ and find the bucket with index h in hash Table 2. The bucket contains a subscription information set $X = \{x_1, x_2, \ldots, x_n\}$. For each subscription information x_i in set X, if w is the last word in x_i, add it to the matched subscription information table and remove it from the bucket in hash Table 2 with index h. For each subscription information x_i in set X, if w is not the last word in x_i, increment the value of N for x_i by 1. Use the new value of N to calculate the new hash value h' $= \text{hash}(w')$, where w' is the $(N + 1)$th word in x_i. Update the hash value of x_i in the corresponding bucket in hash Table 2 with index $h' \leftarrow h$.

The algorithm flowchart is shown in Fig. 7.

4.3 Experimental Results and Analysis

To evaluate the performance of our algorithm, we compared it with the AC algorithm. The experiments were conducted on a 64-bit Windows 11 operating system, using an Intel Core i7-12700H 2.70 GHz CPU and 16 GB DDR4 3200 MHz system memory. The experimental dataset, named "ahocorapy," consists of a list of 50,000 keywords and an input text of 34,199 characters. A comparison of the execution times between our algorithm and the AC algorithm is presented in Table 8. The experimental results indicate that for small-scale inputs, our algorithm exhibits slightly slower performance than the AC algorithm, mainly due to the additional overhead of sorting and hashing. However, this difference is not significant. As the input scale increases, our algorithm outperforms the AC algorithm, demonstrating a clear speed advantage when processing large-scale data and exhibiting better overall performance.

5 Conclusion

We proposes a large-scale information subscription and publication process model for distribution efficiency. A distribution efficiency evaluation index system is constructed, and the impact of each index on information distribution is analyzed.

Fig. 7 The flowchart of the information distribution algorithm

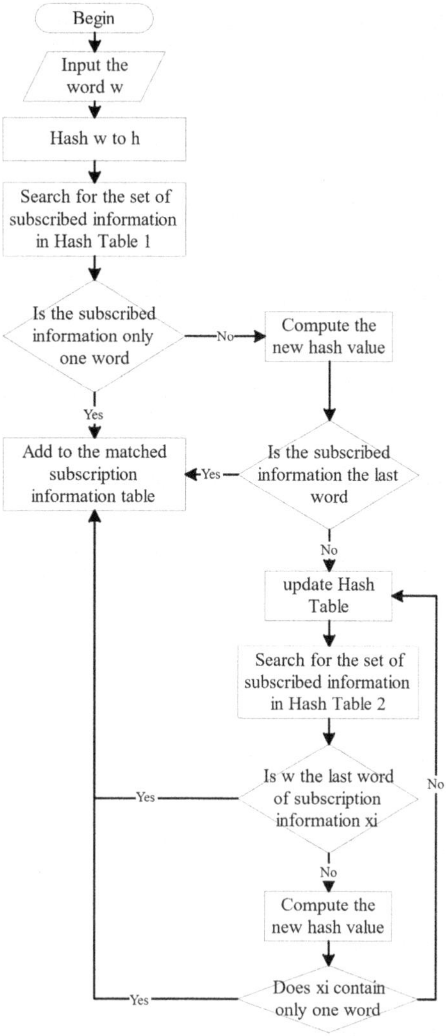

Table 8 Comparative analysis of algorithm execution time

Input size	Time complexity of AC algorithm (ms)	Time complexity of this algorithm (ms)	Speedup ratio
100	1.02	1.99	0.51
1000	16.01	8.01	2.00
10,000	592.03	92.98	6.37
50,000	8234.26	548.35	15.02

Finally, a distribution algorithm suitable for large-scale information is designed. Experimental results demonstrate that the algorithm exhibits high efficiency and reliability in processing large-scale data.

References

1. Zhu W, Deng Y, Qian S, Cao J, Xue G (2022) PEM: a parallel ensemble matching framework for content-based publish/subscribe systems. SEKE, pp 106–111. https://doi.org/10.18293/SEKE2022-051
2. Opyrchal L, Astley M, Auerbach J, Banavar G, Strom R, Sturman D (2000) Exploiting IP multicast in content-based publish-subscribe systems. In: Middleware 2000. Springer, New York, NY, USA, pp 69–88. https://doi.org/10.1007/3-540-45559-0_10
3. Pereira JLM, Fabret F, Llirbat F, Shasha D (2000) Efficient matching for web-based publish/subscribe systems. In: International conference on cooperative information systems, pp 162–173. https://doi.org/10.1007/10722620_17
4. Zhang H, Zhang X, Ding K, Liu M, Liu L, Yuan Z (2022) A fuzzy matching with reasoning publish/subscribe system based on ontology. In: 2nd International conference on consumer electronics and computer engineering (ICCECE). Guangzhou, China, pp 150–156. https://doi.org/10.1109/ICCECE54139.2022.9712661
5. Raiciu C, Rosenblum DS (2006) Enabling confidentiality in content-based publish/subscribe infrastructures. In: Securecomm and workshops, pp 1–11. https://api.semanticscholar.org/CorpusID:1323334
6. Nabeel M, Shang N, Bertino E (2012) Efficient privacy preserving content based publish subscribe systems. In: ACM symposium on access control models and technologies. https://api.semanticscholar.org/CorpusID:14752252
7. Zhang Z (2022) Survey of event matching algorithms in pub/sub systems. Command Inf Syst Technol 13:46–55. https://doi.org/10.15908/j.cnki.cist.2022.02.009
8. Liu X, Lei W, Liu G, Sun J, Zuo Q (2020) Research on efficient fuzzy matching algorithm in battlefield information distribution system. Fire Control Command Control 45:109–113+118
9. Knuth DE, Morris JH Jr, Pratt VR (1977) Fast pattern matching in strings. SIAM J Comput 6(2):323–350. https://doi.org/10.1137/0206024
10. Aho AV, Corasick MJ (1975) Efficient string matching: an aid to bibliographic search. Commun ACM 18(6):333–340. https://doi.org/10.1145/360825.360855

Smoke Detection in Complex Scenery Using Improved Faster-RCNN

Zijun Nie, Hualin Zhan, Lei Luo, and Ziqian Zhang

Abstract Aiming at the complexity of smoke detection scenarios, high target smoke undetect rate, poor detection effect, and so on. This paper propose an object detection algorithm based on the improved Faster-RCNN; firstly, in order to improve the accuracy of the model smoke detection accuracy, we use the Resnet50 feature extraction network to replace the VGG-16 network in original model. The experimental results show that after replacing the network with Resnet50, the model Map50 is improved by 2.4%; secondly, we conduct an experiment using the swin-transformer as the backbone, and the experimental results on the same smoke dataset show that the detection accuracy of Map50 using the swin-transformer is improved by 3.0% compared with the baseline. It can be conclude that the improved Faster-RCNN detection algorithm obtain a better performance and a good detection results.

1 Introduction

Smoke is one of the early signs of fire, and the presence of fire can be detect early before it breaks out through smoke detection, this helps to increase the early alarm rate of fire, providing valuable time for timely evacuation of people, alarming and extinguishing the fire, and reducing the damage caused by fire. In the field of smoke detection, many domestic and foreign scholars have proposed a variety of detection schemes based on traditional algorithms. These methods mainly relies on extracting texture, color, edge, and optical flow features to detect smoke. Cao [1] proposed STCNet, which involves the spatial path of texture feature extraction and the temporal path of capturing smoke motion information. Yuan et al. [2] proposed that Gabor

Z. Nie · L. Luo · Z. Zhang
School of Information and Electromechanical Engineering, Jiangxi Science and Technology Normal University, Nanchang, China

H. Zhan (✉)
Jiangxi Provincial Key Laboratory of Optoelectronics and Communications, Jiangxi Science and Technology Normal University, Nanchang, China
e-mail: james392@163.com

S. Yadav et al. (eds.), *Proceedings of 3rd International Conference on Artificial Intelligence, Robotics, and Communication*, Lecture Notes in Electrical Engineering 1172, https://doi.org/10.1007/978-981-97-2200-6_12

network enables the output of Gabor convolution unit to calculate the histogram characteristics of the image. Ho [3] proposed an SVM classifier-based classifier for smoke detection algorithm and tested it under night time conditions, proving that the algorithm can perform well in complex scenes that the algorithm can show good performance in complex scenes. However, traditional methods face some challenges when dealing with smoke detection tasks in complex scenes and varying lighting conditions, such as a greater reliance on feature selection and susceptibility to interference from noise. To solve these limitations, the rise of deep learning techniques in recent years has brought new opportunities for smoke detection. Deep learning models, especially Convolutional Neural Network (CNN)-based approaches, are capable of learning more complex feature representations, thereby improving the accuracy and robustness of smoke detection. Advanced deep learning architectures such as Faster-RCNN [4], YOLO [5], Retinanet [6], and SSD [7] have also been successfully applied to smoke target detection tasks with remarkable results.

This article will dive in the application of traditional and deep learning methods in the field of smoke detection, analyze their advantages and limitations, and propose directions for improvement and future research in order to promote the continuous progress of smoke detection technology.

2 Methods and Improvements

Faster-RCNN is the object detection algorithm used in this paper, which further improves Fast RCNN by making the target detection process more end-to-end by introducing the Region Proposal Network (RPN), which improves the detection speed.

Faster-RCNN is widely used in target detection tasks, including object detection, vehicle detection, and other scenarios have good application results, the object of study in this paper is based on Faster-RCNN smoke detection task as well, which belongs to one kind of target detection tasks; the feature extraction network used in the original Faster-RCNN algorithm is VGG-16 [8], VGG-16 network structure is shown in Fig. 1, VGG-16 consists of 13 conv layers which use small size 3×3 conv kernels for feature extraction and nonlinear activation by ReLU [9] activation function. These conv layers are stacked at different depths and number of channels to gradually increase the complexity and abstraction capability of the network. After each convolutional layer, VGG-16 performs spatial downsampling using the Maxpooling layer. The operation of pooling helps to extract the main features and reduce the number of parameters. However, since VGG-16 has multiple convolutional and fully connected layers, this makes the number of parameters of the network relatively large, it leads to higher computational cost and require more time for training and inference, in addition, due to the apply of multiple pooling layers, the feature maps after pooling may lose some detail information, which affects the accuracy of detection. Therefore, we introduce the Resnet50 feature extraction network, as shown in Fig. 3, Resnet50 [10] has a deeper network structure compared to VGG-16.

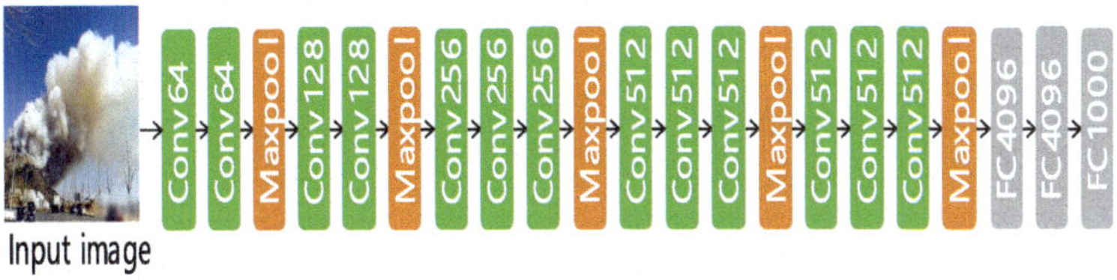

Fig. 1 The structure of VGG-16

A deeper network usually helps to learn more complex features and express more complex patterns. And due to the presence of residual connections, Resnet50 has a smaller number of parameters at greater depths.

The structure of the residual connection module is shown in Fig. 2, and from a mathematical point of view the residual connection can be analyze and express as Eq. (1). Residual joining makes it easier for gradients to be posted back by adding features from the previous layer directly to the desired layer. In addition, the residual connection can also perform multi-scale fusion of features in the network layer, and connect and refuse feature maps of different scales, so that the final feature map can be used for richer semantic information.

$$y_1 = h(x_1) + F(x_1, W_1), \quad x_{l+1} = f(y_1), \quad x_L = x_1 + \sum_{i=1}^{l-1} F(x_i, W_i) \tag{1}$$

x_1 and x_{l+1} are denotes the input and output of the residual unit respectively, F is the learned residual, $h(x_1) = x_1$ denotes the constant mapping, and f is the relu activation function. The features learned from layer 1 to layer L are then denoted as Eq. (1).

Another work of ours is introduce swin-transformer [11] into Faster-RCNN as backbone, the structure of swin-transformer is shown in Fig. 4, compared to VGG-16 which uses simple convolutional layers with a small sense field and weak ability to model global information, swin-transformer introduces a self-attention mechanism

Fig. 2 Residual block

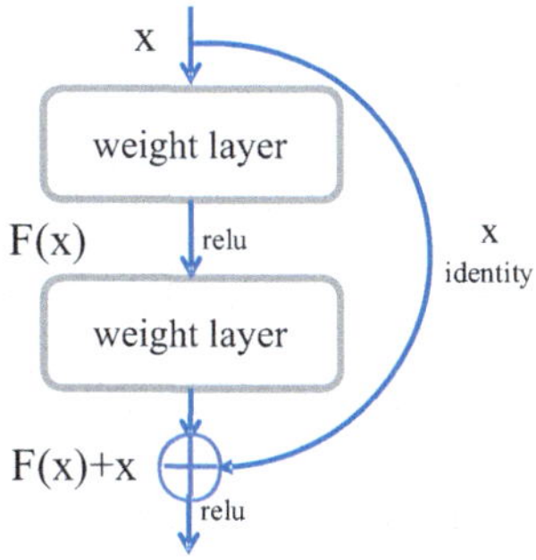

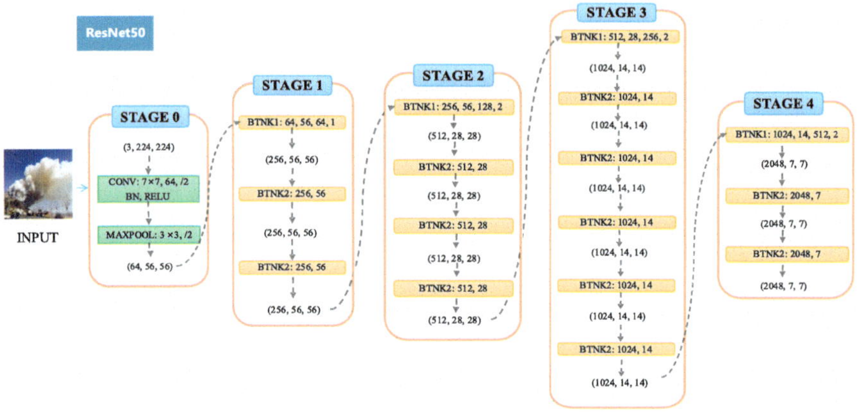

Fig. 3 The structure of Resnet50

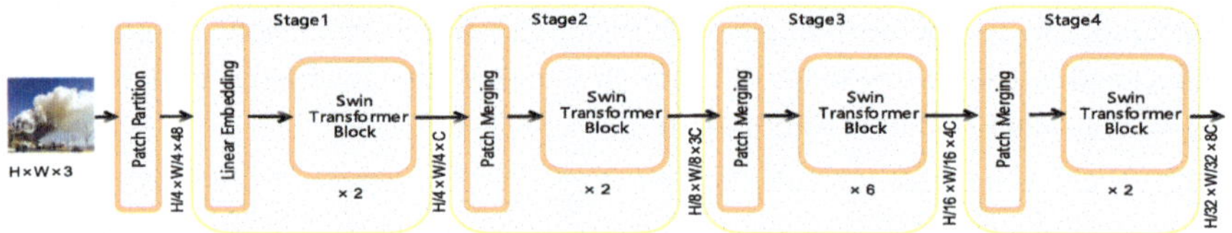

Fig. 4 The structure of swin-transformer

[12], which allows the network to better capture global contextual information and helps to deal with long-range dependencies in images.

The basic component of swin-transformer is swin-block, which is implemented by window-based self-attention mechanism and the mathematical expression of swin-block is Eqs. (2) and (3).

$$\text{Attention}(\mathbf{X}_i) = \text{Softmax}\left(\frac{\mathbf{X}_i \cdot (\mathbf{X}_i)^T}{\sqrt{d_k}}\right) \cdot (\mathbf{X}_i) \tag{2}$$

$$Y_i = \text{Conv}(\text{LayerNorm}(\text{Attention}(X_i) + X_i)) \tag{3}$$

The input denote as $\mathbf{X} \in \mathbb{R}^{H \times W \times C}$, where H is the height of the input, W is the width of the input, and C is the number of channels, and the input tensor is partitioned into window blocks for the purpose of the self-attention mechanism, noting that the window blocks are $\mathbf{X}_i$, and the self-attention mechanism is used for each $\mathbf{X}_i$ in Eq. (4), denotes the matrix multiplication, $\sqrt{d_k}$ is the scaling factor, and d_k is the dimension of the attention mechanism. The output of the attention mechanism is then fused. LayerNorm is the normalization layer. Overall, the mathematical expression of swin-block mainly consists of the steps of local self-attention mechanism, feature fusion,

(a) Training set images (b) Test set images

Fig. 5 The images of SM-dataset

and combination of window blocks. This window self-attention mechanism helps in processing large size images while reducing the computation complexity.

In swin-block, the first step divides the input image into overlapping patches. With this strategy, attention interaction mechanism can be done on each patch, rather than on the entire image. For each patch, a classic conv layer is used to extract local features. And then, after local feature extraction, a cross-patch interaction module is introduced for information transfer and global interaction between different patches. What's more, in order to capture information at different scales, swin-block also introduces a multi-level resolution pyramid structure. Through the above steps, swin-block combines local feature extraction, cross-tile interaction and multi-scale information fusion to achieve better utilization of receptive field and multi-scale information.

3 Experimental

3.1 *Dataset*

In our study, the dataset used for the experiments is SM-dataset. The images in SM-dataset are taken manually and collected from the Internet. The dataset contains images at different times and under different lighting conditions. SM-dataset contains a total of 2500 images, out of which 1950 images in the training set, and 550 images in the test set. These images have no duplicate images in the training set and test set. Some of the smoke images are shown in Fig. 5.

3.2 *Evaluation Metrics*

Mean Average Precision [13] (MAP) and Precision P (Precision) were used as evaluation metrics for the evaluation of the experimental results, respectively. Calculated

Table 1 Algorithm comparison results

Methods	Map50	Improve
Vgg16	0.389	Baseline
Resnet50	0.413	+ 0.024
swin_t	0.378	– 0.011
swin_s	0.401	+ 0.012
swin_v2_s	0.402	+ 0.013
swin_v2_b	0.424	+ 0.035

as the Eq. (4).

$$P = \frac{\mathrm{TP}}{\mathrm{TP} + \mathrm{FP}}, \quad \mathrm{AP} = \int_0^1 P d_R, \quad \mathrm{mAP} = \frac{1}{N}\sum_{i=1}^{N} \mathrm{AP}_i \tag{4}$$

3.3 *Experimental Results Analysis*

In the research experiments of this paper, our model training platform is NVIDIA GeForce GTX 1660 Ti, the operating system is ubuntu18.20. We compare the experimental results of the backbone network improved to Resnet50 with the baseline, engaged in the experimental results can be seen the average detection accuracy Map50 improved by 2.4%. In addition to further experiments, we added the swin-transformer to the Faster-RCNN, and through the experiments and comparisons of several versions of the swin-transformer, we came up with the experimental results showing that the best detection effect and the baseline compared to our algorithm was obtained with the swin_v2_b version of the backbone network improved by 3.5%. It shows that the addition of swin-transformer effectively improves the detection performance of Faster-RCNN (Table 1).

4 Conclusion

In this paper, we improved the Faster-RCNN network, firstly, we replaced the feature extraction network of the original model with the Resnet50 network which has fewer parameters and has a deeper level, and the experimental results show that comparing with the baseline, this operation makes the detection accuracy of the model map50 improved by 2.4%; on the basis of our experiments, we carried out experiments using the swin-transformer as backbone, the swin-transformer algorithm introduce a self-attention mechanism, which can better capture the global contextual information than the original model, and the experimental results show that our improved method improves the map50 value by 3.5% compared to the baseline. Compared with the original method, the improved algorithm has higher smoke detection accuracy.

References

1. Cao Y, Tang Q, Lu X et al (2020) STCNet: spatio-temporal cross network for industrial smoke detection
2. Yuan FN, Xia X, Li G et al (2019) GaborNet for smoke recognition and texture classification. J Image Graph 24(2):269–281
3. Ho CC (2013) Nighttime fire/smoke detection system based on a support vector machine. Math Probl Eng 2013
4. Ren S et al (2017) Faster R-CNN: towards real-time object detection with region proposal networks. IEEE Trans Pattern Anal Mach Intell 1137–49, https://doi.org/10.1109/tpami.2016.2577031
5. Redmon J et al (2016) You only look once: unified, real-time object detection. In: 2016 IEEE conference on computer vision and pattern recognition (CVPR). https://doi.org/10.1109/cvpr.2016.91
6. Lin TY, Goyal P, Girshick R et al (2017) Focal loss for dense object detection. In: Proceedings of the IEEE international conference on computer vision, pp 2980–2988
7. Liu W et al (2016) SSD: single shot multibox detector. In: Computer vision—ECCV 2016. Lecture notes in computer science, pp 21–37. https://doi.org/10.1007/978-3-319-46448-0_2
8. Simonyan K, Zisserman A (2015) Very deep convolutional networks for large-scale image recognition. In: International conference on learning representations
9. Glorot X et al (2011) Deep sparse rectifier neural networks. J Mach Learn Res
10. He K, Zhang X, Ren S et al (2016) Deep residual learning for image recognition. In: Proceedings of the IEEE conference on computer vision and pattern recognition, pp 770–778
11. Liu Z et al (2021) Swin transformer: hierarchical vision transformer using shifted windows. In: 2021 IEEE/CVF International conference on computer vision (ICCV). https://doi.org/10.1109/iccv48922.2021.00986
12. Vaswani A et al (2017) Attention is all you need. In: Neural information processing systems
13. Padilla R et al (2020) A survey on performance metrics for object-detection algorithms. In: 2020 International conference on systems, signals and image processing (IWSSIP). https://doi.org/10.1109/iwssip48289.2020.9145130

CoT-Enhanced CNN-ViT Fusion with Adaptive Token Learning for Precise Facial Expression Recognition

Lingxin Xiong, Jicun Zhang, Xiaojia Zheng, and Yuxin Wang

Abstract Facial expression recognition is a popular research direction in computer vision. Traditional CNN models suffer from spatial localization problems and is difficult to learn the dependencies between different facial regions, thus limiting the global understanding of facial expressions. ViT performs excellently in image categorization and provides a new way of global information perception for the FER task, but lacks inductive bias and is prone to focus on noise. We propose an innovative CoT_FER hybrid architecture, which utilizes the CoT enhancement module to connect CNN and ViT, capturing local regional correlations to improve the global understanding of facial expressions, and adopts an adaptive learning method in the self-attention module of Transformer Encoder to dynamically retain tokens that are most relevant to the expression categorization, to overcome the problem of focusing noise and to improve the computational efficiency. Experiments show that the accuracy of the model on CK+, RAF-DB, and FERPlus is 99.2%, 91.07%, and 90.57%, respectively, demonstrating strong generalization ability and adaptability.

L. Xiong · X. Zheng · Y. Wang
School of Computer Science and Technology, Dalian University of Technology, Dalian, China
e-mail: xionglingxin24@163.com

X. Zheng
e-mail: zhengxiaojia91@163.com

Y. Wang
e-mail: wyx@dlut.edu.cn

J. Zhang (✉)
Neusoft Reach Automotive Technology (Dalian) Co., Ltd., Dalian, China
e-mail: zhangjicun89@163.com

S. Yadav et al. (eds.), *Proceedings of 3rd International Conference on Artificial Intelligence, Robotics, and Communication*, Lecture Notes in Electrical Engineering 1172, https://doi.org/10.1007/978-981-97-2200-6_13

1 Introduction

Facial expression recognition is widely used in medical, lie detection, and safe driving fields, but accurate recognition still faces significant challenges due to the complexity of facial expressions. Traditional manual feature extraction methods perform well on laboratory-controlled datasets, but CNN-based methods [1] are more dominant for field environments with light, angle, and occlusion. Convolutional networks learn deeper features by capturing information from localized regions through convolutional operations, but they also lack an understanding of global facial expressions. In contrast, Vision Transformer (ViT) [2] has advantages in global feature understanding, but its large number of parameters and lack of inductive bias, coupled with the small size of the FER dataset, may lead to model bias toward occluded and noisy regions when trained directly. In this context, we propose the CoT_FER model to address the above issues.

Our main contributions are as follows: (1) For the first time, the CoT module is used to connect the CNN and ViT models, which makes the model more sensitive to local correlations and deepens the model's understanding of the intrinsic connections between different regions of the face. (2) In the self-attention module of the Transformer Encoder, an adaptive learning method is used to dynamically retain the top K tokens that are most relevant to expression categorization, removing noise, and occlusion effects. (3) Satisfactory results are obtained on the datasets CK+, RAF-DB, and FERPlus, which demonstrate the generalization and adaptability of our proposed method.

2 Relate Work

2.1 Convolutional Neural Network on FER

Convolutional networks are the dominant framework in the field of FER. Qian et al. [3] used ventral channel attention and dilated spatial attention mechanisms to enhance feature extraction on CNNs. Xiao et al. [4] used a multi-task co-learning network to efficiently fuse information from different features. Wen et al. [5] paid attention to the features of multiple facial regions at the same time through the mechanism of attention of multiple associates. Inspired by this, we believe that face expression involves the synergistic effect of various facial parts, so we introduce the CoT [6] module after the CNN to better capture the local region association information and thus improve the overall performance.

2.2 Transformer Method on FER

The Transformer architecture has become a research hotspot in the FER field with its excellent global feature capture capability. Xue et al. [7] solved the problem of occlusion by manually setting the parameters to pool the attention weights. Yao et al. [5] fine-tuned the channel space attention transformer to allow the model to focus more on expression-related information. Feng et al. [8] used weight optimization to avoid the influence of unfavorable factors. Unlike the previous methods, we use an adaptive learning method in the Transformer encoder to retain the first K tokens that are most relevant to expression categorization, removing noise, and occlusion to improve performance.

3 The Proposed Method

3.1 Framework of CoT_FER

The CoT_FER model framework is shown in Fig. 1. The framework mainly consists of CNN, a CoT module, and an improved Transformer encoder. Among them, CNN is used to extract local features, CoT is used as a connection module to learn the correlation of features in different regions, and the Transformer encoder is used to learn global features, in its self-attention mechanism, the adaptive learning method is used to retain the first K tokens most relevant to the expression categorization, removing noise and occlusion interferences to focus on the most discriminative region.

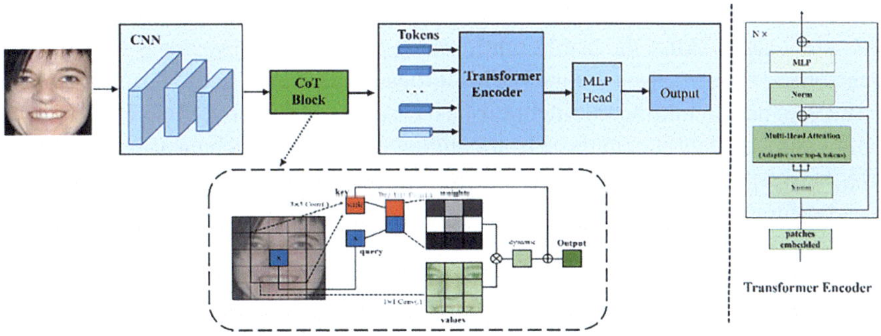

Fig. 1 Structure of CoT_FER model

3.2 CoT Module Injection

The CoT module improves the correlation between local regions and feature expression by learning the contextual information of different regions in the feature map. Using CoT to connect the CNN and Transformer encoder ensures better integration of local features into the overall model, which improves the performance of the facial expression recognition model.

Specifically, the implementation steps of CoT are as follows: for the input feature map $X \in \mathbb{R}^{H \times W \times C}$, define $K = X$, $Q = X$, and $V = XW_v$, first extract the contextual information by a $k \times k$ convolution kernel, regarded as the static contextual expression of X, denoted as K^1. Then, conditional on the fact that K^1 and Q are concatenated in tandem, extract the contextual information by two consecutive 1×1 convolutions (one with activation function, denoted as W_θ, and one without activation function, denoted as W_δ, to obtain the attention matrix $A = \left[K^1, Q\right] W_\theta W_\delta$). The local attention matrix of A at each spatial location is learned based on the query features and context keys, rather than on isolated query key pairs. This enhanced self-attention learning by mining the additional guidance of the static context K^1 is obtained by aggregating all V and A to get the enhanced self-attention, denoted as $K^2 = \text{softmax}(A) * V$, as the dynamic context representation of the input X. Finally, K^2 and K^1 are fused as the output of the CoT block. Overall, $\text{CoT}_{\text{output}(\text{X}_{\text{CNN}})} = K^1 + K^2$.

3.3 Adaptive Token Learning

According to [2], the input image is first chunked and flattened into vectors, and then class_token is added to help learn the mapping relationship between features and class predictions. After the position information is added, the vector is fed into the Transformer encoder, which contains Multi-head Attention, Layer Norm, and MLP modules, where Multi-head Attention has Z self-attention modules. Dot product computation is a key step in the self-attention mechanism, which measures the similarity between Q and K and determines how much attention each token pays to the other positions. The dot product calculation is shown in (1).

$$\left(QK^T\right)_{ij} = \sum_{k=1}^{d} Q_{ik} \cdot K_{jk} \tag{1}$$

After calculating the similarity score, it is converted into a set of attention weights that sum to 1 using the softmax function. These weights are multiplied by V to get the final attention output. As shown in Eq. (2), K and V are query vectors, key vectors, and value vectors, respectively. d denotes the dimension of the K vector.

$$\text{Attention}(Q, K, V) = \text{Softmax}\left(\frac{QK^T}{\sqrt{d}}\right)V \tag{2}$$

Finally, the output is obtained by processing the class_token through MLP Head. The stronger the correlation between the remaining tokens and the class token, the more important the information they carry to contribute to the expression recognition result.

With a set of learnable parameters $[K_1, K_2, K_3, \cdots, K_N]$, the first K tokens dynamically retained during the training process are pooled against the attention weight matrix to achieve denoising and masking.

4 Experiments and Comparison

4.1 FER Datasets

CK+ [9]: laboratory-controlled benchmark dataset (extended Cohn-Kanade). Seven basic expressions were selected and the last three frames of the labeling sequence were extracted for experiments.

RAF-DB [10]: The Real-world Affective Faces Database, contains 29,672 images of facial expressions of different ages and individuals. In the experiment, 12,271 images are used for training and 3068 images are used for testing.

FERPlus: Provides more accurate labeling results compared to the FER2013 dataset [11]. The votes of 10 individuals obtain emotion labels. In the experiment, 25,373 images were used for training and 3225 for testing.

4.2 Experiment Setup

The experiments use the IR50 network pre-trained on the Ms-Celeb-1M dataset and the vit-small [2] model pre-trained on the ImageNet dataset. The image size is uniformly scaled to 112 × 112. The adaptive learning K value is initially defaulted to 1. Cross entropy is used as the loss function and SGD is used as the optimizer. The learning rate is fixed to 1e-3 with a batch size of 48 and an epoch of 60 and implemented on NVIDIA GeForce RTX 3080Ti.

Table 1 Ablation study on CoT block and adaptive method

CoT Block	Adaptive method	CK+	RAF-DB	FERPlus
×	×	97.968	89.569	89.764
×	√	98.981	90.61	90.446
√	×	98.705	90.775	90.322
√	√	99.200	91.07	90.570

4.3 Ablation Study

4.3.1 Ablation Experiments on the Proposed Module

To validate the effectiveness of the proposed modules, ablation experiments were performed on three datasets. Table 1 shows that both individual modules effectively improve the accuracy, and the combination of the two further improves the performance by 1.232, 1.501, and 0.806% in the CK+, RAF-DB, and FERPlus datasets, respectively.

4.3.2 Visualization and Analysis

Figure 2 demonstrates the confusion matrices on the CK+, RAF-DB, and FERPlus datasets, and overall, the model performs best in the categories of happy, surprised, and angry with high discrimination. The two expressions, surprise and happy, are difficult to distinguish in local regions, and the local relevance learning of the CoT module strengthens the global perception ability of the model with significant results.

To further explore the role of the CoT module at the connection, we visualize the feature maps before and after the CoT module in the inference stage, and randomly select the images in the RAF-DB test set, the results used for visualization are shown in Fig. 3. It can be seen that after the processing of the CoT module, the regions

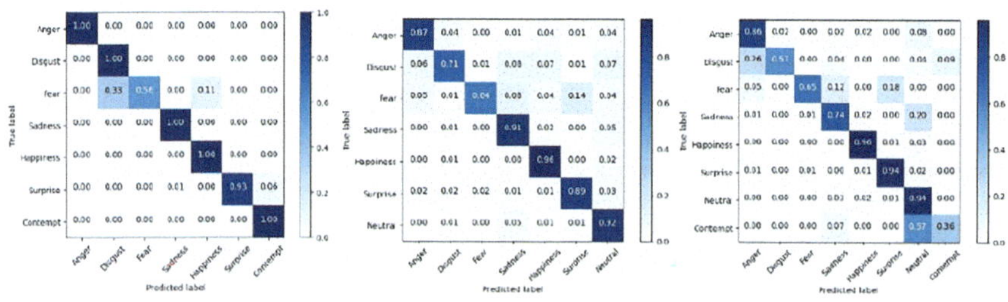

(a) (b) (c)

Fig. 2 Confusion matrix of CoT_FER on CK+ (**a**), RAF-DB (**b**), FERPlus (**c**)

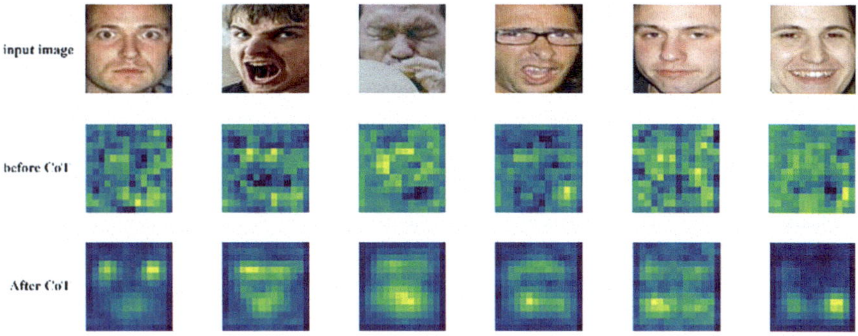

Fig. 3 Visualization of feature maps before and after the CoT module

of interest of the model are more concentrated and the local correlation of different regions is stronger.

4.4 *Comparison Study*

It was evaluated on three datasets and compared with state-of-the-art methods. Table 2 showed that 99.20% was achieved on CK+ (tenfold cross-validation method), 91.07% on RAF-DB, and 90.57% on FERPlus, which were superior to other models. Compared to fusing global and local features to learn feature information [12] comprehensively and fine-tuning the channel space attention transformer [5] to allow the model to focus more on expression-related information, CoT_FER utilizes the CoT module to understand the correlation between local features, enhances the comprehensive understanding of expressions, uses adaptive learning to preserve topK tokens, removes noise and occlusion, and improves the overall expression recognition performance.

Table 2 Performance comparison on CK+, RAF-DB, FERPlus

CK+		RAF-DB		FERPlus	
Method	Accuracy (%)	Method	Accuracy (%)	Method	Accuracy (%)
ZFER[13]	98.74	DAN[1]	89.70	FT-CSAT[5]	89.26
VDAC[3]	98.98	CFNet[4]	89.99	AGT[12]	89.40
CFNet[4]	99.07	FST[8]	90.38	FST[8]	90.41
CoT_FER (Ours)	99.20	CoT_FER (Ours)	91.07	CoT_FER (Ours)	90.57

5 Conclusion

In this work, we propose the CoT_FER model that uses the CoT module to enhance fine-grained local feature relationships for enhanced local and global expression understanding and an adaptive approach to remove noise interference. It performs competitively on CK+, RAF-DB, and FER+ datasets. Plans are to extend the model to handle video-based expression recognition problems.

References

1. Wen Z, Lin W, Wang T et al (2023) Distract your attention: multi-head cross attention network for facial expression recognition. Biomimetics 8(2):199
2. Dosovitskiy A, Beyer L, Kolesnikov A et al (2020) An image is worth 16x16 words: transformers for image recognition at scale. arXiv preprint arXiv:2010.11929
3. Qian Z, Mu J, Tian F (2023) Ventral-Dorsal attention capsule network for facial expression recognition. Digital Signal Process 136:103978
4. Xiao J, Gan C, Zhu Q et al (2023) CFNet: Facial expression recognition via constraint fusion under multi-task joint learning network. Appl Soft Comput 141:110312
5. Yao H, Yang X, Chen D et al (2023) Facial expression recognition based on fine-tuned channel-spatial attention transformer. Sensors 23(15):6799
6. Li Y, Yao T, Pan Y et al (2022) Contextual transformer networks for visual recognition. IEEE Trans Pattern Anal Mach Intell 45(2):1489–1500
7. Xue F, Wang Q, Tan Z et al (2022) Vision transformer with attentive pooling for robust facial expression recognition. IEEE Trans Affect Comput
8. Feng H, Huang W, Zhang D et al (2023) Fine-tuning Swin transformer and multiple weights optimality-seeking for facial expression recognition. IEEE Access 11:9995–10003
9. Lucey P, Cohn JF, Kanade T et al (2010) The extended Cohn-Kanade dataset (CK+): a complete dataset for action unit and emotion-specified expression. In: 2010 IEEE computer society conference on computer vision and pattern recognition-workshops. IEEE, pp 94–101
10. Li S, Deng W, Du J (2017) Reliable crowdsourcing and deep locality-preserving learning for expression recognition in the wild. In: CVPR, vol. 28. IEEE, pp 2584–2593
11. Goodfellow IJ, Erhan D, Carrier PL et al (2013) Challenges in representation learning: a report on three machine learning contests. In: Neural information processing: 20th international conference, ICONIP 2013. Daegu, Korea, November 3–7, 2013. Proceedings, Part III 20. Springer Berlin Heidelberg, pp 117–124
12. Sun N, Song Y, Liu J et al (2023) Appearance and geometry transformer for facial expression recognition in the wild. Comput Electr Eng 107:108583
13. Shahzad T, Iqbal K, Khan MA et al (2023) Role of zoning in facial expression using deep learning. IEEE Access 11:16493–16508

Evaluation Method of Cognitive Level Based on Machine Learning and Wearable Device

Shuai Tao, Hongbin Hu, Jie Zhao, and Liwen Kong

Abstract Dementia is an incurable syndrome, but early detection and treatment of dementia can help slow down the progress of dementia. Recent studies have highlighted the association between overall cognitive function, motor function, and gait abnormalities. In this study, we enrolled 302 participants from the Rehabilitation Hospital Affiliated to the National Rehabilitation Aids Research Center, screening 193 according to set criteria, including 137 patients with Mild Cognitive Impairment (MCI) and 56 healthy controls (HC). Gait parameters were collected using a wearable device while participants performed single and dual tasks. Focusing on gait cycle, kinematics parameters, and time–space parameters, we employed recursive feature elimination to identify crucial features. Utilizing the participant's Moca score as the response variable, we established a machine learning model that quantitatively evaluates cognitive levels based on gait features. The temporal and spatial parameters of toe-off angle and heel strike angle emerged as pivotal markers with significant clinical relevance for evaluating cognitive levels. These parameters hold substantial clinical application value in the prevention or delayed onset of dementia.

1 Introduction

Dementia is a syndrome characterized by a decline in memory, cognition, behavioral ability, and intelligence. Although dementia is considered incurable, there is evidence to suggest that early treatment for dementia can slow down the progression of dementia. With the development of technology, using artificial intelligence for assistance has become a reality [1]. In recent years, studies have shown a relationship between overall cognitive function, motor function, and gait abnormalities [2]. The emergence of artificial intelligence and wearable sensing technology [3] has

S. Tao (✉) · H. Hu · L. Kong
Dalian University, Dalian 116622, China
e-mail: taoshuai@dlu.edu.cn

J. Zhao
Zhongshan Hospital Affiliated to Dalian University, Dalian 116001, China

S. Yadav et al. (eds.), *Proceedings of 3rd International Conference on Artificial Intelligence, Robotics, and Communication*, Lecture Notes in Electrical Engineering 1172, https://doi.org/10.1007/978-981-97-2200-6_14

made precise gait analysis possible. Wu [4] using the lightGBM algorithm to select patients with low cognitive levels. This study mainly investigated the relationship between the responsiveness and cognitive level of participants. Jung [5] used gait sequence features and long-term short-term memory networks to predict the risk of cognitive impairment. Saho [6] used SVM to classify patients into high cognitive risk and low cognitive risk. This study only conducted a classification study and did not establish a quantitative evaluation model. Noh [7] used machine learning models to predict patient MMSE scores based on features from gait parameters and physical fitness tests. However, the gait feature dimensions used in this study are limited. Matsuura [8] used single and dual-task gait parameters combined with statistical algorithms to predict the cognitive level of patients. Ricciardi [9] identified PD patients with MCI using decision tree (DT), random forest (RF), and k-nearest neighbor (KNN) algorithms. These studies focus on features other than gait features, without delving into the relationship between various gait features and cognitive levels. This research focuses on MCI patients and normal individuals, with gait parameters such as gait cycle, kinematic parameters, and spatiotemporal parameters as the research focus. Important feature combinations are selected and a machine learning model that can accurately predict cognitive level is established. The relationship between these features and cognitive level is also studied.

2 Materials and Methods

2.1 Subjects of Study

Between September 2018 and December 2022, 302 participants were enlisted from the Rehabilitation Hospital, an institution affiliated with the National Rehabilitation Assistive Device Research Center. Upon rigorous adherence to the pre-set inclusion and exclusion parameters, a final cohort of 193 participants was established. This cohort comprised 137 individuals diagnosed with MCI and 56 who acted as HC. The research study received ethical clearance from the Research Ethics Committee of the Rehabilitation Hospital affiliated with the National Rehabilitation Assistive Device Research Center (Approval Number: S20220226), and written informed consent was secured from all participants upon their enrollment. This study employed the Moca [10] to evaluate the overall cognitive functions of the participants.

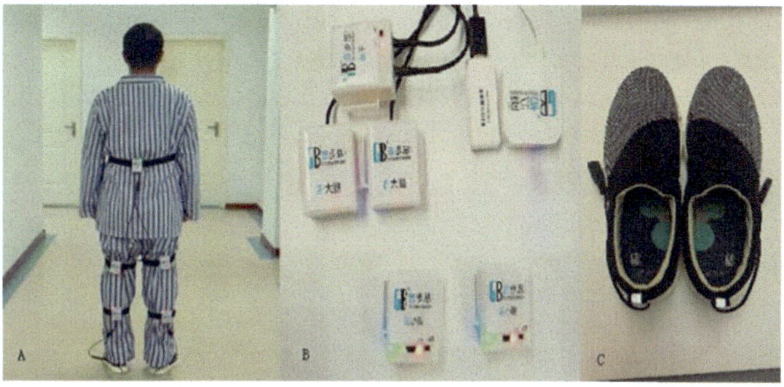

Fig. 1 JiBuEn equipment and correct wearing

2.2 Equipment and Performance Test

All gait tests were conducted inside the spacious corridor of the rehabilitation hospital affiliated with the National Rehabilitation Auxiliary Research Center. This experiment uses the JiBuEn [11] system. The JiBuEn device and correct wearing are shown in Fig. 1.

2.3 Performance Test and Gait Parameters

Participants carried out two gait tests (one single-task test, one dual-task test) under guidance. In the single-task test, free walking, participants walked at a comfortable speed without any additional activities. The dual-task test was the countdown-from-100 test, which required participants to start counting down from 100 (like 100, 99, 98…etc.) while walking. The gait parameters collected in this study were consistent for both single and dual tasks, including Gait cycle, Pace, Cadence, Stride length, Stride time, Swing time, Support time, Toe-off angle, Heel strike angle, Asymmetric coefficient [12], Gait consumption [13].

2.4 Feature Selection and Model

The data comprised 51 independent variables including the sex of the participant, body information, gait variability, and gait parameters. All data was standardized before input. We use SVM, RF, Lasso, Elastic Net, Adaboost, and GBDT to establish regression models, respectively. The dataset was randomly divided 100 times. The training set comprised 70% of the samples, and the test set comprised 30% of the samples. Then, we used smote to adjust the balance of training set. The Linear Recursive Feature Elimination (RFE) technique [14] was used for feature selection to reduce the input variables. The best feature combinations were used as inputs of model and the optimal hyperparameters were determined using grid search. Ultimately, the RMSE from each model's 100 experiments was compared to select the model with the smallest error as the optimal model.

3 Result

3.1 Feature Selection Results

The accumulated number of feature combinations chosen by different machine learning models is illustrated in Fig. 2. From the graph, it can be inferred that the degree of variation in the two linear regression models, Lasso and ElasticNet, is relatively small. The number of features does not prominently influence them. SVM selects fewer optimal features. This suggests that SVM may not be ideal for handling high-dimensional data. On the other hand, the ensemble learning models RF, Adaboost, and GBDT exhibit relatively similar behavior. On a broader note, their RMSE tends to decrease as the number of features increases, indicating their capability to consider more dimensional information, thereby possessing better potential to generate accurate predictions.

3.2 Training Results

We used every model's optimal feature combination as the input, and conducted a grid search to select the best parameters. Table 1 lists the RMSE statistical data of each machine learning model and the number of features in the model's best feature combination. From the table, it can be seen that the RF, Adaboost, and GBDT models all opted for a large number of features, with GBDT selecting the most features. RF has the smallest mean value.

We decided to select RF for further research. Figure 3 presents the ranking of features according to their importance to the model, with the x-axis indicating the importance of a feature to the prediction outcome. The top three features are single-task heel strike angle, single-task pace, and single-task toe-off angle. This conclusion is similar to our previous research [15], where the heel strike angle and toe-off angle do play a significant role.

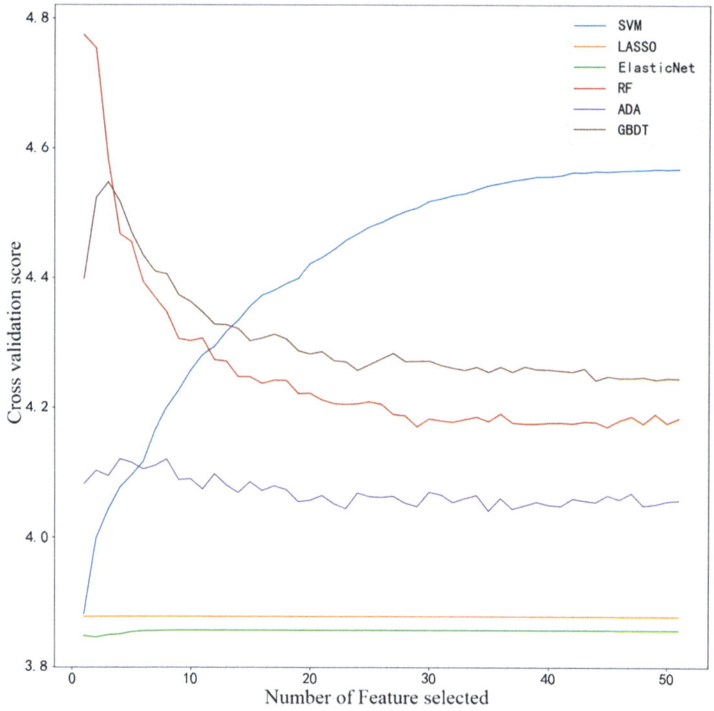

Fig. 2 Cumulative variable count

Table 1 Results of machine learning model

ML	Mean	Median	Min	Max	Variable
Lasso	3.8369	3.8749	2.5276	4.6934	1
ElasticNet	3.8369	3.8605	2.4418	4.6482	2
SVM	3.7764	3.8013	2.6584	4.6616	1
Adaboost	3.9891	4.0203	2.9300	4.9155	34
RF	3.7640	3.7510	2.9140	5.0530	40
GBDT	3.9373	3.9712	2.8637	4.8745	44

4 Conclusions

Previous research has demonstrated statistically significant differences in gait parameters and kinematic parameters between individuals with cognitive impairment and healthy individuals. However, limited research has considered time–space parameters (toe-off angle, heel strike angle) as factors in assessing cognition. In this study, we expanded the scope to include gait cycles, kinematic parameters, and time–space parameters, with a specific emphasis on the latter. The analysis revealed significant

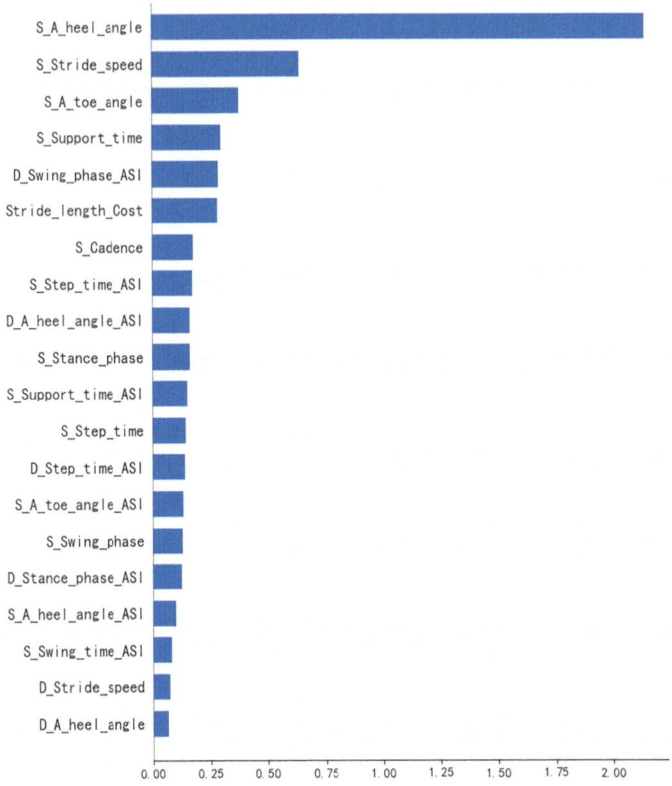

Fig. 3 Importance of features

differences in time–space parameters among subjects with different cognitive levels. Additionally, time–space parameters ranked high in importance according to machine learning algorithms. Recent studies have identified toe-off angle and heel strike angle as crucial features for distinguishing between healthy individuals and those with gait disorders [16]. These parameters may also serve as typical gait features for differentiating post-stroke dementia from Alzheimer's disease (AD) [17]. Reduced heel strike angle is indicative of impaired foot clearance, leading to an increased risk of falls, which significantly impacts the quality of life for individuals with MCI and AD [18]. Contrary to previous research highlighting the significance of dual-task walking in influencing gait parameters among MCI patients, our experiment found that single-task heel strike angle and single-task toe-off angle ranked higher in importance compared to dual-task features. We discovered that the significance of single and dual-task heel strike angles differed significantly between HC and MCI groups but was less distinct in the AD group. This may be attributed to AD subjects being more prone to "shuffling" during walking [19]. Furthermore, many previous studies focused on classification tasks without the need for precise predictions of cognitive levels. In contrast, our experiment aimed to predict subjects' cognitive levels continuously. Consequently, the machine learning model paid greater attention to features exhibiting continuous changes across different cognitive scores. The use of the Smote algorithm balanced the dataset, ensuring an equal number of samples for each cognitive score, resulting in uniform attention from the machine learning algorithm to subjects with varying cognitive levels.

In summary, this study found that: (1) gait parameters collected by wearable devices can serve as useful clinical tools for evaluating cognitive level; (2) time–space parameters (toe-off angle, heel strike angle) are important features for quantitatively evaluating cognitive level, and have important clinical application value in preventing or delaying the occurrence of MCI in the future; (3) the current experimental paradigm needs to be improved for AD patients with "shuffling" during walking and future research should further explore other effective experimental paradigms.

Acknowledgements The research funded by Interdisciplinary project of Dalian University (Project Number: DLUXK-2023-QN-004).

References

1. Lo F, Su F, Chen S, Qiu J, Du J (2021) Artificial intelligence aided innovation education based on multiple intelligence. In: 2021 IEEE International conference on artificial intelligence, robotics, and communication (ICAIRC). Fuzhou, pp 12–15. https://doi.org/10.1109/ICAIRC52191.2021.9544874
2. Tao S, Han X, Kong LW (2022) Machine learning-based gait analysis for recognition of amnestic mild cognitive impairment and Alzheimer's disease. Chin Gen Pract 25(31):3857–3865. https://link.cnki.net/urlid/13.1222.R.20220919.1557.015
3. Zhao X, Ma X, Shan Y, Gong X (2021) A wearable foot pressure sensor. In: 2021 IEEE International conference on artificial intelligence, robotics, and communication (ICAIRC). Fuzhou. pp 8–11. https://doi.org/10.1109/ICAIRC52191.2021.9544992
4. Wu S, Matsuura T, Okura F (2021) Detecting lower MMSE scores in older adults using cross-trial features from a dual-task with gait and arithmetic. IEEE Access 9:150268–150282. https://doi.org/10.1109/ACCESS.2021.3126067
5. Jung D, Kim J, Kim M (2021) Classifying the risk of cognitive impairment using sequential gait characteristics and long short-term memory networks. IEEE J Biomed Health Inf 25(10):4029–4040. https://doi.org/10.1109/JBHI.2021.3073372
6. Saho K, Uemura K, Sugano K (2019) Using micro-Doppler radar to measure gait features associated with cognitive functions in elderly adults. IEEE Access 7:24122–24131. https://doi.org/10.1109/ACCESS.2019.2900303
7. Noh B, Yoon H, Youm C (2021) Prediction of decline in global cognitive function using machine learning with feature ranking of gait and physical fitness outcomes in older adults. Int J Environ Res Public Health 18(21):11347. https://doi.org/10.3390/ijerph182111347
8. Matsuura T, Sakashita K, Grushnikov A (2019) Statistical analysis of dual-task gait characteristics for cognitive score estimation. Sci Rep 9(1):1–12. https://doi.org/10.1038/s41598-019-56485-w
9. Ricciardi C, Amboni M, de Santis C (2020) Machine learning can detect the presence of mild cognitive impairment in patients affected by Parkinson's disease. In: 2020 IEEE International symposium on medical measurements and applications (MeMeA). Bari, pp 1–6. https://doi.org/10.1109/MeMeA49120.2020.9137301.

10. Nasreddine ZS, Phillips NA, Bédirian V (2005) The Montreal cognitive assessment, Moca: a brief screening tool for mild cognitive impairment. J Am Geriatrics Soc 53(4):695–699. https://doi.org/10.1111/j.1532-5415.2005.53221.x
11. Tao S, Zhang X, Cai H (2018) Gait based biometric personal authentication by using MEMS inertial sensors. J Ambient Intell Hum Comput 9(5):1705–1712. https://doi.org/10.1007/s12652-018-0880-6
12. White R, Agouris I, Fletcher E (2005) Harmonic analysis of force platform data in normal and cerebral palsy gait. Clin Biomech 20(5):508–516. https://doi.org/10.1016/j.clinbiomech.2005.01.001
13. Muir-Hunter SW, Montero-Odasso M (2016) Gait cost of using a mobility aid in older adults with Alzheimer's disease. J Am Geriatrics Soc 64(2):437–438. https://doi.org/10.1111/jgs.13973
14. Pullanagari R, Kereszturi G, Yule I (2018) Integrating airborne hyperspectral, topographic and soil data for estimating pasture quality using recursive feature elimination with random forest regression. Remote Sens 10(7):1117. https://doi.org/10.3390/rs10071117
15. Rao H, Shi X, Rodrigue AK (2019) Feature selection based on artificial bee colony and gradient boosting decision tree. Appl Soft Comput 74:634–642. https://doi.org/10.1016/j.asoc.2018.10.036
16. Lindh-rengifo M, Jonasson SB, Ullén S (2022) Components of gait in people with and without mild cognitive impairment. Gait Posture 93:83–89. https://doi.org/10.1016/j.gaitpost.2022.01.012
17. Ni L, Lv W, Sun D (2021) Pathological gait signatures of post stroke dementia with toe-off and heel-to-ground angles discriminate from Alzheimer's disease. Front Neurosc 13:766884. https://doi.org/10.3389/fnagi.2021.766884
18. Alcock L, Galna B, Lord S (2016) Characterisation of foot clearance during gait in people with early Parkinson's disease: deficits associated with a dual task. J Biomech 49(13):2763–2769. https://doi.org/10.1016/j.jbiomech.2016.06.007
19. Ginis P, Pirani R, Basaia S (2017) Focusing on heel strike improves toe clearance in people with Parkinson's disease: an observational pilot study. J. Physiotherapy 103(4):485–490. https://doi.org/10.1016/j.jbiomech.2016.06.007

Ship Image Segmentation Based on Multi-dimensional Information and Feature Fusion

Yanxiang Cai, Haicheng Xie, Jianfeng Yang, and Tianqi Yu

Abstract Ship images in inland waterway scenarios have the characteristics of fixed background and continuous image. Therefore, motion detection is a feasible preprocessing method to mine the dataset information. For the feature extraction problem of RGB image and motion information, an asymmetric two-branch structure is adopted to perform convolution operation, respectively. For the problem of the scale gap between ship targets, we design a multi-scale convolutional attention module, which is used to capture the features of targets of different scales by setting up different receptive fields. To fuse features of the two branches, an adaptive feature fusion module is designed and inserted into both the encoder and decoder, which can effectively fuse the features of two branches. Experimental results show that the segmentation method proposed in this work can reach 94.88% of mIoU in the ship image dataset, compared to 92.88% of U-Net.

1 Introduction

The inland navigation system has always played a major role in transportation due to its economic advantages [1], but the continuous growth in the number of vessels has also put higher demands on the inland navigation regulatory system. Deep learning-based image segmentation methods have flourished in recent years and are widely used in the fields of medical images, automated driving, pedestrian detection, etc. Therefore, deep learning-based semantic segmentation methods have also

---

Y. Cai · H. Xie · J. Yang · T. Yu (✉)
School of Electronic and Information Engineering, Soochow University, Suzhou 215006, China
e-mail: tqyu@suda.edu.cn

Y. Cai
e-mail: 20214228046@stu.suda.edu.cn

J. Yang
e-mail: jfyang@suda.edu.cn

S. Yadav et al. (eds.), *Proceedings of 3rd International Conference on Artificial Intelligence, Robotics, and Communication*, Lecture Notes in Electrical Engineering 1172, https://doi.org/10.1007/978-981-97-2200-6_15

been introduced to the task of regulating waterways and ships in inland waterway scenarios.

Yupeng Ma [2] used a joint approach of AlexNet and SSA-SVM classifier for ship target recognition to solve the problem of ship target recognition in small datasets. Wenkang Zhou [3] proposed a ship image segmentation method based on the attention mechanism and depth-separable convolution, which optimized the network's edge localization ability for multi-scale targets. However, there is still a lack of publicly available ship image datasets for inland waterway scenarios, and the above-mentioned methods are limited to using single RGB image information for model training, lacking more effective preprocessing of the dataset information. In addition, there is a large disparity in the scale of the ship targets in images.

To resolve the above issues, a ship image segmentation method is proposed in this paper, based on the multi-dimensional information and feature fusion. Firstly, the dataset is preprocessed to extract the motion features as a priori information to participate in the model training. Secondly, two branches are adopted to process the motion features and the fused features, respectively, and then the attention mechanism-based adaptive feature fusion module to implement the fusion of the multi-dimensional information at the intermediate level of the network. Experiments show that the method proposed in this paper can achieve 94.88% of mIoU, and the segmentation effect is significantly improved compared to the baseline.

The rest of the paper is organized as follows. Classical approaches to semantic segmentation and feature fusion are described in Sect. 2. In Sect. 3, the segmentation method for ship image is proposed in detail. The experimental evaluation is conducted in Sect. 4. Finally, the work of this paper is concluded in Sect. 5.

2 Related Work

This section briefly analyzes the image segmentation from two aspects, namely, semantic segmentation and feature fusion.

2.1 Semantic Segmentation

Semantic segmentation has undergone a vigorous development process, representing significant advances in semantic segmentation from fully convolutional networks represented by FCN [4], U-Net [5], Deeplab [6], and in recent years, the transformer structure evolved from natural language processing. Jonathan Long et al. [4] proposed the Full Convolutional Network (FCN), which implements an end-to-end semantic segmentation method. Olaf Ronneberger [5] proposed the U-Net, which uses a classical codec structure and skip connection to achieve good segmentation performance in several tasks. SegNeXt [7] believes that successful semantic segmentation models have the following three key attributes, namely, a powerful encoder, interaction of multi-scale features, and spatial attention. While these works have proven effective on mainstream image segmentation benchmarks, they are still RGB-only approaches that lack attention to other valid information about the dataset. Our method is designed considering the dataset characteristics.

2.2 Feature Fusion

Feature fusion is an important part of the network structure, and the common feature fusion methods in deep networks can be divided into the following two types. (a) Linear approaches. The shortcut connection method used in ResNet [8] to solve the degradation is essentially addition, while skip connection is used in U-Net to fuse the feature maps of shallow and deep networks. (b) Integration based on attention mechanisms. The channel attention module is used in SENet [9] for feature fusion. SA-NET [10] proposes to use shuffle attention in the network. Dai [11] proposes a multi-scale channel attention module to fuse features of inconsistent semantics and scales. However, linear methods allocate the features with fixed weights regardless of the variance of contents, and approaches based on attention mechanisms only use one input feature to generate weight. Our method is aware of both features and uses high-level features as guidance to generate weights.

3 Method

In this section, our proposed method for ship image segmentation is provided in detail, based on the multi-dimensional features extraction and adaptive feature fusion. Basically, U-Net [5] is adopted as the baseline, which is a classical encoder–decoder network.

3.1 Dataset Preprocessing

This dataset is composed of ship images in inland waterway scenarios. Considering that the images are continuous image frames intercepted from surveillance videos, motion detection is adopted in this work to mine the dataset information.

The motion detection method used in this paper is background subtraction, but due to the complexity of the water surface, the more obvious noise such as the near water splash cannot be removed by morphological processing, so the background subtraction method is used in conjunction with SVM classifiers as well as morphological operations for the detection of the motion features, to reduce the noise to a certain extent.

3.2 Network Structure

In order to better learn the texture, spatial location, and other information in RGB images and motion features, a two-branch structure is used to encode the motion information and RGB images separately. Aiming at the problem of large disparity of ship target scales in images, the attention strategy of cascaded atrous convolution is used to capture target features of different scales more efficiently with minimal increase in parameter cost. Finally, for the feature fusion problem, an intermediate level fusion strategy is adopted, i.e., the adaptive feature fusion module (AFFM) is designed based on the attention mechanism, which is utilized to perform the fusion at the encoder and decoder stages, respectively. The overall structure of the network is shown in Fig. 1.

Motion Branch. The Motion Branch is designed as an auxiliary branch with reference to the ResNet [8], and a 7×7 initial convolutional layer is used first, because a large convolutional kernel provides a large receptive field to capture the global information of the image more efficiently. This is followed by a standard residual module that further extracts and passes features and implements a twofold down-sampling to better preserve spatial information.

RGB Branch. The RGB Branch basically follows the encoder–decoder structure of the baseline, where multiple convolutional blocks with the same structure are stacked. Aiming at the problem of the scale gap of the ship target in the image, we design a multi-scale convolutional attention (MS-CA) module, and the structure is shown in Fig. 2. This module cascades multi-branch atrous convolution, sets different dilations to obtain different receptive fields, and finally obtains the spatial attention map Att $\in R^{H \times W}$ by 1×1 convolution and sigmoid function. It can be written as:

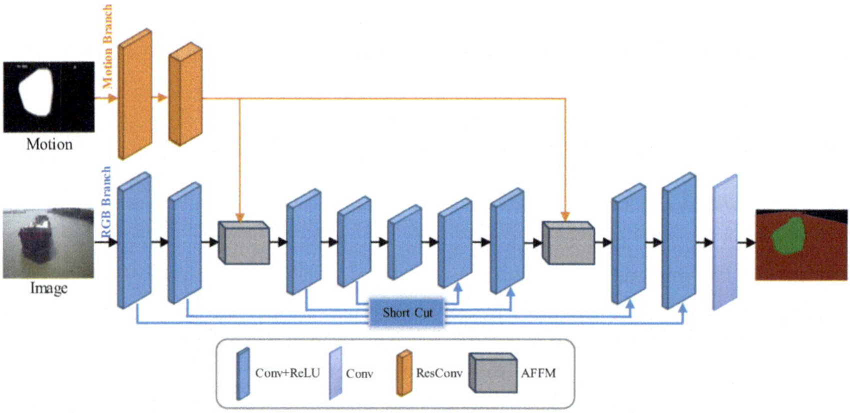

Fig. 1 Model architecture. Motion Branch (orange rectangle) and RGB Branch (blue rectangle) receive motion and RGB information, respectively. The AFFM is used for feature fusion

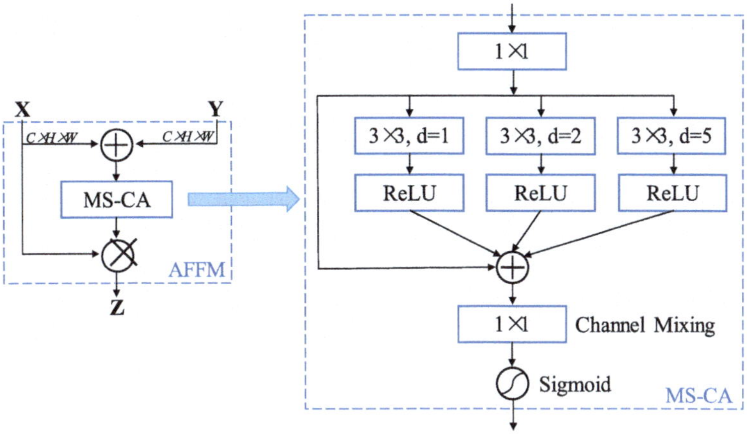

Fig. 2 Illustration of the proposed MS-CA and AFFM

$$\mathrm{Att} = \sigma\left(f^{1\times 1}\left(\sum_{i=0}^{2} f^{3\times 3}(F, d) + F\right)\right), \tag{1}$$

where the $F \in R^{C\times H\times W}$ is the input feature map, σ denotes the sigmoid function, $f^{n\times n}$ means the convolution operation with a kernel size of $n \times n$, and d denotes the dilation parameter of the atrous convolution. Considering that there is a pixel coverage problem with the atrous convolution, the hybrid dilated convolution (HDC) [12] is used here to make the dilation convolution cover all the pixels of the feature map by setting reasonable dilations.

Based on the MS-CA module, the adaptive feature fusion module (AFFM) is designed, and the structure is shown in Fig. 2, where $X, Y \in R^{C\times H\times W}$ denote the feature maps from RGB Branch and Motion Branch, and $Z \in R^{C\times H\times W}$ is the fused feature. To reduce the computational complexity, addition is used as the initial input fusion of the module. It is formalized as:

$$Z = \mathrm{Att}(X + Y) \otimes X, \tag{2}$$

where ⊗ denotes the element-wise multiplication.

4 Performance Evaluation

For experimental evaluation, the following experiments are conducted in this section.

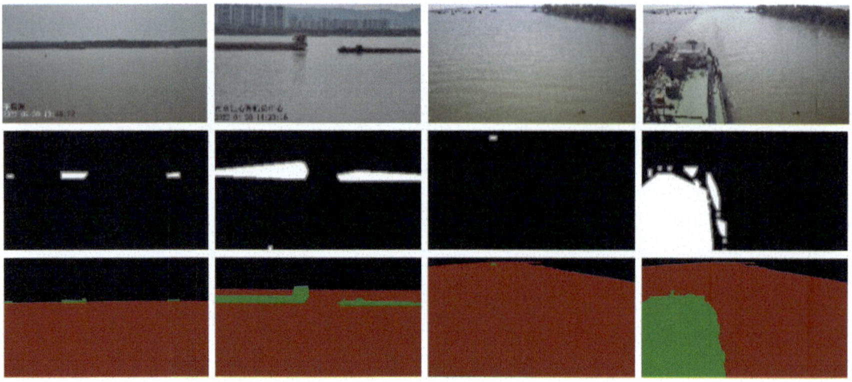

Fig. 3 Custom ship image dataset (SID)

4.1 Experimental Dataset

This experiment is conducted by using a custom ship image dataset (SID) with a total of 1440 samples, of which 1240 samples are used for training and 200 samples for validation. The dataset was preprocessed and the results obtained are shown in Fig. 3.

4.2 Experimental Setting

We conduct experiments by using PyTorch and use the mean Intersection over Union (mIoU) as the metric of segmentation. The batch size is set to 8, the training epoch is 50, and the Adam is applied to train the model. We set the initial learning rate to 0.005 and use the poly-learning rate decay policy. Considering the problem of category imbalance in the dataset, the experiment chooses weight cross-entropy as the loss function. It can be expressed by the mathematical formula as:

$$L = -\sum_{i=0}^{k-1} w_i y_i \log(p_i), \tag{3}$$

where w_i denotes the category weight, y_i means true category, p_i denotes the predicted probability of the model for category as i, and k denotes the total number of categories.

4.3 Experimental Results

In order to evaluate the performance of the proposed method, we compare our method with the baseline, and the experimental results are shown in Table 1, where the results are ignored for the background class and only the metric of the foreground classes is counted. In the experiments, we investigate different ways of fusing features, where addition and concatenate connection are used to fuse feature in input stage, and middle-stage fusion methods include addition and AFF module. AFF × 2 means inserting the AFF module in the encoder and decoder at the same time.

The experimental results show that feature fusion using addition in the original feature space does not give a boost to the training results of the network, while con has a good effect. Inserting two AFFMs in the middle stage leads to better training results for the network. Figure 4 shows a qualitative comparison with the baseline on the SID. We can see that our method recognizes well when processing object details.

Table 1 Comparison with baseline and classical segmentation networks

Method	Motion branch	Feature fusion method		SID		
		Input stage	Middle stage	Category 1	Category 2	mIoU
SegNet	×	×	×	97.45	84.89	91.17
DeepLabv3	×	×	×	97.07	88.42	92.75
U-Net	×	×	×	97.20	88.56	92.88
U-Net	×	Add	×	96.54	77.87	87.21
U-Net	×	Cat	×	97.58	89.67	93.63
U-Net	✓	×	Add	86.44	75.52	80.98
U-Net	✓	×	AFF	97.57	90.52	94.05
U-Net	✓	Cat	AFF	94.50	77.49	86.00
Ours	✓	×	AFF × 2	97.50	92.26	94.88

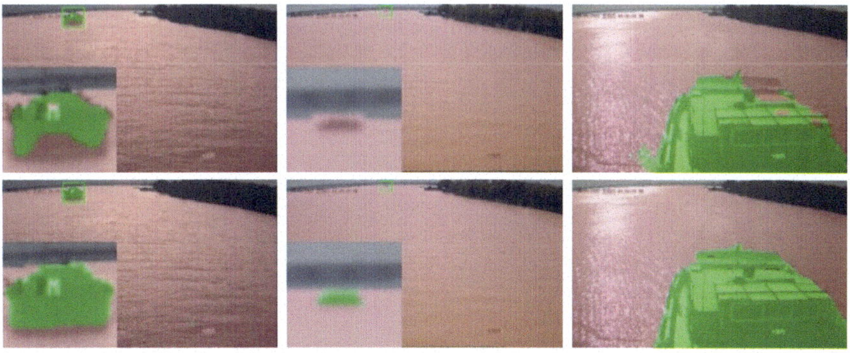

Fig. 4 Qualitative comparison of U-Net (Line 1) and the method we proposed (Line 2)

5 Conclusions

In this paper, we analyzed the less involved inland waterway scenario and the data characteristics in this scenario. Subsequently, we proposed a feasible dataset preprocessing method for the above scenario and designed an asymmetric two-branch convolutional structure, a multi-scale convolutional attention (MS-CA) module, and an adaptive feature fusion module (AFFM). Experiments show that the ship image segmentation method proposed for inland waterway scenarios improves the mIoU metrics by 2% compared to the baseline.

Current semantic segmentation methods are mostly limited to using a single RGB image for training. Instead, this paper shows that it is possible to discover the information of the dataset and design a reasonable feature fusion approach to achieve the purpose of using auxiliary information to enhance the segmentation results.

References

1. Chen D, Fan T, Yuan H et al (2022) Review and prospect on system operation supervision technology of inland river navigation system 22(06):1–14. https://doi.org/10.16097/j.cnki.1009-6744.2022.06.001
2. Ma Y, Zheng M, Wu Y et al (2023) Based on neural network feature extraction research on SSA-SVM ship target identification method. Ship Sci Technol 45(08):158–164
3. Zhou W (2021) Research on multi-scale semantic segmentation technology based on ship images. Soochow University. https://doi.org/10.27351/d.cnki.gszhu.2021.000833
4. Shelhamer E, Long J, Darrell T (2017) Fully convolutional networks for semantic segmentation. IEEE Trans Pattern Anal Mach Intell 39(2017):640–651. https://doi.org/10.1109/TPAMI.2016.2572683
5. Ronneberger P, Fischer T (2015) Brox.U-net: convolutional networks for biomedical image segmentation. In: 2015 International conference on medical image computing and computer-assisted intervention (MICCAI). Springer, pp 234–241. https://doi.org/10.1007/978-3-319-24574-4_28
6. Chen L, Papandreou G, Schroff F, Adam H (2017) Rethinking atrous convolution for semantic image segmentation. arXiv preprint. https://doi.org/10.48550/arXiv.1706.05587
7. Guo M, Lu C, Hou Q et al (2022) SegNeXt: rethinking convolutional attention design for semantic segmentation. arXiv preprint. https://doi.org/10.48550/arXiv.2209.08575
8. He K, Zhang X, Ren S, Sun J (2016) Deep residual learning for image recognition. In: 2016 IEEE conference on computer vision and pattern recognition (CVPR). Las Vegas, NV, USA, pp 770–778. https://doi.org/10.1109/CVPR.2016.90
9. Hu J, Shen L, Sun G (2018) Squeeze-and-excitation networks. In: 2018 IEEE/CVF conference on computer vision and pattern recognition. Salt Lake City, UT, USA, pp 7132–7141. https://doi.org/10.1109/CVPR.2018.00745
10. Zhang Q, Yang Y (2021) SA-Net: shuffle attention for deep convolutional neural networks. In: ICASSP 2021–2021 IEEE international conference on acoustics, speech and signal processing (ICASSP). Toronto, ON, Canada, pp 2235–2239. https://doi.org/10.1109/ICASSP39728.2021.9414568

11. Dai Y, Gieseke F, Oehmcke S, Wu Y, Barnard K (2021) Attentional feature fusion. In: 2021 IEEE winter conference on applications of computer vision (WACV). Waikoloa, HI, USA, pp 3559–3568. https://doi.org/10.1109/WACV48630.2021.00360
12. Wang P et al (2018) Understanding convolution for semantic segmentation. In: 2018 IEEE winter conference on applications of computer vision (WACV). Lake Tahoe, NV, USA, pp 1451–1460. https://doi.org/10.1109/WACV.2018.00163

Real-Time 6-DoF Object Pose Estimation Network Based on 2D-3D Coordinate Correspondence

Shuai Li, Jinlong Chen, Minghao Yang, and Jianhua Su

Abstract Object 6D pose estimation is essential for a variety of applications, including autonomous driving, robotic manipulation, and automated harvesting. Due to the influence of different lighting conditions and occlusions, objects with different poses may exhibit significant variations in appearance in different views. This makes the estimation of the 6D pose of objects from a single RGB image a challenging task. This paper proposes a deep learning network capable of simultaneously detecting and estimating the 6D pose (3D rotation + 3D translation) of target objects. As establishing a large-scale dataset using real data is extremely challenging, this paper introduces an automatic collection scheme based on virtual data generated from real images. We conducted experiments on the generated astronaut dataset using our data collection scheme for 6D pose estimation. The experimental results show that the accuracy of 6D pose estimation based on the ADD metric is 96%. Therefore, this method has significant potential and can be extended to other downstream tasks.

S. Li
Guangxi Key Laboratory of Trusted Software, Guilin University of Electronic Technology, Guilin 541004, Guangxi, China

J. Chen
Key Laboratory of Intelligent Processing of Computer Image and Graphics, Guilin University of Electronic Technology, Guilin 541004, Guangxi, China

M. Yang (✉)
Laboratory of Brain Atlas and Brain-Inspired Intelligence, Institute of Automation, Chinese Academy of Sciences (CASIA), Beijing 100190, China
e-mail: mhyang@nlpr.ia.ac.cn

Transdisciplinary Platform of Functional Connectone and Brain-Inspired Intelligence, Institute of Automation, Chinese Academy of Sciences (CASIA), HuaiRou, Beijing 101400, China

J. Su
National Key Laboratory of Multimodal Artificial Intelligence, Institute of Automation, Chinese Academy of Sciences (CASIA), Beijing 100190, China
e-mail: jianhua.su@ia.ac.cn

S. Yadav et al. (eds.), *Proceedings of 3rd International Conference on Artificial Intelligence, Robotics, and Communication*, Lecture Notes in Electrical Engineering 1172, https://doi.org/10.1007/978-981-97-2200-6_16

1 Introduction

Real-time object detection and 6D pose estimation are crucial for technologies such as autonomous driving, automated harvesting, and robotics. However, challenges arise in real-world conditions due to variations in lighting, differences in object appearances, and occlusions. Developing a practical object 6D pose estimation algorithm is highly challenging.

To address this issue, this paper proposes a 2D-3D coordinate correspondence-based object 6D pose estimation network. The network decomposes the 6D pose into 3D rotation and 3D translation, predicting each separately. For 3D translation, a direct prediction method is employed, while for 3D rotation, the method predicts not only the two-dimensional projection positions of the target object at each pixel in the image but also the three-dimensional positions of each point in the target object in space. This information is used to solve the 3D rotation of the target object through 2D-3D correspondence. Since collecting relevant datasets for target object 6D pose is challenging, the paper utilizes a method based on real images to generate virtual data for training and testing.

The contributions of this paper can be summarized as follows: (1) proposing a 2D-3D coordinate correspondence-based object 6D pose estimation network. The learning pattern of the deep neural network is guided to recover the 2D projection coordinates of the object at each pixel in the image and the spatial coordinates of each point in the 3D space. The information pertaining to 2D projection and 3D spatial data is fused to estimate the 6D pose of the target object. (2) Using a method based on real images to generate a large-scale 6D pose dataset, which can be utilized for estimating the 6D pose of objects.

The subsequent sections of the paper are structured as follows: Sect. 2 provides an overview of related work, while Sect. 3 elucidates the proposed method, offering detailed insights into the recovery of 2D and 3D information for the target object. Section 4 introduces the dataset constructed in this paper, along with relevant details and experimental results. Section 5 concludes the discussion.

2 The Related Work

In the past decade, object 6D pose estimation has been a promising technology. Various research efforts have primarily focused on developing visual perception algorithms for object 6D pose estimation, utilizing both traditional and deep learning-based approaches.

Recent studies indicate that leveraging deep learning with only RGB data for object 6D pose estimation demonstrates outstanding performance. These investigations can be classified into two main categories: methods that directly regress the 6D pose and those that incorporate post-processing steps subsequent to 2D detection. Methods directly regressing the 6D pose do not require any post-processing steps, enabling faster inference speed but at the cost of some loss in accuracy. Yin et al. [1] utilized a graph neural network for detecting the 6D pose of objects, integrating appearance information from RGB inputs and geometric information from depth images. Kehl et al. [2] and Sundermeyer et al. [3] employed an approach that divides rotation into specific regions with continuous ranges, addressing it as a classification problem thereafter. Xiang et al. [4] introduced a shape matching loss to tackle the problem of rotation blurriness arising from symmetric objects, given that symmetric objects can exhibit multiple 3D rotations.

Approaches that incorporate 2D detection and post-processing achieve the final 6D pose information by addressing the Perspective-n-Point (PnP) problem, relying on keypoint detection from the identified object in 2D images. Prior research has primarily concentrated on keypoint extraction, encompassing tasks such as predicting the eight 2D corner points of the bounding box for the 3D model [5–7]. However, keypoint extraction becomes challenging if the target object is occluded by another object. Recent studies propose a composite method using pixel voting and RANSAC to treat the point set of uncovered object surfaces as keypoints [8, 9]. This approach presupposes the precise detection of keypoint positions in two-dimensional images for three-dimensional objects by deep learning networks. Utilizing these keypoints in 2D images enables accurate retrieval of the pose information for the corresponding 3D object.

3 Method

This paper proposes a 2D-3D coordinate correspondence-based object 6D pose estimation method, as illustrated in Fig. 1. The approach involves using a target detector (DetectNet) to obtain the position of the target object in the image. Subsequently, the region of interest is extracted and fed into both the rotation estimation network (RotNet) and the translation estimation network (TransNet). The translation estimation network directly yields the translation matrix (T) for the target object, while the rotation network provides correspondences between 2D points and 3D points for the target object. This 2D and 3D information is then fed into the perspective point-solving network (PnPNet) to estimate the rotation matrix (R).

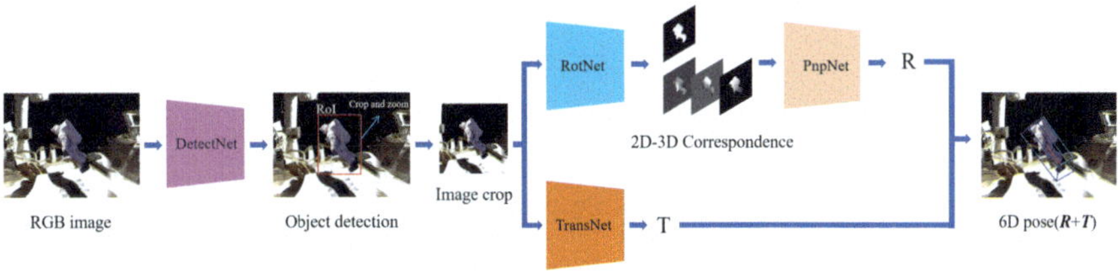

Fig. 1 Real-time object 6D pose estimation method based on 2D-3D coordinate correspondence

4 2D-3D Coordinate Correspondence Regression

We adopt a method based on densely corresponding coordinate points to estimate rotation, which exhibits robustness against occlusion and noise. To establish 2D-3D correspondence points, the network concurrently forecasts the 3D coordinates for all designated pixels and assigns a confidence value to each pixel, determining its association with the target object. Instead of using additional network branches, we integrate the predicted pixel confidence into the 3D coordinate regression because they share the same output dimensions, and their values have precise positional correspondence. The specific implementation steps are as follows: first, extract features from the target region using the backbone network. Then, introduce a rotation network composed of convolutional and deconvolutional layers to process the features and upscale them into a four-channel coordinate–confidence map. The four channels include three channels for the coordinate map (M_{coor}) and one single channel for the confidence map (M_{conf}), sharing all features within the network. In M_{coor}, individual pixel values denote a 3D coordinate, with each channel representing a coordinate axis within the designated 3D coordinate system.

(1) **Constructing the Ground Truth for M_{coor}**

To train the rotation estimation network, we need to use the ground truth of rotation matrices and translation matrices to project the three-dimensional object model onto the image plane, creating the ground truth. First, the points of the three-dimensional object model are transformed from the world coordinate system to the camera coordinate system, as calculated by Eq. (1).

$$P_C = P_W R_{gt} + T_{gt}, \tag{1}$$

where $P_W = (X_W, Y_W, Z_W)^T$ represents the point of the three-dimensional object model in the camera coordinate system, $P_C = (X_C, Y_C, Z_C)^T$ represents the point of the three-dimensional object model in the world coordinate system, and R_{gt} and T_{gt} are the ground-truth rotation matrix and translation matrix. Then, the points of the three-dimensional object model in the camera coordinate system are mapped to the image plane, with the mapping relationship shown in Eq. (2).

$$\begin{cases} i = \frac{X_C f_x}{Z_C} + c_x \\ j = \frac{Y_C f_y}{Z_C} + c_y \end{cases}, \tag{2}$$

where (i, j) is the pixel coordinate on the image plane, and f_x, f_y, c_x, and c_y are the camera intrinsic parameters. The ground truth of M_{coor} can be represented as $f(i, j) = \mathrm{P_W}$. The ground-truth coordinates label as shown in Fig. 2.

(2) **Build 2D-3D Coordinate Correspondence**

By setting a threshold, we can extract target pixels from the confidence map. However, due to the downsampling in neural networks, the size of the object in the

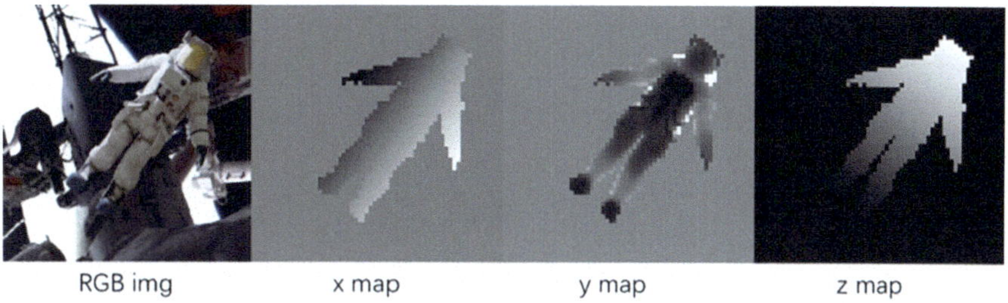

Fig. 2 Ground-truth coordinates' labels

RGB image is often different from the size of the object in the coordinate map. To establish a 2D-3D coordinate correspondence, we accurately map pixels from the coordinate map to the RGB image while maintaining precision. We specify the size and center of the object in the RGB image as $\left(\tilde{S}_x, \tilde{S}_y\right)$ and (c_u, c_v), while in the coordinate map, they are $\left(S_x, S_y\right)$ and $\left(c_i, c_j\right)$. Given a pixel (i, j) in the coordinate map, the associated pixel (u, v) in the RGB imagecan be determined using the formula presented in Eq. (3).

$$\begin{cases} u = c_u + \frac{S_x}{\tilde{S}_x}(i - c_i) \\ v = c_v + \frac{S_x}{\tilde{S}_x}\left(j - c_j\right) \end{cases}. \tag{3}$$

With the correspondence between PnP and RANSAC, solving the rotation matrix R can be easily achieved.

4.1 Scale-Invariant Translation Estimation

In pursuit of a more robust and accurate estimation, our approach involves the direct prediction of translation from the image, bypassing the need for 2D-3D coordinate correspondences. This mitigates the potential impact of scale errors associated with predicted 3D coordinates. Our proposition involves a scale-invariant approach to translation estimation, aiming for high precision and efficiency by focusing on local image blocks. Firstly, calculate the information of the locally sampled image blocks in the global image, denoted as T_G (including position $C_{x,y}$ and size (h, w)). Then, introduce an additional translation estimation network in the backbone network to predict the scale-invariant translation $T_S = \left(\Delta_x, \Delta_y, t_z\right)$. Here, Δ_x and Δ_y represent the offset from the bounding box center to the object center. The network is trained to regress not absolute offsets, but relative offsets, as shown in Eq. (4).

$$\begin{cases} \Delta_x = \frac{O_x - C_x}{w} \\ \Delta_y = \frac{O_y - C_y}{h} \\ t_z = \frac{T_z}{r} \end{cases}, \tag{4}$$

where t_z is the scaled depth, and this relative offset remains constant for sampling (i.e., scale invariant). Finally, the translation $T = \left(T_x, T_y, T_z\right)$ can be solved by combining T_S with T_G, as shown in Eq. (5).

$$\begin{cases} T_x = (\Delta_x w + C_x)\frac{T_z}{f_x} \\ T_y = \left(\Delta_y h + C_y\right)\frac{T_z}{f_y} \\ T_z = t_z r \end{cases}. \tag{5}$$

4.2 Network Architecture

This paper utilizes intermediate supervisory learning to guide the neural network in learning the 2D-3D point correspondence of the target object, achieving 6D pose estimation by analyzing the 2D-3D point correspondence. The target object pose estimation network based on 2D-3D correspondence mainly consists of four modules: DetectNet, backbone, RotNet, and TransNet, responsible for target detection, feature extraction of the region of interest (RoI), target object rotation estimation, and target object translation estimation. Figure 3 shows the network architecture of the proposed method.

Object Detector (DetectNet): This paper employs the pre-existing object detector YOLOV5 for initial target object detection to obtain the position of the target object in the image, referred to as the region of interest (RoI). The corresponding RoI is then augmented into a square, scaled to 256 × 256 × 3, and sent to the backbone for feature extraction. To enhance robustness, during training, the position of the target object in the RoI is uniformly translated. During testing, the 256 × 256 × 3 RoI is input into the backbone to extract RoI features.

This paper uses ResNet34 as the backbone, performing initial feature extraction on the 256 × 256 × 3 RoI outputted by DetectNet to obtain a feature map of size 512 × 8 × 8. After feature extraction, the network splits into two branches: RotNet and TransNet. In the RotNet branch, after three upsampling operations, the feature map is mapped to a size of 4 × 64 × 64 as the RotNet branch. The output of RotNet consists of two parts. The first part is a 1 × 64 × 64 prediction mask, penalizing incorrectly

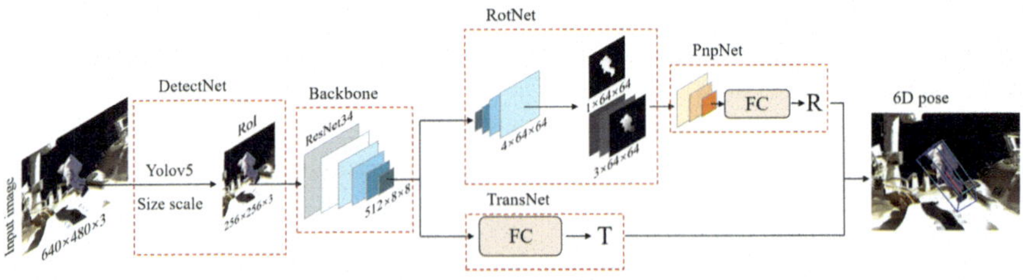

Fig. 3 Network architecture

detected points when calculating the loss. The second part is a $3 \times 64 \times 64$ prediction of a 3D coordinate map, where each pixel corresponds to a 3D coordinate point of the target object. Once the predicted 3D coordinate map is obtained and the 2D-3D correspondence is established, it serves as input to PnPNet. PnPNet resolves the 2D-3D correspondence to obtain the target object's rotation matrix R. PnPNet includes three convolutional layers and three fully connected layers, ultimately outputting the rotation matrix R. In the TransNet branch, features extracted by the backbone go through three convolutional layers and four fully connected layers to directly output the offset of the target object's center relative to the optical center. The translation matrix T can be obtained from Eq. (5).

Because our method directly regresses continuous coordinate maps, and the predicted coordinate map of the object is not continuous on the boundaries, it can make the model difficult to converge. To address this issue, we propose the Masked Confidence–Coordinates' Loss (MCC Loss). Specifically, this loss function only calculates the loss for the foreground region of the coordinate map. For the confidence map, the loss is calculated for all regions, as shown in Eq. (6). This mechanism avoids the influence of non-target regions, aiding the network in providing more accurate coordinates.

$$\mathcal{L} = L_1\left(\sum_{i=1}^{n} M_{\text{conf}}\left(M_{\text{coor}_j} - \tilde{M}_{\text{coor}_j}\right)\right) + L_1\left(M_{\text{conf}} - \tilde{M}_{\text{conf}}\right). \tag{6}$$

5 Experiments

5.1 Dataset

Since generating accurate labels for the 6D pose of the target object solely from 2D images is challenging, this paper proposes a method for generating synthetic data based on real images to establish a 6D pose dataset for the target object. The dataset includes 20 scenes, and based on the pose distribution in the training set, 1000 images are randomly rendered for each scene. For each target in the training set, the rotation range E is calculated, and a random rotation R is generated within E. The translation vector (T) is randomly generated, utilizing the mean and variance computed from the training set. Synthetic images can be generated by applying this translation vector along with the rotation matrix (R).

5.2 Training Strategy

The proposed method is trained and tested using the PyTorch framework, running on a single Tesla V100 GPU with 16 GB of VRAM. The initial learning rate is configured at 10^{-4}, and a cosine scheduler is implemented for dynamic adjustment of the learning rate. During the training process, the RMSprop optimizer is used for gradient descent. To achieve better training results, the training is divided into three stages: first, separately train RotNet and TransNet; then, train them together. Each stage involves training for 160 epochs.

5.3 Evaluation Metrics

In this paper, we utilize three standard estimation metrics to assess the effectiveness of the proposed method: 2D projection, 5° 5 cm, and ADD. The Average 3D Error (ADD) metric quantifies the average discrepancy in 3D coordinates between keypoints transformed by the estimated pose (R and T) and the authentic 3D coordinates of those keypoints. The calculation formula for ADD is shown in Eq. (7):

$$e_{\mathrm{ADD}} = \mathrm{avg}(Rx + T) - \left(R'x + T'\right)_2. \tag{7}$$

Here, R' and T' represent the true annotations of rotation and translation. In the context of 5°5 cm, the pose is deemed accurate if the errors in rotation and translation are both below 5° and 5 cm, respectively. The 2D projection metric computes the distance between the 2D projection coordinates of keypoints, determined by the estimated pose (R and T), and the 2D projection coordinates of the keypoints' ground truth. The formula for 2D projection is given by Eq. (8).

$$e_{\mathrm{2Dproj}} = \frac{1}{N}\sum K(Rx + T) - K\left(R^{*}x + T^{*}\right). \tag{8}$$

Here, K represents the camera intrinsic parameters.

5.4 Comparison with State-of-the-Art Methods

To showcase the efficacy of the proposed method, we conducted a performance comparison against state-of-the-art approaches, including YOLO6D [6], PVNet [9], Single Stage [10], segmentation driven [11], and Wide Depth Range [12]. As depicted in Table 1, bold text highlights the superior performance. Relative to the Wide Depth-Range method, the proposed 2D-3D coordinate correspondence-based approach exhibits a 0.48% lower performance in the 2D projection aspect. However, for ADD and the 5° 5 cm metrics, the proposed method outperforms Wide Depth Range by

Table 1 Performance comparison on astronaut 6D datasets

Method	Category	5° 5 cm	ADD	2D proj	Runtime (ms)
YOLO6D	Two stage	0.89	0.31	34.01	22.68
Single stage	One stage	3.21	0.53	0.25	27.23
Segmentation driven	Two stage	18.62	5.38	43.86	40.34
PVNet	Two stage	59.31	32.93	96.71	38.68
Wide depth range	One stage	73.42	39.89	**98.92**	30.22
Proposed	Two stage	**94.05**	**78.43**	98.44	**9.3**

38.54% and 20.63%, respectively. Furthermore, the proposed method demonstrates significantly faster processing speed compared to other methods, accomplishing a time of 9.30 ms. This represents a 59.0% improvement over YOLO6D's processing time of 22.68 ms.

Two-stage methods based on PnP/RANSAC algorithms (YOLO6D and PVNet) exhibit high accuracy in rotation matrix estimation but have suboptimal performance in translation matrix estimation. This is attributed to the significant variations in the object's position within the perspective of a single camera. Single Stage demonstrates poor performance in metric-based pose estimation because it solely employs multiple 1D convolutional layers to replace the PnP algorithm, resulting in limited improvement over Single-Stage methods in essence. In contrast to these methods, the proposed approach utilizes scale-invariant translation for translation matrix estimation, significantly enhancing the accuracy of 6D pose estimation.

5.5 *Ablation Study*

This section comprises a set of ablation experiments designed to investigate the influence of scale-invariant translation and the Masked Confidence–Coordinates' Loss on the estimation of object 6D pose. We conducted a thorough assessment of scale-invariant translation and Masked Confidence–Coordinates' Loss using the generated astronaut dataset. Table 2 showcases the outcomes derived from the coordinate confidence map utilizing PnP (excluding scale-invariant translation estimation). Trained exclusively on real images, our method demonstrates a substantial enhancement in performance on the 5°5 cm metric, rising from 61.76 to 82.18%. In the case of real + synthetic images, our method surpasses the baseline with a considerable margin (our 95.34% compared to 88.67% for 5°5 cm). Table 2 includes an assessment of the contribution from each individual component. Using YOLOV5 for detection during testing, the performance is nearly identical to YOLOV3. On Tiny YOLOV3, there is only a marginal decrease in accuracy. MCC Loss imparts high modularity and flexibility to our pose network, demonstrating robust performance with a variety of detectors without retraining.

Table 2 Ablation study on dynamic masked confidence–coordinates' loss

Row	Methods			5°5 cm	ADD	2D proj
	Syn	MCCLoss	Detector			
1	×	×	YOLOV5	61.76	50.71	87.04
2	×	✓	YOLOV5	82.18	70.24	94.94
3	✓	×	YOLOV5	88.67	73.94	95.34
4	✓	✓	YOLOV5	**94.05**	**78.43**	**98.44**
5	✓	✓	YOLOV3	93.82	75.03	97.67
6	✓	✓	Tiny YOLOV3	92.96	74.81	95.93

Table 3 Ablation study on SITE

Row	Methods		5°5 cm	ADD	2D proj
	Site	Detector			
1	×	YOLOV5	93.54	74.05	95.86
2	✓	YOLOV5	**94.05**	**78.43**	**98.44**
3	✓	YOLOV3	93.82	75.03	97.67

The evaluation results using SITE are shown in Table 3. The utilization of SITE for estimating translation, rather than coordinates, leads to a significant enhancement in ADD accuracy, rising from 74.05 to 78.43%. SITE not only attains superior performance across all objects but also showcases a more balanced overall performance. In terms of translation, our SITE delivers highly accurate estimates for all objects, highlighting the advantages of our decoupled pose estimation strategy.

6 Conclusions

In this work, a method for object 6D pose estimation based on 2D-3D correspondence was developed by decomposing rotation and translation. To achieve highly accurate, efficient, and robust pose estimation, we introduced the Masked Confidence–Coordinates' Loss and scale-invariant translation estimation techniques. To generate a large amount of training data, we employed a method for virtual data generation based on real images, creating a dataset for training and testing object 6D poses. Our approach exhibits proficiency in handling textureless and occluded objects, attaining state-of-the-art results across all common metrics on the generated astronaut dataset. For the future, we anticipate the application of our method in other related fields, such as agricultural robotics.

Acknowledgements This work is supported by the Science and Technology on Aerospace Flight Dynamics Laboratory (No.KGJ61422102103 11), the Guangxi Key Research and Development Program (AB21220038, AB23026048), the National Natural Science Foundation of China (NSFC)

(No.61873269), the Beijing Natural Science Foundation (J210012, L192005), Hebei Natural Science Foundation (F2021205014), the Natural Science Foundation of Guangxi of China (No: 2020GXNSF AA297061, 2019GXNSFDA185006, 2019GXNSFDA185007).

References

1. Yin P, Ye J, Lin G, Wu Q (2021) Graph neural network for 6D object pose estimation [formula presented]. Knowledge-Based Syst 218:106839
2. Kehl W, Manhardt F, Tombari F, Ilic S, Navab N (2017) SSD-6D: making RGB-based 3D detection and 6D pose estimation great again. In: Proceedings of the IEEE international conference on computer vision 2017, pp 1530–1538
3. Sundermeyer M, Marton ZC, Durner M, Brucker M, Triebel R (2018) Implicit 3D orientation learning for 6D object detection from RGB images. In: Lecture notes computing science (including subseries lecture notes artificial intelligence lecture notes bioinformatics) 11210 LNCS, pp 712–729
4. Xiang Y, Schmidt T, Narayanan V, Fox D (2017) PoseCNN: a convolutional neural network for 6D object pose estimation in cluttered scenes. arXiv Prepr. arXiv1711.00199
5. Rad M, Lepetit V (2017) BB8: a scalable, accurate, robust to partial occlusion method for predicting the 3D poses of challenging objects without using depth. In: Proceedings of IEEE international conference on computer vision 2017, pp 3848–3856
6. Tekin B, Sinha SN, Fua P (2018) Real-time seamless single shot 6D object pose prediction. In: Proceedings of IEEE computer society conference on computer vision and pattern recognition, pp 292–301
7. Tremblay J, To T, Sundaralingam B, Xiang Y, Fox D, Birchfield S (2018) Deep object pose estimation for semantic robotic grasping of household objects, pp 1–11
8. Song C, Song J, Huang Q (2020) HybridPose: 6D object pose estimation under hybrid representations. In: Proceedings of IEEE computer society conference on computer vision and pattern recognition, pp 428–437
9. Peng S, Liu Y, Huang Q, Zhou X, Bao H (2019) PVNet: pixel-wise voting network for 6dof pose estimation. In: Proceedings of IEEE computer society conference on computer vision and pattern recognition
10. Hu Y, Fua P, Wang W, et al (2020) Single-stage 6D object pose estimation. In: 2020 IEEE/CVF conference on computer vision and pattern recognition (CVPR); 2020 Jun 13–19. IEEE Press, Seattle, USA, Piscataway, pp 2930–3299
11. Hu Y, Hugonot J, Fua P, et al (2019) Segmentation-driven 6D object pose estimation. In: 2019 IEEE/CVF conference on computer vision and pattern recognition (CVPR), 2019 Jun 15–20. IEEE Press, Long Beach, CA, USA, Piscataway, pp 3385–3394
12. Hu Y, Speierer S, Jakob W et al (2021) Wide-depth-range 6D object pose estimation in space. In: 2021 IEEE/CVF conference on computer vision and pattern recognition (CVPR), 2021 Jun 21–25. IEEE Press, Nashville, USA, Piscataway, pp 15870–15879

Traffic Light Detection Based on Deep Learning

Yuelian Gu, Liying Wu, Tongqiang Xu, and Lin Chen

Abstract This article proposes a traffic signal recognition algorithm based on deep learning, addressing prominent issues in traffic signal light detection, such as low detection accuracy, slow speed, and large volume making it difficult to install on mobile devices. The algorithm is built upon the YOLOv5S network module and employs Convolutional Neural Network (CNN) for feature extraction and classification. Two image enhancement techniques, adaptive anchor boxes and adaptive image scaling, are used to improve detection speed and accuracy. Additionally, the Mosaic data augmentation technique is applied to enhance the accuracy and robustness of traffic signal light recognition. Extensive comparative experiments were conducted using a self-collected traffic signal light database. The results indicate that the proposed method, compared to traditional approaches, effectively handles interference factors such as lighting conditions, background variations, and changes in viewing angles. It achieves high accuracy and practicality in traffic signal light recognition, with an average precision rate of 93%. This method enables rapid and accurate detection of traffic signal lights and can be widely applied in traffic signal analysis research.

1 Introduction

Traffic signal lights are vital components of urban road traffic management. They guide the movement of pedestrians and vehicles, ensuring both safety and efficiency in traffic flow. Correct identification of traffic signal lights holds significant importance in fields such as autonomous driving, traffic monitoring, and intelligent traffic management.

In the realm of traffic signal light detection and recognition, various methods have been proposed. Zhang et al. [1] introduced an traditional two-stage target detection algorithm, that used a network to locate possible object regions in an image, and

Y. Gu · L. Wu (✉) · T. Xu · L. Chen
Linyi University, Linyi 276000, Shandong, China
e-mail: wuliying@lyu.edu.cn

S. Yadav et al. (eds.), *Proceedings of 3rd International Conference on Artificial Intelligence, Robotics, and Communication*, Lecture Notes in Electrical Engineering 1172, https://doi.org/10.1007/978-981-97-2200-6_17

then used another network to classify the objects. Bi et al. [2] proposed an improved algorithm named as YOLO-RFB based on YOLOv4, by pruning the trunk and introducing worms hole convolution layer and other methods have successfully improved the detection of small and medium targets in traffic signs sample effect. Bochkovskiy et al. [3] operated YOLOv4 on commonly adopted GPUs of Maxwell, Pascal, and Volta architectures to achieve state-of-the-art results. However, these methods are primarily tailored for circular signal lights, rendering them less effective for signals of other shapes. They are also sensitive to color variations, leading to higher false detection rates and limited applicability.

To address these challenges, this article proposes a traffic signal light recognition algorithm based on deep learning. Leveraging the YOLOv5S network module and Convolutional Neural Networks (CNN) for feature extraction and classification, the algorithm enhances detection speed and accuracy through adaptive anchor boxes and adaptive image scaling, two image enhancement techniques. Additionally, the algorithm incorporates Mosaic data augmentation techniques, elevating the accuracy and robustness of traffic signal light recognition.

2 Convolutional Neural Network

2.1 A Subsection Sample

The Convolutional Neural Network (CNN) [4] is a type of multilayer supervised learning neural network designed for processing data with grid-like structures, such as images and videos. It consists of multiple layers, including convolutional layers, downsampling layers, and fully connected layers, among others. These layers work together to extract and enhance features from images to facilitate classification tasks. The convolutional layers employ convolutional operations on input images using local receptive fields and weight sharing, allowing them to extract features. Downsampling layers reduce the size of feature maps, decreasing computational complexity and model intricacy. Simultaneously, they enhance feature invariance, reduce noise, and process data volume, offering stability against deformations and distortions. The fully connected layers categorize the features obtained from convolutional and downsampling layers. The activation values obtained through SoftMax in the fully connected layers represent the image features extracted by the Convolutional Neural Network [5]. The architecture of the Convolutional Neural Network is shown in Fig. 1.

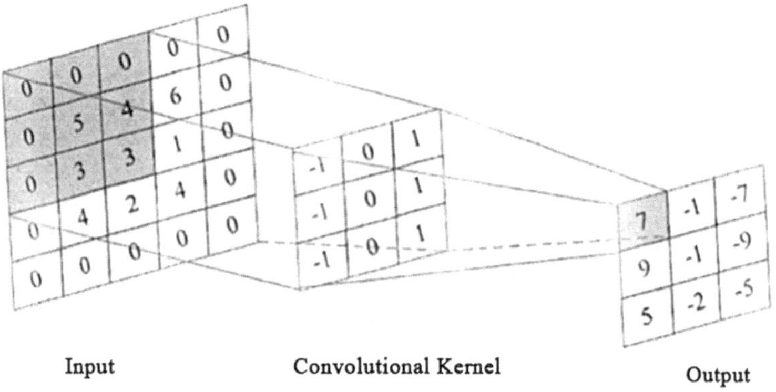

Fig. 1 Convolutional neural network architecture

3 Traffic Signal Recognition Method Based on YOLOv5s

In deep learning-based methods, the YOLO series algorithms have faster detection speed and relatively higher recognition accuracy compared to other one-stage algorithms. Additionally, their network model weights' files are smaller. Therefore, this paper decides to use YOLOv5 for detecting and recognizing traffic signals.

3.1 The YOLOv5s Network Architecture

YOLOv5 is a deep learning model widely used for object detection, including four models: YOLOv5s, YOLOv5m, YOLOv5l, and YOLOv5x. In order to strike a balance between speed and accuracy, this study opts for the YOLOv5s network, which is small, fast, and easily portable to in-vehicle systems, for traffic signal detection. YOLOv5s consists of input layer, backbone network, neck network, and head network. Its network architecture is illustrated in Fig. 2.

Its network architecture consists of four parts: input, backbone, neck, and prediction. The backbone serves as the main network to extract information from the images for later use by the network. The neck's role is to better fuse or extract features provided by the backbone, thereby improving the network's performance.

The basic components include CBL, Res unit, CSPX, and SPP. CBL is a structure composed of convolutional layers, batch normalization, and the leaky ReLU activation function. Res unit allows the construction of very deep neural networks [6]. CSPX is a Convolutional Neural Network structure used for object detection, which further enhances the performance and efficiency of object detection by introducing a multi-scale feature fusion mechanism. SPP utilizes different scales of max-pooling to perform multi-scale feature fusion.

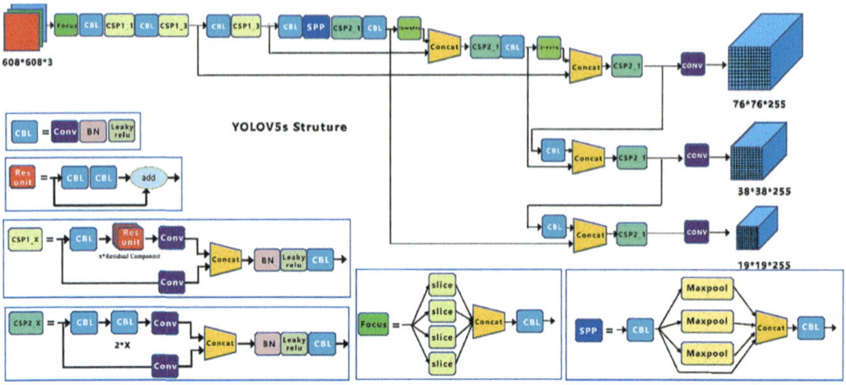

Fig. 2 YOLOv5s network architecture

3.2 *The Core Fundamentals of YOLOv5s*

Input Layer

(1) Mosaic Data Augmentation

YOLOv5s [7] utilizes the Mosaic data augmentation technique to enhance its performance. Mosaic combines four images through random scaling, random cropping, and random arrangement, as illustrated in Fig. 3. This method effectively increases the number of samples in the training set and enhances the network's generalization ability. In YOLOv5s, Mosaic data augmentation works exceptionally well, particularly in detecting small objects.

(2) Adaptive Anchor Box Computation

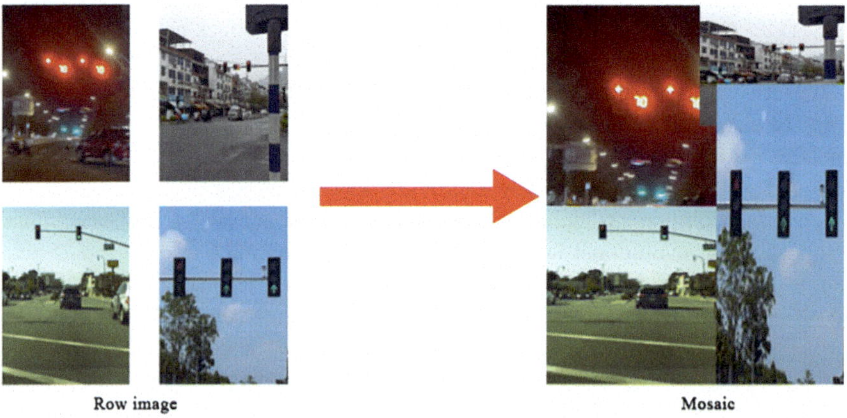

Fig. 3 Data augmentation illustration

In adaptive anchor box computation, first, the receptive field size corresponding to each pixel in the original image is calculated based on the input image size and the size of the feature map. Then, multiple anchor boxes of different sizes and ratios are generated according to the receptive field size. To further improve the adaptability of the anchor boxes, resizing and adjusting the aspect ratios of the boxes can be done to better match the shape and size of the target objects. Adaptive anchor box computation effectively addresses the issue of fixed sizes in traditional anchor boxes, making them more responsive to changes in the targets, thereby enhancing the accuracy and stability of object detection.

(3) Adaptive Image Scaling

To mitigate the impact of varying input image aspect ratios on object detection, the YOLOv5s algorithm uniformly resizes the original images to standard dimensions such as 416 × 416 or 608 × 608, as shown in Fig. 4. The resized images are then input into the network for object detection. This approach benefits the network's ability to effectively learn and extract features from objects of different scales, enhancing detection accuracy. Additionally, it reduces the computational workload for both training and inference, thus improving algorithm efficiency.

However, there are some drawbacks, such as poor adaptability to objects of different sizes and aspect ratios, as well as information redundancy. To address this issue, YOLOv5s employs an adaptive method of adding black borders. Based on the actual size and aspect ratio of the image, the algorithm calculates the minimum black borders needed to adjust the image to a fixed size, reducing information redundancy and enhancing adaptability, as illustrated in Fig. 5.

Backbone

(1) Focus Structure

The focus structure is a neural network architecture used in computer vision tasks. It can be considered as a specialized Convolutional Neural Network structure comprising two crucial components: the focus layer and the channel shuffle layer. The focus layer transforms high-resolution input images into low-resolution feature

Fig. 4 Adaptive image scaling in YOLOv5s

Fig. 5 Adaptive adding black borders in YOLOv5s

maps, as shown in Fig. 6. This transformation reduces computational complexity and memory usage while enhancing detection accuracy. The channel shuffle layer rearranges channel dimensions, enabling different feature channels to collaborate, thus improving feature expression capabilities and robustness.

(2) The CSP1_X structure is applied to the backbone main network, while another CSP2_X structure is applied to the neck

Cross-Stage Partial Network (CSP) is a network architecture, as shown in Fig. 7. It mainly consists of two parts: CSPDarknet53 and CSPX. CSPDarknet53 is a ResNet structure composed of numerous ResUnit and CBL modules [8]. CSPX, on the other hand, is composed of convolutional layers and X ResUnit modules concatenated together. Each ResUnit module includes a cross-stage structure, used to divide and merge feature maps into two parts. This design allows the network to reduce computational load without sacrificing accuracy, thereby enhancing inference speed in practical applications [9].

Neck

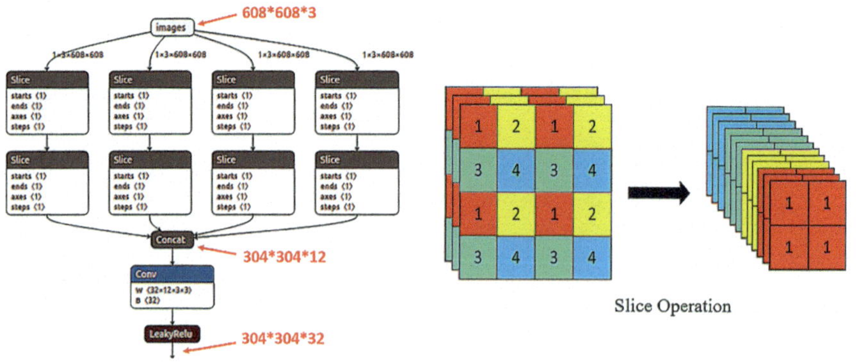

Fig. 6 Focus structure diagram

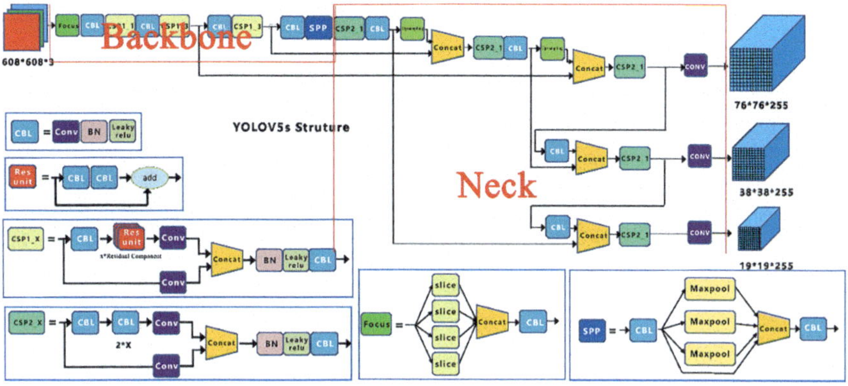

Fig. 7 Diagram of CSP structure

(1) Bounding Box Loss Function

In YOLOv5s, CIOU_Loss is used as the bounding box loss function.

$$\text{CIOU\_ Loss} = 1 - \text{CIOU} = 1 - \left(\text{IOU} - \frac{\text{Distance\_2}^2}{\text{Distance\_}C^2} - \frac{\nu^2}{(1 - \text{IOU}) + \nu}\right), \tag{1}$$

$$\nu = \frac{4}{\pi}\left(\arctan\frac{W^{gt}}{h^{gt}} - \arctan\frac{W^{p}}{h^{p}}\right)^2. \tag{2}$$

(2) Non-maximum Suppression (NMS)

Post-processing is a crucial step in object detection, often involving NMS operations to remove redundant bounding boxes. YOLOv5s utilizes weighted NMS to handle duplicate bounding boxes. The main idea is to score all boxes and then filter them based on these scores, thereby enhancing the accuracy and efficiency of NMS. By employing DIOU_NMS, as shown in the yellow arrowed portion (see Fig. 8), even occluded motorcycles that were previously obscured can be detected.

4 YOLOv5s Analysis of Experimental Results

In order to evaluate the improved YOLOv5s model proposed in this paper, a total of 13,195 experimental datasets were used. Among them, there are 3053 traffic light detection datasets and 10,142 foreign datasets. Both datasets have been annotated with red, green, and yellow lights and can be directly used for training object detection models. Examples of dataset samples are shown in Fig. 9.

Fig. 8 NMS example

Fig. 9 Sample dataset image

4.1 Experimental Evaluation Metrics

(1) mAP

Mean Average Precision, this is the most commonly used evaluation metric in the field of object detection [10].

(2) PR curve

In the PR curve, 'P' represents precision, and 'R' represents recall. It illustrates the relationship between precision and recall. Typically, recall is set as the x-axis, and precision is set as the y-axis. Table 1 shows the confusion matrix.

Table 1 Confusion matrix

Real situation	Prediction result	
	Positive example	Counterexample
Positive example	TP	FN
Counterexample	FP	TN

Where TP represents the number of true positive cases (correctly identified positive instances); FN represents the number of false negative cases (positive instances incorrectly classified as negative); TN represents the number of true negative cases (correctly identified negative instances); FP represents the number of false positive cases (negative instances incorrectly classified as positive).

$$\text{Precision} = \frac{\text{TP}}{\text{TP} + \text{FP}}, \tag{3}$$

$$\text{Recall} = \frac{\text{TP}}{\text{TP} + \text{FN}}. \tag{4}$$

(3) F1-Score

F1-Score, also known as balanced F Score, is defined as the harmonic mean of precision and recall.

$$\text{F1} - \text{Score} = \frac{2}{\frac{1}{\text{precision}} + \frac{1}{\text{recall}}} = 2 \times \frac{\text{precision} \times \text{recall}}{\text{precision} + \text{recall}} \tag{5}$$

The F1-Score metric combines the results of both precision and recall. F1-Score values range from 0 to 1, where 1 indicates the best model output and 0 represents the poorest model output.

4.2 Experimental Results

The images in the training set are randomly divided according to the ratio of 7:2:1. Approximately 9000 images are used for the training set, 2600 images for the test set, and the remaining images are used for the validation set. The initial learning rate of the network is set to 0.01, and the number of training epochs is fixed at 300. After setting the network parameters, the network model is trained using a GPU.

Using the visualization tool TensorBoard, we can observe the visualization graph of training epochs. It can be seen that mAP gradually increases with the training epochs (see Fig. 10). After the model convergence, the mean Average Precision (mAP) metric for YOLOv5s traffic signal detection is approximately mAP_0.5 = 0.93.

Figure 11 displays the F1-Score curve and the precision–recall curve.

The trained YOLOv5s model was tested in real traffic scenarios, and the results indicate that the model performs well. It can effectively recognize traffic signals and meets real-time requirements. The recognition time is approximately 30 ms with CPU (4 threads) and around 25 ms with GPU.

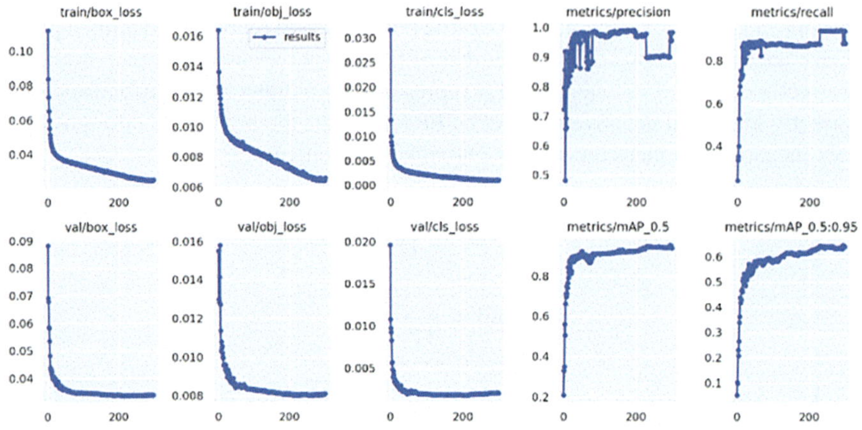

Fig. 10 Visualization of training epochs

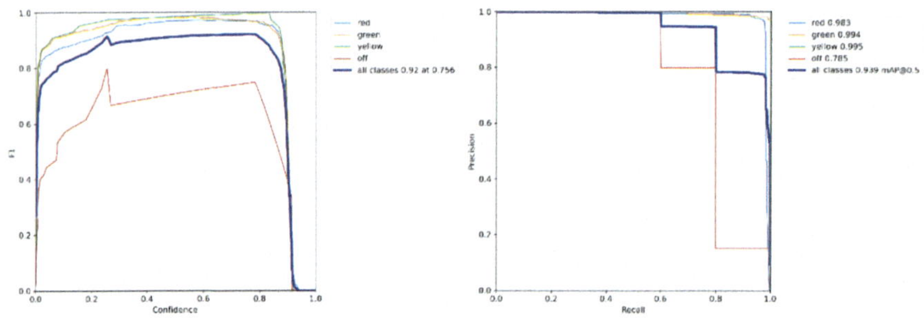

Fig. 11 F1-Score curve (left) and PR curve (right)

It displays sample traffic signal recognition in well-light daytime conditions (see Fig. 12). It is evident that the recognition rate of traffic signals is high in well-lit daytime scenarios. In Fig. 13, samples of traffic signal recognition in low-light nighttime conditions are shown. Compared to daytime, the nighttime scenes suffer from insufficient illumination and increased interference from vehicle lights and streetlights. However, the accuracy of traffic signal recognition remains 93%.

5 Summary

To address the challenges posed by traditional traffic signal detection methods, such as low accuracy, slow speed, and large size that hampers deployment on mobile devices, this study proposes a traffic signal recognition algorithm based on deep learning. The algorithm incorporates a series of optimization strategies, including data augmentation and model architecture design, resulting in significantly improved

Fig. 12 Partial experimental results in well-light scenes

Fig. 13 Experimental results in dimly lit scenes

execution speed. When accelerated with GPU, the algorithm achieves an average recognition time of 25 ms. Compared to traditional image processing-based methods, this represents a speed increase of over 200%, making it highly suitable for practical applications. Moreover, the proposed algorithm takes into account the variations in traffic signal appearance across different scenarios, demonstrating robustness against external factors such as lighting and weather conditions. Its practical applicability is therefore substantial. In future research, the algorithm will explore incorporating various information such as signal color, shape, and position into signal recognition. By fusing multimodal information, the aim is to further enhance the accuracy and robustness of traffic signal recognition.

References

1. Zhang HY, Wang Y, Dayoub F (2021) VarifocalNet: an Iouaware dense object detector. In: Proceedings of the 2021 IEEE/CVF conference on computer vision and pattern recognition. Nashville, pp 8514–8523. https://doi.org/10.48550/arXiv.2008.13367
2. Bi ZQ, Xu FQ, Shan MJ (2021) YOLO-RFB: an improved traffic sign detection model. In: Proceedingsofthe12th international conference on mobile computing, applications, and services. Springer, pp 3–18. https://doi.org/10.1007/978-3-030-99203-3_1
3. Bochkovskiy A, Wang CY, Liao HYM (2020) YOLOv4: optimal speed and accuracy of object detection. arXiv.2004.10934. https://doi.org/10.48550/arXiv.2004.10934
4. Li XM, Xie ZJ, Deng X (2022) Traffic sign detection based on improved faster R-CNN for autonomous driving. J Super Comput 78:7982–8002. https://doi.org/10.1007/s11227-021-04230-4
5. Huang QY, Cai ZC, Lan T (2021) A single neural network for mixed style license plate detection and recognition. IEEE Access 9:21777–21785. https://doi.org/10.1109/ACCESS.2021.3055243
6. Tan MX, Quoc VLe (2019) EfficientNet: rethinkingmodel scaling for convolutional neural networks. In: Proceedings of international conference on machine learning (ICML). https://doi.org/10.48550/arXiv.1905.11946
7. Zhou, AL, Tan GX (2023) Traffic signal detection algorithm based on YOLOv5s. J Guangxi Univ Sci Technol 34(04):69–76. https://doi.org/10.16375/j.cnki.cn45-1395/t.2023.04.010
8. Li ZW, Zhang W, Yang XL (2023) An enhanced deep learning model for obstacle and traffic light detection based on YOLOv5. Electronics 12(10). https://doi.org/10.3390/ELECTRONICS12102228
9. Wang QY, Zhang Q, Liang XT (2021) Traffic lights detection and recognition method based on the improved YOLOv4 algorithm. Sensors 22:200–200. https://doi.org/10.3390/S22010200
10. Ehtesham H, Yasser K, Imtiaz A (2023) Learning deep feature fusion for traffic light detection. J Eng Res 11(3). https://doi.org/10.1016/J.JER.2023.100128

Big Data Algorithm and Robot Modeling

Robot Navigation Based on Reinforcement Learning: An Overview

Yunfeng Wan and Jiaying Zhang

Abstract In recent years, the market for real-world robots has become more and more extensive; robots in warehousing and logistics, public services, home management, and other fields are becoming more and more popular. The market demand for better navigation algorithms is becoming stronger, and algorithms based on reinforcement learning are the key focus of future research. In this overview, we analyze the existing three main types of problems in this field and summarize the existing cutting-edge algorithms. Then, we reveal the remaining problems and optimizations in this area and give possible solutions and future research directions.

1 Introduction

Robot navigation is an important research direction in the field of robotics and artificial intelligence, which aims to enable robots to autonomously plan reasonable paths, avoid obstacles on the path, move from the starting location to the target location, and be able to get rid of the constraints of a single map, so that they can adapt to new environments quickly. In recent decades, more and more experts and scholars have devoted themselves to this field, constantly exploring more powerful and better algorithms.

Traditional robot navigation algorithms are mainly based on path planning after collecting information. Often, it is necessary to first acquire environmental information through sensors such as lasers or cameras as a way to achieve map construction and robot localization. The most typical algorithms are SLAM algorithms and many

Y. Wan (✉)
College of Artificial Intelligence and Automation, Beijing University of Technology, Beijing 100021, China
e-mail: wyf359977120@gmail.com

J. Zhang
College of Information and Communication Engineering, Beijing University of Technology, Beijing 100021, China

S. Yadav et al. (eds.), *Proceedings of 3rd International Conference on Artificial Intelligence, Robotics, and Communication*, Lecture Notes in Electrical Engineering 1172, https://doi.org/10.1007/978-981-97-2200-6_18

of their derivatives, such as Visual SLAM, LSD-SLAM, ORB-SLAM, GMapping and Hector SLAM, and so on.

The path planning after information collection is based on the constructed environment map and the computed position of the robot, and the robot's motion path is calculated. Path planning algorithms are mainly categorized into global path planning and local path planning; the former is mainly applicable to static or semi-static environments, while the latter is mainly applied to dynamic or unknown environments. Global path planning needs to grasp all the information of the environment and perform path planning based on all the information of the environment map, and common algorithms include e A-star, colony optimization, rapid exploration random tree, etc. Local path planning only requires real-time collection of environmental information by sensors to understand the environmental map information and then determine the location of the location map and its local distribution of obstacles, so that the optimal path from the current node to a sub-target node can be selected, and common algorithms are Artificial Potential Field and dynamic window approach.

Although traditional robot navigation algorithms can quickly find the optimal or suboptimal paths, they also have some drawbacks, such as the need to model the system, the need to geometrically divide the search area, the low coverage in the search space, the narrow scope of the search, and it is difficult to explore the unknown region. Although there are some improved algorithms such as RRT that can solve some of the problems, they still cannot fully meet the navigation needs in complex environments.

In recent years, the development of deep learning technology has made robot navigation algorithms more efficient and intelligent. Reinforcement algorithms allow robots to adapt to different environments through learning and achieve more flexible action decisions. Reinforcement learning algorithms do not rely on manually labeled trajectories, but only need to specify the planning goals (no collision, reach the destination, shortest path, etc.), and let the robot continuously try and iteratively update the network in the simulated environment. This allows the robot to adapt to dynamic or unknown environments and improves the robustness and flexibility of path planning.

Reinforcement learning-based navigation algorithms and hybrid algorithms of reinforcement learning and traditional path planning have become the future development trend and mainstream in the field of robot navigation. In this paper, we analyze the current status of the robot navigation field, summarize the three most important subfields of this field at present, including indoor navigation, navigation in dynamic environments and multi-robot navigation, and compare the differences and performance between the relevant algorithms for each problem that provide the advantages and disadvantages of these algorithms in dealing with the corresponding problems. In addition, the problems and optimizations of these algorithms during the research process are analyzed in this paper and collect commonly used simulation tools and datasets in this type of research. Finally, possible solutions and research directions are proposed to address the shortcomings of the existing algorithms.

2 RL-Based Navigation

Navigation, i.e., controlling the robot to move from the current position to the target position, is one of the core functions of mobile robots. As mobile robots are more and more widely used in production and life, the environments they face are more complex and changeable, and the efficiency and accuracy of the navigation algorithms are required to be more efficient, which pose a great challenge to the autonomous robot navigation technology.

In recent years, many studies, both nationally and internationally, have shown that reinforcement learning is well suited to the field of robot navigation, which is essentially a large-scale, sparse reward task dealing with high-dimensional observations. The application of reinforcement learning to robot navigation systems involves the use of fuzzy logic, neural networks, and other techniques to achieve flexibility, resilience, and adaptivity. Such systems provide robots with a motion planning framework that is capable of automatic learning and self-adjustment, thus enabling the robot to more efficiently and consistently find the optimal path, avoid obstacles, and better cope with changes in the environment so that the robot can function well in new environments.

Point-to-point complex spatial navigation is now quite well developed, and there is a large amount of well-developed research work on this problem, with many algorithms that are highly accurate, efficient, and applicable to new environments. There are also superior algorithms for anti-jamming, obstacle avoidance, localization point recovery, extrication, fault recovery, and rapid training of reinforcement learning agents.

In addition, human–computer interaction and distributed motion planning applied to multi-robot cooperative navigation have become the key research direction in the field of robot navigation nowadays, and this field has stronger practicability and broader market prospect, which is an important driving force for the future mobile robots to better integrate with people's production and life. Human–computer interaction is currently focused on semantic navigation and natural language navigation, but the existing algorithms all have low accuracy, poor adaptability to new environments, and other problems and need to be supported by a larger dataset to update and iterate more perfect and newer algorithms, in the expectation that this problem can be put into products in the future. Meanwhile, there is also multi-robot motion planning, and although the single-robot navigation problem has been quite perfect, many robot navigation algorithms, such as FMP-DRL, still have the problem of discrete outputs and incoherent robot motion. This problem is also crucial for future applications of robots and requires deeper research based on this algorithm.

2.1 Indoor Navigation

2.1.1 Overview

The indoor navigation problem is one of the most important subfields in the field of robot navigation, which focuses on how to realize the localization, planning, and control of robots in indoor environments. Compared with outdoor navigation, indoor navigation faces more challenges because indoor environments are more complex and often subject to many limitations, such as signal interference and narrow space.

Indoor navigation algorithms often contain two parts: collision-free path planning in C-space generation and robot motion control. Robot motion control is a challenging problem because the robot cannot just follow the collision-free paths in C-space, but also needs to face many complex problems in the process, such as satisfying task constraints, dealing with changes in the environment, compensating for sensor noise, measurement errors, unmodeled system dynamics, and so on. Reinforcement Learning Control (RL) can be a good solution to the above problems compared to the complex and numerous traditional robot motion control algorithms. However, RL algorithms have a core problem: Rewards tend to be scarce when performing long-distance indoor navigation, which makes it difficult to train good agents. Therefore, in many recent studies of reinforcement learning-based indoor navigation algorithms, researchers have focused on investigating how to compensate for the drawbacks of reinforcement learning by adopting, e.g., adopting specific path planning algorithms, to achieve better results.

2.1.2 Development

Faust et al. [1] proposed PRM-RL in 2018, which is an indoor navigation algorithm that combines sampling-based path planning with reinforcement learning. In this algorithm, AutoRL as the RL method [2], the trained RL agent focuses only on the local task, performs local point-to-point motion control, learns sensor noise compensation, and meets task constraints, while the global long-distance navigation task is realized by the PRM assisted by the RL agent. Its framework is shown in Fig. 1 [2].

The traditional probabilistic roadmap (PRM) algorithm is mainly divided into two stages: randomly sampling a large number of robot positional points in C-space, searching for neighboring nodes for each node, establishing linear connections, and constructing a roadmap; according to the information of the start point, the goal point, and the roadmap, a feasible path is searched from the roadmap by using heuristic search algorithms. PRM-RL cleverly uses the agent to determine the connectivity and further construct the roadmap, instead of simply establishing linear connections between neighboring nodes in C-space in the traditional algorithm. This algorithm makes up for the respective shortcomings of PRM and RL, effectively improving the problem of RL's poor operation in long-distance, sparse reward environments; and the problem of PRM's inefficiency in dense and narrow environments, and its

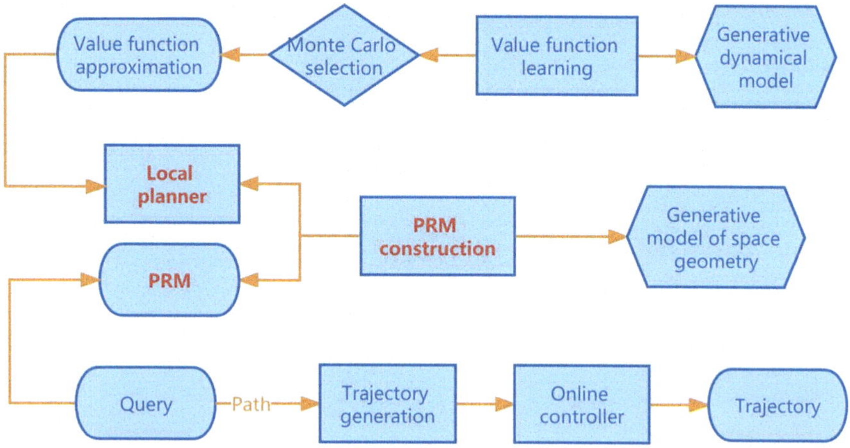

Fig. 1 PRM-RL framework

poor adaptability to environmental changes. Chiang et al. [3] took a similar idea to the PRM-RL algorithm and proposed RL-RRT, an algorithm in which agent also focuses on LOCAL motion control and obstacle avoidance and acts as a local planner during the path planning process. The difference is that RRT-RL employs the Rapidly Exploring Randomized Tree (RRT), which is a sampling-based path planning algorithm, same as PRM. In order to grow the tree toward a more optimal region, a reachability estimator trained by means of supervised learning is introduced in this algorithm, which predicts the time for the RL policy to reach a predetermined state in the face of an obstacle. Gao et al. [4] proposed PRM-TD3, which combines a deep reinforcement learning algorithm win Delayed Deep Deterministic policy Gradient (TD3) combined with PRM, which significantly reduces the algorithm training and testing time, improves the development efficiency, and has a strong generalization ability. Its framework is shown in Fig. 2 [4].

Unlike the above algorithms that use the agent as a local planner and robot motion controller, which are fused with path planning algorithms to realize long-distance navigation, search on the replay buffer (SORB) proposed by Eysenbach et al. [5] in 2019 tries to address this problem from another perspective. This algorithm decomposes a long-distance navigation task into many simpler subtasks, each with its own sub-target point. The role of RL in this algorithm is to provide an appropriate abstraction of the environment, i.e., to construct a graph consisting of nodes and edges. On this basis, a traditional planning algorithm can find the sub-target points of these small tasks in turn, thus realizing long-distance navigation (Table 1).

In these studies, the clever combination of reinforcement learning and traditional path planning compensates for the respective shortcomings of the two and demonstrates in experiments their high accuracy, high efficiency, and fast adaptation to new environments, which is a major development direction for future indoor navigation algorithms.

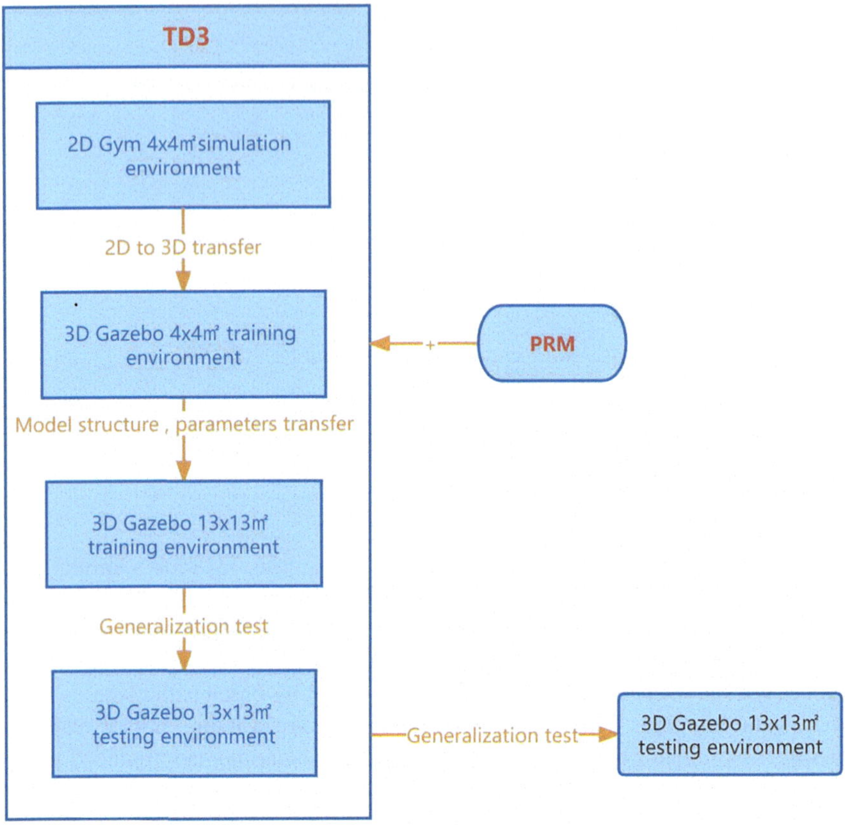

Fig. 2 PRM-TD3 framework

2.2 Navigation in Dynamic Environments

2.2.1 Overview

Indoor navigation algorithms perform well in complex spaces with fixed obstacles, but in robot application scenarios filled with dynamic obstacles, these algorithms are often not sensitive enough to make obstacle avoidance and further motion planning. Nowadays, robots are increasingly used in daily life, with service robots operating in indoor public places, and food delivery and courier robots operating in outdoor streets, which need to deal with dense crowds and various moving obstacles. Obstacle avoidance and navigation among dynamic obstacles are a key capability for robots to move in real-world environments.

Current algorithms face three main challenges; firstly, when facing obstacles like pedestrians with unpredictable and incoherent movements, the robot needs to react more robustly and efficiently to changes in these obstacles compared to stationary

Table 1 Indoor navigation

	Advantages	Disadvantages	Dataset
PRM-RL/ PRM-AutoRL [1, 2]	The ability to adapt better to new environments, generate long-range and extensive proxies, improves the poor performance in sparse reward environments. AutoRL can search for deep RL rewards and neural network architecture through large-scale hyperparameter optimization	There may be a catastrophic forgetting problem in deep reinforcement learning algorithms. AutoRL helps to address this issue	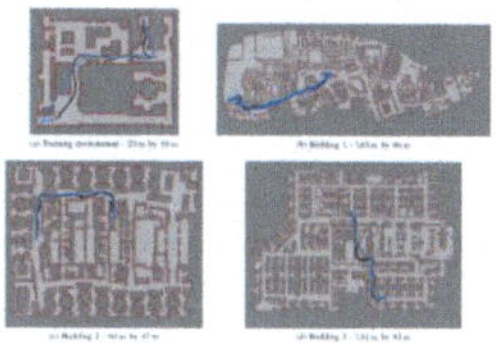 Four real-world maps, with the first one used for training (22.7 × 18.0 m). The next three maps are 12 and 52 times larger than the first map and are used for evaluation
RRT-RL [3]	Improved the problem of high computational complexity and inability to detect obstacles in TTR calculations. It allows for quick adaptation to new environments, low operational costs, and high efficiency	Although the obstacle perception accessibility estimator correctly predicts TTR, its performance is poor in environments where true accessibility is often significantly influenced by large obstacle structures. This is because the estimator only uses local information to approximate accessibility	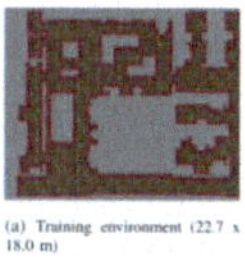 The training environment is the same as PRM-RL [1]. (22.7 × 18.0 m) 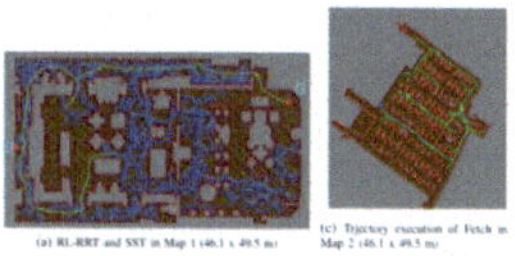 Maps that are 12 and 18 times larger than the training set are used for evaluation. Left figure shows a cluttered environment, while right figure represents a narrow corridor

(continued)

Table 1 (continued)

	Advantages	Disadvantages	Dataset
PRM-TD3 [4]	Design the algorithm based on DRL in a 2D environment and then transfer the designed algorithm to a simple 3D environment for retraining to significantly reduce algorithm training and testing time, improve development efficiency, and possess strong generalization capabilities	PRM is essentially a sampling-based algorithm, and each planned result is different, which means that it may not always yield the optimal path	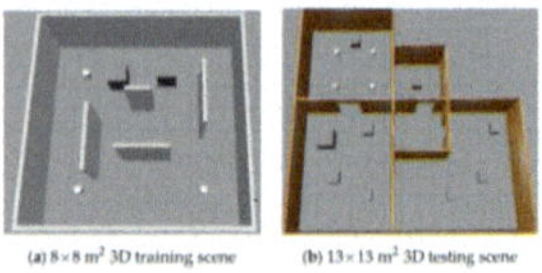 Using Gazebo in Robot Operating System (ROS) (a widely used robotics platform) as a 3D simulation environment, build their own 3D models. Left figure for training and right figure for evaluation purposes
SORB [5]	Perform broad-scale planning of tasks in high-dimensional observations. Overcome the drawbacks of planning and reinforcement learning. The design is simple (with humans only providing an example target state) and avoids reward shaping, which can cause the agent to favor suboptimal solutions and be more flexible	Highly reliant on goal-oriented reinforcement learning, thus performing poorly when dealing with overly complex tasks. Moreover, the planning algorithm used is quite simple	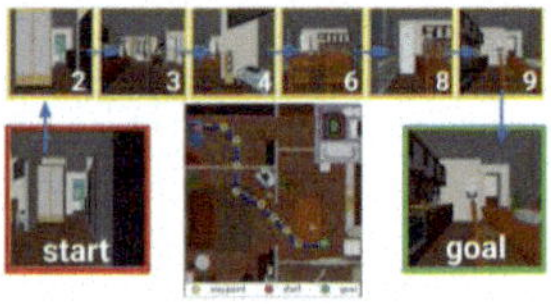 Using the 3D house dataset from the SUNCG dataset [6], studied extending the algorithm from 2 to 3D and stitching together four images from the east, west, south, and north into one panoramic image

obstacles to avoid collisions. Second, when there are a large number of dense dynamic obstacles like pedestrians in the environment, the robot can experience freezing phenomenon and navigation loss. Freezing phenomenon means that the robot will consider any movement as unsafe. Loss-of-navigation phenomenon is when the robot loses its localization information and is unable to find its position on the map due to the occlusion of features required for localization in complex dynamic environments. These two phenomena can cause the robot to stop in place or oscillate between several directions, preventing further navigation. Deep reinforcement learning has been shown in many studies to be suitable for solving such problems posed by dynamic obstacle avoidance navigation. It is worth noting that in this field, due to the complex and dynamic nature of the environment, researchers tend to prefer autonomous modeling through ROS for conducting simulation experiments rather than relying on existing datasets.

2.2.2 Development

Chen et al. [7] proposed a new algorithm GA3C-CADRL based on deep reinforcement learning in 2018, which is different from the previous works in which key assumptions are often required for the motion rules of dynamic obstacles, thus lacking generalization to new environments and new types of dynamic obstacles in real application scenarios. This algorithm learns obstacle avoidance among various types of dynamic obstacle agents through deep reinforcement learning, which no longer requires assumptions about the subject behavior of dynamic obstacles, and is a major breakthrough in the field of dynamic obstacle avoidance.

Reinforcement learning has also been effective in solving the problem of localization loss and freezing in robots. Liang et al. [8] proposed the DenseCAvoid algorithm in 2020. The algorithm is mainly based on the environment of dense crowds, but I believe that the algorithm can also be generalized to dynamic environments in a broader sense. The DenseCAvoid algorithm is also a hybrid algorithm based on reinforcement learning. In order to generate robust and safe trajectories in dense crowds, the algorithm combines pedestrian trajectory prediction with a dynamic obstacle avoidance algorithm based on deep reinforcement learning; in order to solve the freezing problem of the robot, the algorithm trains policies using the input data from the visual sensors and pedestrian trajectory prediction data using the deep reinforcement learning algorithm Proximal Policy Optimization (PPO) to train the policy so that the possibility of both problems is greatly reduced. Sathyamoorthy et al. [9] optimized DenseCAvoid for the robot freezing problem and proposed Frozone, a new algorithm to deal with the robot freezing problem. The algorithm constructs a potential freezing zone in front of the robot by predicting the pedestrian trajectory and modifies the robot's motion speed and direction according to this zone, so that the probability of freezing decreases dramatically again compared to DenseCAvoid. The framework of this algorithm is shown in Fig. 3 [9].

Unlike previous algorithms that only minimize the likelihood of freezing and localization loss problems by improving the accuracy of mobile agent predictions, utilizing sensor data, etc., Fan et al. [10] proposed a new navigation framework in 2018 to help the robot resume normal operation even when it is unfortunate enough to fall into these two problems. The core idea is to move the robot to a location with a more adequate landmark feature by making the robot move to a more adequate location, which in turn allows the robot to resume normal operation. This structure consists of two main parts: a global navigation algorithm based on reinforcement learning and a local algorithm that helps the robot recover its own localization point in the global map. In this framework, the robot switches between two modes depending on the real-time situation. When the robot is in normal mode, the agent performs smooth global path navigation. However, when the robot loses its own localization information, it switches to recovery mode. The robot samples

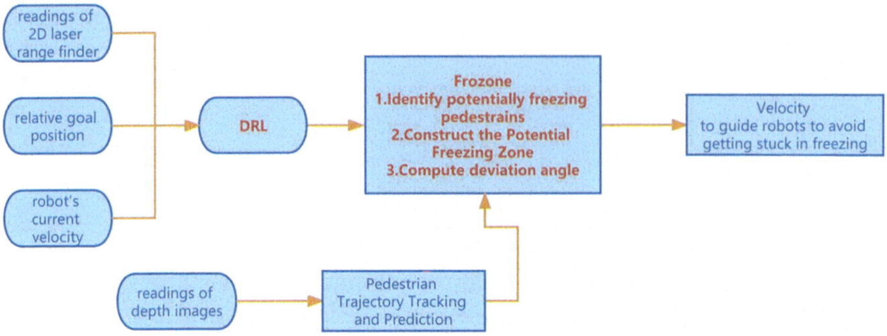

Fig. 3 Frozone framework

a set of locations with sufficient landmark feature, called recovery points, in the 2D map generated during navigation. Then, the reachability of the recovery points is optimized by reinforcement learning to select the best recovery points. Finally, the robot will dynamically select an optimal recovery point based on the real-time situation and keep approaching it to reposition itself in the map. The framework is shown in Fig. 4 [10].

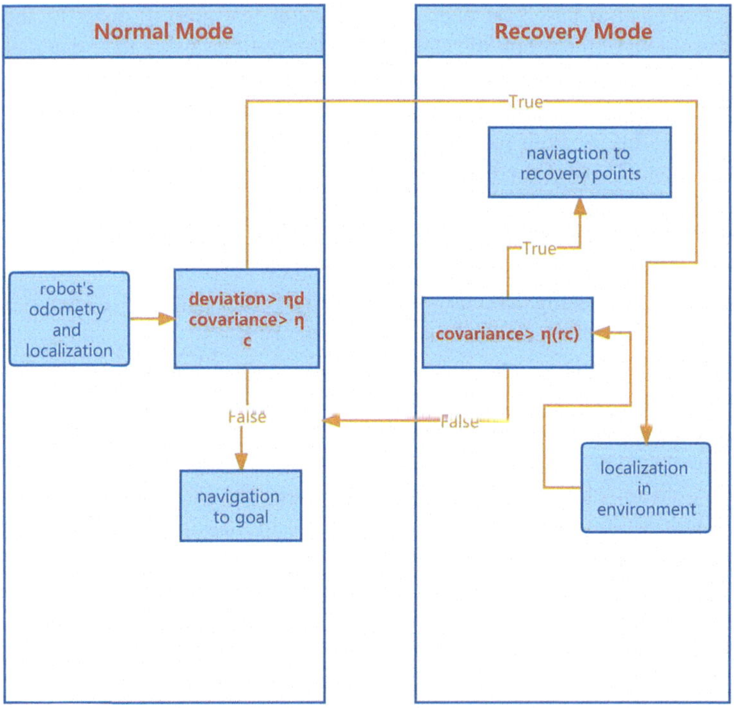

Fig. 4 Escape algorithm framework of Fan et al. [10]

Compared with indoor navigation, navigation in dynamic environments is more complex, and robots are more likely to collide, freeze, and lose positioning. The current research has used reinforcement learning to avoid accidents as much as possible or to help robots recover from accidents, but the current algorithm still has room for further improvement, and the field still needs more in-depth research (Table 2).

2.3 Multi-robot Navigation

2.3.1 Overview

In fact, in many of today's robot application scenarios, a single robot is no longer sufficient to fulfill some special work requirements. In the fields of warehousing and logistics, autonomous driving, UAV formation, etc., multiple robots are required to work together in the same space, which requires the robots in each space to navigate and generate collision-free global path planning, especially avoiding collisions between robots compared to single-robot navigation. Multi-robot navigation systems are categorized into two main types: (1) Centralized navigation: A central node is needed to obtain comprehensive information about the positions, velocities, and goal positions of all robot agents and to plan the paths of all robots. (2) Decentralized navigation: The movements of individual robots are planned completely autonomously, and there is no centralized controller. Decentralized navigation is simpler to compute than centralized navigation and therefore has better scalability in terms of the number of agents and computation. At the same time, decentralized navigation does not require a robust communication network between multiple robots and a central node as in centralized navigation, which can better cope with communication constraints and node loss. However, decentralized navigation also faces challenges such as how to ensure that all robots complete the overall task.

As mentioned in the previous section, navigation algorithms require real-time modeling and analysis of obstacles and surroundings. In past research, many classical multi-agent pathfinding algorithms [11] have been proposed, and these algorithms often use heuristics. The computational complexity of traditional navigation algorithms for multi-robot systems is exponentially related to the number of agents in the multi-robot system, which leads to the fact that as more and more robots are required to join in the modern robotic work environment, many of the traditional algorithms are no longer suitable for solving the multi-robot navigation problem. Recently, reinforcement learning has been demonstrated in a number of studies to be suitable not only for solving single-robot navigation, but also as a solution for multi-robot navigation, especially in the field of decentralized navigation, which is becoming more and more widely viewed and applied.

Table 2 Navigation in dynamic environments

	Advantages	Disadvantages	Dataset
GA3C-CADRL [7]	There is no longer a need for prior assumptions and predictions about the motion behavior of dynamic obstacles, as it demonstrates unprecedented adaptability to new environments and new types of dynamic obstacles. Furthermore, it achieves excellent performance while using only 2D radar and image information, making it cost-effective	The current performance of the algorithm is excellent, but it lacks sufficient testing and requires more comprehensive analysis in more complex environments to quantify the relationship between the algorithm and sensor performance	Not explicitly mentioned
DenseCAvoid [8]	Similar to GA3C-CADRL [7], no necessary assumptions are made about the behavior of dynamic obstacles (pedestrians). A novel network architecture and reward function are used to generate robust motion trajectories in dynamic environments. This is achieved through the utilization of multi-sensor data and pedestrian trajectory prediction data	The accuracy of this algorithm is limited by the precision of pedestrian trajectory prediction. The overall performance is mainly influenced by the synthetic datasets used in the training phase and the limitations of the sim-to-real paradigm. Additionally, there is a lack of sufficient consideration for the dynamic constraints in real-world scenarios	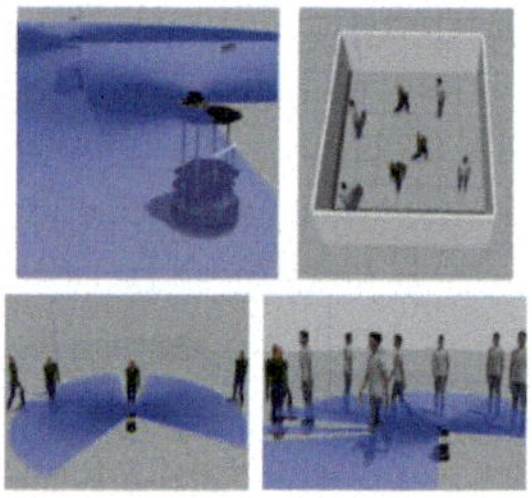 Using Gazebo in ROS as a 3D simulation environment, build their own 3D models. Utilizing the TurtleBot 2 robot for field testing in real-world scenarios

(continued)

Table 2 (continued)

	Advantages	Disadvantages	Dataset
Frozone [9]	To address the issues of robot freezing and localization loss, a potential freezing zone is constructed ahead of the robot by predicting pedestrian trajectories. This allows for controlled robot movement, significantly reducing the probability of freezing compared to DenseCAvoid [8]	Similar to DenseCAvoid [8], the performance of this algorithm is constrained by the pedestrian trajectory prediction algorithm. There is a lack of consideration for the dynamics constraints of the robot	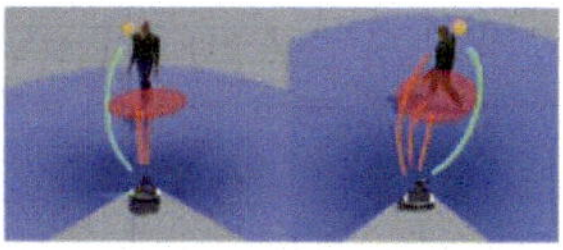 An autonomous 3D scene is created using ROS Kinetic and Gazebo 8.6. The simulation model of a Clearpath Jackal robot is employed, which is connected to Gazebo's Hokuyo 2D laser scanner model and Orbbec Astra depth camera model
Recovery points [10]	Formulate a novel framework to handle the navigation lost and the robot freezing problem simultaneously by switching between the normal mode and the recovery mode during navigation. Propose a reinforcement learning-based recovery algorithm that enables the robot to regain the localization by approaching recovery points adaptively		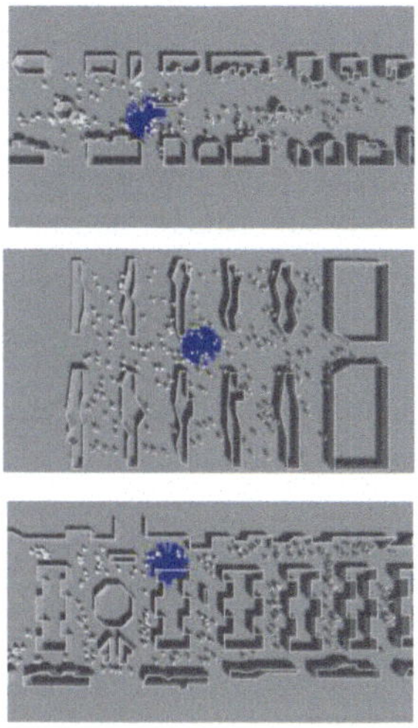 A combination of two ROS platforms, Gazebo and Menge, is used to build the simulation environment. The Gazebo simulator is responsible for simulating the robot system, while Menge acts as the driver, simulating the movement of crowds within the Gazebo environment

2.3.2 Development

Semnani et al. [12] proposed Forced-Based Planning (FMP) in 2020, a real-time decentralized navigation algorithm based on distributed reinforcement learning applied to the motion of a large number of robot agents. This algorithm allows each robot to generate its own independent motion trajectory based on force-based control using only one piece of information, the relative positions of neighboring robot agents. However, one problem with this algorithm is that it often fails to generate time-optimal paths due to the limitations of decentralized navigation. To remedy this problem, Semnani et al. [13] subsequently combined the FMP algorithm with deep reinforcement learning to propose a hybrid FMP-RL algorithm by taking advantage of the fact that the RL navigation algorithm always produces optimal paths and utilizing FMP to compensate for the disadvantage that the RL algorithm performs poorly in dense obstacles and dynamic environments. In this framework, each robot in the multi-robot system enables FMP when the robot faces freezing scenario, high-risk scenario, or simple scenario and adopts GA3C-CADRL [7] when FMP fails to generate optimal paths. DRL-FMP overall control framework is shown in Fig. 5 [14].

Liu et al. [15] proposed a decentralized navigation algorithm for multi-robot systems based on evolutionary reinforcement learning (MAPPER) in 2020. The algorithm adopts a similar direction to SORB [5], which decomposes the long-distance navigation task into many subtasks to improve the shortcomings of reinforcement learning algorithms, and an evolutionary training method, which makes the algorithm extremely scalable and can be easily scaled up to large-scale and complex multi-robot environments.

Damani et al. [13] improved their PRIMAL proposed in 2019 and in 2021 [16] proposed PRIMAL2. PRIMAL is also a distributed RL-based multi-robot decentralized navigation algorithm, and while the algorithm performs well in terms of number

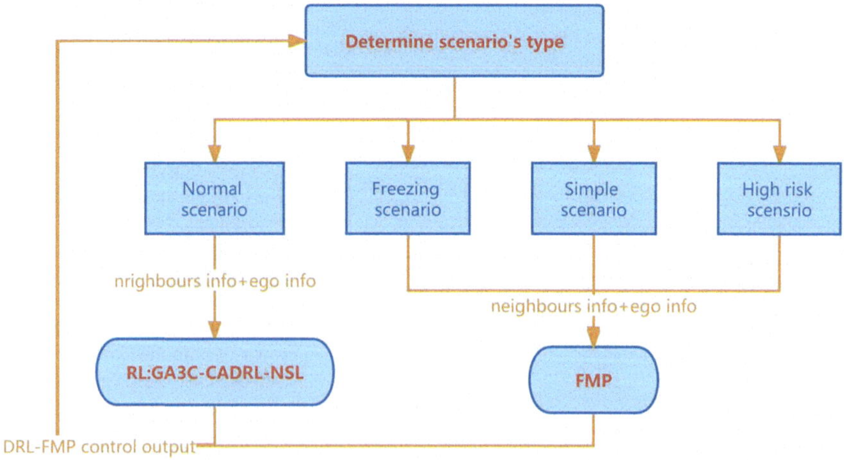

Fig. 5 DRL-FMP control framework

of robots scalability, it does not perform well in structured, obstacle, and robot agent-dense environments that require extensive agent coordination. PRIMAL2 remedies this problem by employing a new distributed training code and relying on centralized navigation for imitation learning. Notably, PRIMAL2 also addresses the problem of lifelong navigation, where the robot needs to reach a series of goal points, being repeatedly assigned new goal locations instead of stopping after reaching the first one, which is promising in current robot application scenarios.

Long et al. [17] proposed RL-based end-to-end decentralized navigation collision avoidance algorithm in 2018. In terms of collision avoidance, centralized algorithms have good performance due to the ability to plan optimal paths better for each robot, but the drawbacks of centralized navigation have also led researchers to look for solutions corresponding to decentralized navigation. The algorithm passes sensor measurements directly to the agent's steering commands and learns the optimal strategy in a multi-scenario, multi-stage training framework, which approximates the performance of centralized navigation with the scalability of decentralized navigation. Lin [18] proposed a decentralized navigation algorithm with an end-to-end strategy also used in 2019, which has the major advantage of not needing to construct an obstacle map and directly mapping the raw sensor data mapping from lidar into continuous control commands to realize navigation.

Decentralized multi-robot navigation is currently becoming a trend, and decentralized navigation is increasingly favored by researchers for its many advantages. Existing research based on reinforcement learning has realized end-to-end navigation, and point-to-point navigation is effective, which can initially meet the needs of lifelong navigation. In view of the broad market application prospect of multi-robot system, this field is worth investing more research (Table 3).

3 Current Challenge and Solution

Although robot navigation algorithms based on reinforcement learning have made significant progress in recent years, there are still some problems in many types of research: (1) Many studies remain in 2D scene simulation, but in addition to the direction in actual navigation, it is also necessary to consider the factors such as terrain and obstacles in half space in 3D scene. (2) Most of the studies' training and evaluation are conducted in scene simulation, lacking the evaluation of the actual robot in the real field. This is necessary because many problems are only exposed in real-world scenarios, e.g., RRT-RL algorithms are significantly affected by the structure of large obstacles, which cannot be accurately modeled in simulation; end-to-end navigation algorithms are also difficult to simulate the actual noise impacts on sensors in simulation scenarios with light, fog, and humidity. (3) Some human–computer interaction (HCMI) navigation algorithms based on semantic commands and natural language have emerged [19–21], and this direction has a good market application prospect, but more kinds of human–computer interaction algorithms with practical application scenarios can be researched. (4) For the sake of safety, obstacle

Table 3 Multi-robot navigation

	Advantages	Disadvantages	Dataset
FMP-DRL [12, 14]	Proposing a collaborative multi-robot training hybrid algorithm that addresses the issue of collision avoidance during multi-robot motion, ensuring that each robot generates collision-free paths	The algorithm generates discrete outputs instead of continuous outputs, resulting in discontinuous motion of the robot agents and a lack of robot agents with more complex kinematics. Additionally, due to the limitations of decentralized navigation, it often fails to produce the time-optimal paths	Not explicitly mentioned
MAPPER [15]	By decomposing long-distance navigation tasks into local subtasks, the limitations of reinforcement learning can be mitigated. Additionally, an evolutionary training approach is utilized, ensuring excellent scalability of the algorithm. This allows for easy extension to large-scale and complex multi-robot environments	High reliance on waypoints during the navigation process. It has poor safety performance and there is a possibility of collision with other robots and obstacles	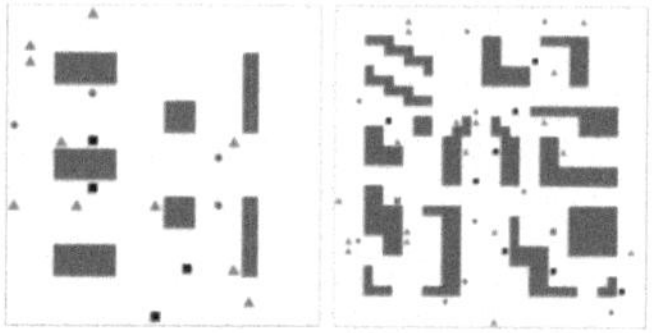 Grid world simulation environment. During the process of evolutionary training, the size of the grid environment, the number of agents, and dynamic obstacles are all continuously increased

(continued)

Table 3 (continued)

	Advantages	Disadvantages	Dataset
PRIMAL2 [13, 16]	As a decentralized navigation algorithm, it relies on centralized navigation for imitation learning, combining the advantages of both types of algorithms to some extent. It also addresses the lifelong navigation problem	More suitable for large-scale multi-robot systems, performance is mediocre when there are few agents in the system. In addition, implicit agent coordination still needs further improvement	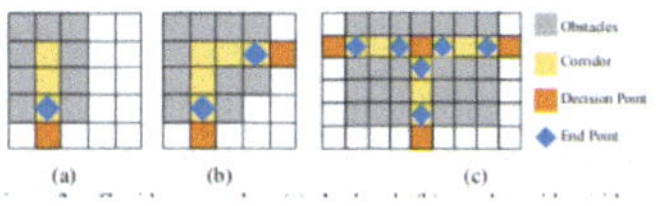 Two-dimensional discrete 4-connected grid worlds where agents, goals, and obstacles occupy one grid cell, respectively
End-to-End multi-robot collision avoidance algorithm [17]	Through multi-scenario and multi-stage training, it narrows the gap between centralized navigation and decentralized navigation in the field of multi-robot obstacle avoidance, surpassing previous decentralized navigation obstacle avoidance algorithms. It also demonstrates good adaptability to new environments	There is still a gap in performance compared to the global planner of centralized navigation. However, considering the many irreplaceable advantages of decentralized navigation, further research is needed for this algorithm	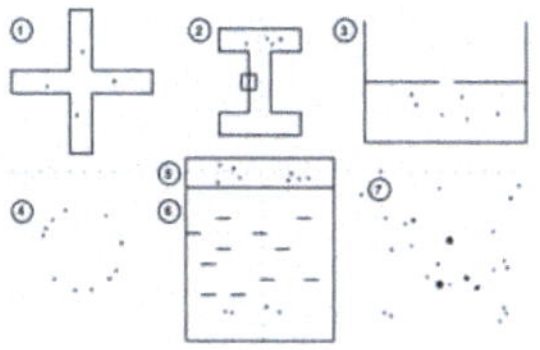 Simulations of large robot groups with laser scanners were conducted in the stage simulator in ROS. The above images depict 2D simulated scenarios used for training the collision avoidance algorithm

avoidance should be prioritized: vehicle, human, and stationary obstacles, and current research has not yet ventured into this area.

In response to these problems, future researchers can: (1) According to the purpose of the study, consider more use of 3D scene simulation. (2) The actual scene evaluation is carried out when conditions permit, and the actual test is carried out in the complex indoor environment, the field environment with bad road conditions, and the human–computer interaction scene with human participation. (3) Research such as fixed-point delivery through face recognition, gesture instruction recognition, target tracking, and other human–computer interaction navigation algorithms. (4) Add low-speed and high-speed moving objects to the environment, and consider the priority output of the reinforcement learning algorithm.

In addition, the existing algorithms in this field still have some optimizations to be made: (1) High computational complexity: The computation time should be further reduced to meet the real-time application of robot navigation. (2) High storage complexity: The storage space requirement should be further reduced to lower the size of the model. (3) Low accuracy of human–robot interaction algorithms: The accuracy of natural language and semantic navigation is still low. (4) Unreasonable assumptions: In many algorithms, some parameters need to be assumed during training and application, or the action pattern of obstacles should be assumed in advance.

To address these shortcomings, future researchers can: (1) model compression, including model distillation, pruning, and quantization. (2) use more complex neural network architectures, increase the size of the dataset using data enhancement techniques, employ parameter aggregation and pooling techniques, use multiple support vector regression, and use more data clustering. (3) Collect better and more training data, and optimize data preprocessing methods, data enhancement; use learning algorithms such as meta-learning, fresh-shot learning. (4) Cancel some assumptions before model training for experimental simulation; compare the difference of experimental results. To find better and more universal algorithms, AutoRL, a reinforcement learning algorithm that has been proved to be quite effective in hyperparameter optimization, can be used to improve the existing algorithms.

4 Conclusion

As more and more application domains begin to replace human labor with robots, more efficient, accurate, and functional navigation algorithms are evolving at a higher rate than ever before. Traditional navigation algorithms are gradually becoming less adapted to the increasingly complex navigation environments and task requirements, and reinforcement learning has been shown in many studies to be an excellent solution to the robot navigation problem. By analyzing the existing literature and tools in this field, I classify the current field of reinforcement learning-based robot navigation into three main areas: Indoor navigation, navigation in dynamic environments, and multi-robot navigation. This paper introduces the research objectives, application scenarios, and existing research progress in these areas and the innovations of these researches. Additionally, to provide convenience for future scholars, this article sorts the datasets and modeling tools used by these researchers in training and evaluating the algorithms. Finally, this paper reveals the current problems and optimizations in this area and provides some possible research directions and improvement ideas that will help researchers to develop more successful robot navigation algorithms in the future.

References

1. Faust A, Oslund K, Ramirez O, Francis A, Tapia L, Fiser M, Davidson J (2018) PRM-RL: long-range robotic navigation tasks by combining reinforcement learning and sampling-based planning. In: IEEE International conference on robotics and automation (ICRA). Brisbane, pp 5113–5120
2. Francis A, Faust A, Chiang HTL, Hsu J, Kew JC, Fiser M, Lee TWE (2020) Long-range indoor navigation with PRM-RL. IEEE Trans Rob 36(4):1115–1134
3. Chiang HTL, Hsu J, Fiser M, Tapia L, Faust A (2019) RL-RRT: Kinodynamic motion planning via learning reachability estimators from RL policies. IEEE Robot Autom Lett 4(4):4298–4305
4. Gao J, Ye W, Guo J, Li Z (2020) Deep reinforcement learning for indoor mobile robot path planning. Sensors 20(19):5493
5. Eysenbach B, Salakhutdinov RR, Levine S (2019) Search on the replay buffer: bridging planning and reinforcement learning. In: Advances in neural information processing systems. Vancouver, p 32
6. Song S, Yu F, Zeng A, Chang AX, Savva M, Funkhouser T (2017) Semantic scene completion from a single depth image. In: IEEE conference on computer vision and pattern recognition, pp 1746–1754
7. Everett M, Chen YF, How JP (2018) Motion planning among dynamic, decision-making agents with deep reinforcement learning. In: IEEE/RSJ International conference on intelligent robots and systems (IROS). Madrid, pp 3052–3059
8. Sathyamoorthy AJ, Liang J, Patel U, Guan T, Chandra R, Manocha D (2020) Densecavoid: real-time navigation in dense crowds using anticipatory behaviors. In: IEEE International conference on robotics and automation (ICRA). Paris, pp 11345–11352
9. Sathyamoorthy AJ, Patel U, Guan T, Manocha D (2020) Frozone: freezing-free, pedestrian-friendly navigation in human crowds. IEEE Robot Autom Lett 5(3):4352–4359
10. Fan T, Cheng X, Pan J, Long P, Liu W, Yang R, Manocha D (2019) Getting robots unfrozen and unlost in dense pedestrian crowds. IEEE Robot Autom Lett 4(2):1178–1185
11. Stern R (2019) Multi-agent path finding–an overview. In: Artificial intelligence: 5th RAAI Summer School, Dolgoprudny, pp 96–115
12. Semnani SH, de Ruter A, Liu H (2020) Force-based algorithm for motion planning of large agent. IEEE Trans Cybern 52(1):654–665
13. Damani M, Luo Z, Wenzel E, Sartoretti G (2021) PRIMAL2: Pathfinding via reinforcement and imitation multi-agent learning-lifelong. IEEE Robot Autom Lett 6(2):2666–2673
14. Semnani SH, de Ruiter AH, Liu HH (2020) Multi-agent motion planning for dense and dynamic environments via deep reinforcement learning. IEEE Robot Autom Lett 5(2):3221–3226
15. Liu Z, Chen B, Zhou H, Koushik G, Hebert M, Zhao D (2020) Mapper: multi-agent path planning with evolutionary reinforcement learning in mixed dynamic environments. In: IEEE/RSJ International conference on intelligent robots and systems (IROS). Las Vegas, pp 11748–11754
16. Sartoretti G, Kerr J, Shi Y, Wagner G, Kumar TS, Koenig S, Choset H (2019) Primal: Pathfinding via reinforcement and imitation multi-agent learning. IEEE Robot Autom Lett 4(3):2378–2385
17. Long P, Fan T, Liao X, Liu W, Zhang H, Pan J (2018) Towards optimally decentralized multi-robot collision avoidance via deep reinforcement learning. In: IEEE international conference on robotics and automation (ICRA). Brisbane, pp 6252–6259
18. Lin J, Yang X, Zheng P, Cheng H (2019) End-to-end decentralized multi-robot navigation in unknown complex environments via deep reinforcement learning. In: IEEE International conference on mechatronics and automation (ICMA). Stockholm, pp 2493–2500
19. Shah P, Fiser M, Faust A, Kew JC, Hakkani-Tur D (2018) Follownet: robot navigation by following natural language directions with deep reinforcement learning. arXiv preprint arXiv:1805.06150

20. Mousavian A, Toshev A, Fišer M, Košecká J, Wahid A, Davidson J (2019) Visual representations for semantic target driven navigation. In: International conference on robotics and automation (ICRA). Montreal, pp 8846–8852
21. Druon R, Yoshiyasu Y, Kanezaki A, Watt A (2020) Visual object search by learning spatial context. IEEE Robot Autom Lett 5(2):1279–1286

Active Control Technology of Supply Chain Intelligent Work Robot Based on Speech Enhancement Technology

Zhanbiao Tang, Sanlin Gu, Lixing Zhang, Jiao He, and Lin Tian

Abstract In order to improve the efficiency and quality of supply chain management, achieve intelligent management of the supply chain, reduce management costs, and provide competitive advantages for enterprises and meet market demands. Propose an active control method for supply chain intelligent operation robots based on speech enhancement technology. By preprocessing the voice signal of the supply chain intelligent operation robot and utilizing the Teager energy operator, the characteristics of the supply chain intelligent operation robot's voice signal are extracted. On this basis, based on speech enhancement technology, the least squares support vector machine is used to output control results and achieve active control of supply chain intelligent operation robots. The experimental results show that the supply chain intelligent operation robot proposed in this paper has high active control accuracy and short time.

1 Introduction

Supply chain management plays a crucial role in enterprise operations. With the further advancement of globalization and the intensification of market competition, the supply chain has become increasingly complex and vast, and the demand for efficient management and flexible response has become increasingly strong [1, 2]. However, traditional supply chain management methods have some problems, such as poor information exchange, long decision-making time, and low accuracy, which are difficult to meet the rapidly changing market demand. In addition, the continuous development of intelligent technology provides new opportunities for improving supply chain management. The rise of technologies such as artificial intelligence,

Z. Tang · S. Gu · L. Zhang · J. He
State Grid Gansu Electric Power Company Materials Company, Lanzhou 730050, Gansu, China

L. Tian (✉)
State Grid Siji Feitian (Lanzhou) Cloud Data Technology Co., LTD, Lanzhou 730050, Gansu, China
e-mail: evelyn09_ok@yeah.net

S. Yadav et al. (eds.), *Proceedings of 3rd International Conference on Artificial Intelligence, Robotics, and Communication*, Lecture Notes in Electrical Engineering 1172, https://doi.org/10.1007/978-981-97-2200-6_19

machine learning, and the Internet of Things has made intelligent robots an important tool for supply chain management. Intelligent robots can autonomously perform job tasks through their ability to perceive, learn, and make decisions, improving the efficiency, quality, and adaptability of the supply chain [3, 4]. Given the above background, the research on supply chain intelligent operation robots has emerged. It aims to develop autonomous and intelligent work robots by combining intelligent technology with supply chain management practices and achieve their active control. In order to improve the efficiency and quality of supply chain management, provide better competitiveness for enterprises, and meet the market demand for efficient, flexible, and intelligent supply chain management.

At present, scholars in relevant fields have conducted research on the control of intelligent operation robots in the supply chain. Yang et al. [5] proposes a remote control method for industrial robot joysticks based on robot operating systems. This study designs a robot remote control system on how to remotely control industrial robots using a joystick, achieving information exchange between the robot operating system and the control cabinet. A new application program interface has been developed to achieve the secondary development of industrial robots. Ethernet connections are used to exchange system data and reduce communication latency. In order to improve the operational efficiency of the robot, the joystick is used as an external input device of the robot operating system to efficiently and reliably achieve the motion operation of the robot. This method has certain feasibility. Lima et al. [6] proposed a user oriented natural robot control method based on thin plate splines and long-term recursive convolutional networks. Nonlinear customizable pose mapping based on thin plate spline curves directly converts human motion into robotic arm motion in a nonlinear manner, allowing for matching of different bodies with different workspace shapes and kinematic constraints. The second type is a deep neural network hand state classifier based on long-term recursive convolutional networks, which utilizes the temporal coherence of the obtained deep data. The results indicate that this method can successfully complete the proposed task and provide more accurate operations in areas closer to the boundary of the workspace. However, the above methods still have problems with low control accuracy and long time. Therefore, an active control method for supply chain intelligent operation robots based on speech enhancement technology is proposed.

2 Active Control Method for Intelligent Operation Robot in Supply Chain

2.1 Preprocessing Supply Chain Intelligent Operation Robot Voice Signals

In the process of inputting voice commands for supply chain intelligent operation robots, various factors such as noise interference, user speaking volume, distance from the voice receiver, and internal current burrs in the system can have a certain impact on the active control results of supply chain intelligent operation robots. Therefore, it is necessary to preprocess the voice signals of supply chain intelligent operation robots, which are mainly divided into the following four parts:

(1) Median filtering: Median filtering can effectively remove noise such as impulse signals and burrs caused by the speech sampling system itself. Assuming that $z(n)$, $z(n-1)$, and $z(n+1)$ represent the current position and adjacent speech data vectors, and median$(\cdot)$ represents the median operation, the median filtering formula is as follows:

$$z(n) = \text{median}[z(n-1), z(n), z(n+1)]. \tag{1}$$

(2) Normalization: Due to the different effects of different sound amplitudes on feature extraction of speech signals, it is necessary to normalize the speech signal and map it to the same interval $[-1, 1]$ to eliminate the impact of sound amplitude differences. Assuming that max$(\cdot)$ represents the maximum operation, the normalized result of the median filtered speech signal is represented as:

$$\tilde{z}(n) = z(n) \big/ \max[|z(n)|]. \tag{2}$$

(3) Pre-emphasis: Because the greater the energy contained in the speech signal, the lower the corresponding frequency. Therefore, when the user speaks, most of the energy is distributed in the bass part. Pre-emphasis operation is to process the high-frequency signal through a high-pass filter to achieve the overall smoothness effect of the speech signal. Assuming that y represents the input speech signal, α represents a constant, and $\alpha \in (0.93, 0.98)$, the high-pass filter transfer function is represented as:

$$G(y) = \tilde{z}(n)\left(1 - \alpha y^{-1}\right). \tag{3}$$

(4) Framing and windowing: Framing achieves a smooth transition between frames by segmenting speech, with a frame shift between adjacent frames to ensure the continuity of the speech signal after framing. The frame shift is usually 0–0.5 times the frame length. Assuming that $y_d(m)$ represents the m frame of the

speech signal output by the high-pass filter, $W(\cdot)$ represents the moving window function with a length of M, l represents the moving distance, and Hamming window is selected as the window function, the m frame $y_w(m)$ of the speech signal after framing is as follows:

$$\begin{cases} y_w(m) = G(y) \cdot y_d(m) \cdot W(m) \cdot l \\ W(m) = 0.54 - 0.46\cos\left(\dfrac{2\pi m}{M-1}\right), m \in (0, M-1) \end{cases}. \tag{4}$$

Through the above process, complete the speech signal preprocessing of the supply chain intelligent operation robot.

2.2 Extracting Speech Signal Features of Smart Operation Robots in the Supply Chain

After preprocessing the speech signal of the supply chain intelligent operation robot, the Teager energy operator [7, 8] is used to extract the speech signal features of the supply chain intelligent operation robot. The process for extracting speech signal features of supply chain intelligent operation robots is as follows:

Step 1: After pre-emphasis processing and framing, the $y'_w(m)$ expression of the supply chain intelligent operation robot voice signal is as follows:

$$y'_w(m) = w[y_w(m) - y_w(m-1)]. \tag{5}$$

In Formula (5), w represents the weight coefficient. Strengthen the high-frequency signals contained in the voice of the supply chain intelligent operation robot through the set weight coefficients. Add the Hamming window to the speech signal after framing, and obtain the speech signal of the supply chain intelligent operation robot after windowing processing.

Step 2: Calculate the Teager energy of the speech signal after windowing processing.

Step 3: Perform a fast Fourier transform on the Teager energy to convert the time-domain speech signal of the supply chain intelligent operation robot into the frequency domain, and obtain the energy spectrum value of each frame of the supply chain intelligent operation robot's speech signal.

Step 4: Use a high-pass filter to filter the obtained energy spectrum, take the logarithm of the filter output value, and obtain the logarithmic form of the energy spectrum.

Step 5: Using the discrete cosine transform method, process the obtained logarithmic energy spectrum, and the obtained results are the speech signal features of the supply chain intelligent operation robot. The formula for extracting the speech signal features of the supply chain intelligent operation robot is as follows:

$$G(i) = \sum_{i=1}^{N} m_i y'_w(m) \cos\left[\frac{1}{N}(i - 0.5)\right]. \tag{6}$$

In Formula (6), m_i and N represent the feature dimensions of the speech signal and the number of high-pass filters, respectively. By using Formula (6), the speech signal features of supply chain intelligent operation robots can be extracted.

2.3 Realize Active Control of Intelligent Operation Robots in the Supply Chain

Hidden Markov models are widely used in fields such as natural language processing and pattern recognition. Utilizing hidden Markov models to enhance the speech signal of supply chain intelligent operation robots, the speech enhancement results [9, 10] are used as inputs for active control of supply chain intelligent operation robots, and the control results are output using least squares support vector machines to achieve active control of supply chain intelligent operation robots.

The active control of intelligent operation robots in the supply chain belongs to a nonlinear regression problem. Assuming that Table $\{(x_1, y_1), (x_2, y_2), \cdots, (x_l, y_l)\}$ shows a set of voice sample datasets, there exists a nonlinear function for active control of supply chain intelligent operation robots as follows:

$$f(x) = G(i)\langle \omega \cdot \varphi(x) \rangle + b. \tag{7}$$

In Formula (7), ω and b represent the weight vector and bias, respectively, while $\varphi(x)$ represents the nonlinear mapping function that maps the input samples to the high-dimensional feature space.

Select the radial basis function as the kernel function of the least squares supports vector machine for active control of supply chain intelligent operation robots, and its expression is as follows:

$$K(x, x_i) = \exp -\left(\frac{\|x - x_i\|^2}{2\sigma^2}\right). \tag{8}$$

In Formula (8), σ represents the width of the kernel function.

The expression of the least squares support vector machine regression model for the active control of the final supply chain intelligent operation robot is as follows:

$$f'(x) = G(i) \sum_{j=1}^{l} \alpha_j K(x, x_i) + b. \tag{9}$$

In Formula (9), α_j represents the Lagrange multiplier.

Table 1 Simulation parameter settings

Parameter	Value	Parameter	Value
Sampling frequency	48 kHz	Digitalizing bit	16 bits
Window size	3 × 3	Moving step size	2
Frame offset	0.2	Weight coefficient	0.01
Penalty parameter	0.2	Width of kernel function	8
Relaxation variable	0.9	Dimensions of eigenvectors	3

Using the least squares support vector machine regression model of Formula (9), output the active control results of the supply chain intelligent operation robot.

3 Experiments and Analysis

3.1 *Experimental Environment Settings*

In order to verify the effectiveness of the active control method for supply chain intelligent job robots based on speech enhancement technology, the experiment was conducted in a MATLAB environment using the Windows 10 operating system, Intel (R) Pentium (R) CPU G645 @ 3.60 GHz processor, and 64 GB of memory. Draw up the test platform, namely Matlab R2019 b, and carry out the simulation test in the environment with the main frequency of 1. Build a real-time simulation platform based on the simulation parameters in Table 1.

Set the speech signal feature quantity of the supply chain intelligent operation robot to 100 GB, and use the method of reference [5] and the method of reference [6] as a comparison method to verify the feasibility of this method.

3.2 *Accuracy of Active Control for Smart Work Robots in the Supply Chain*

Smart operation robots play an important role in the supply chain and can perform tasks autonomously. Accurate control ability can ensure the smooth completion of tasks and avoid mistakes or misoperation, thus improving the overall production efficiency. The accurate control ability of smart operation robots can ensure that there are no mistakes in the processing, inspection, and packaging in the production process. Maintain product consistency and compliance with standards, and ensure the quality of final products. In order to verify the accuracy of the active control of the supply chain intelligent work robot proposed in this article, three methods were

Table 2 Accuracy of active control for supply chain intelligent work robots using different methods

Supply chain intelligent operation robot voice signal feature quantity/GB	The proposed method/%	The method of Yang et al. [5]/%	The method of Lima et al. [6]/%
20	98.6	89.6	85.2
40	97.8	88.5	86.4
60	98.3	87.9	86.9
80	98.1	88.7	84.8
100	96.9	87.6	85.1

compared and the accuracy of the active control of the supply chain intelligent work robot using different methods was obtained as shown in Table 2.

As can be seen from Table 2, the method proposed in this paper has high accuracy for the active control of intelligent working robots in supply chain, and when the number of voice signal features of smart operation robots in supply chain is 100 GB, it can reach 96.9%. However, the active control accuracy of the method in reference [5] and the method in reference [6] for the smart operation robots in supply chain is 87.6 and 85.1%. This is because this method uses Teager energy operator to extract the features of the speech signal of intelligent working robots in supply chain. These features can capture the instantaneous changes, rhythms, and important contents in the speech signal, better distinguish different speech instructions or signals, and increase the adaptability of the model to noise, interference, and signal changes, thus improving the accuracy of active control.

3.3 *Active Control of Time for Smart Operation Robots in the Supply Chain*

Smart operation robots are an important production asset in the supply chain. Effectively controlling its running time can improve the efficiency of resource utilization, enable it to complete tasks more efficiently, avoid unnecessary idleness or excessive operation, and thus maximize the utilization of resources. Further validate the active control time of the supply chain intelligent operation robot proposed in this article, and compare three methods to obtain the active control time of the supply chain intelligent operation robot using different methods, as shown in Table 3.

As can be seen from Table 3, when the number of speech signal features of the intelligent operation robot in the supply chain is 100 GB, the active control time of the intelligent operation robot in the supply chain is 4.3 ms, while the method in Yang et al. [5] is 5.8 ms and the method in Lima et al. [6] is 6.9 ms. This is because this method uses the output control results of least squares support vector machine to map the data in high-dimensional feature space to low-dimensional feature space and

Table 3 Active control time of supply chain intelligent operation robots using different methods

Supply chain intelligent operation robot voice signal feature quantity/GB	The proposed method/ms	The method of Yang et al. [5]/ms	The method of Lima et al. [6]/ms
20	0.5	1.2	2.4
40	1.0	2.4	3.7
60	2.2	3.3	4.9
80	3.1	4.6	5.8
100	4.3	5.8	6.9

greatly reduces the complexity of model construction and calculation by reducing the dimension of features. Through the hyperplane, the regression result of new input data can be predicted quickly and accurately, and the corresponding actions can be executed quickly in the active control process of intelligent operation robot in supply chain, thus reducing the active control time.

4 Conclusion

This article proposes an active control method for supply chain intelligent operation robots based on speech enhancement technology. By preprocessing the voice signals of supply chain intelligent operation robots, extract the features of the supply chain intelligent operation robot's voice signals. Based on speech enhancement technology and least squares support vector machine, active control of supply chain intelligent operation robots is achieved, effectively improving control accuracy and shortening control time.

References

1. Sun H, Mao W, Dang Y et al (2022) Optimum path for overcoming barriers of green construction supply chain management: a grey possibility DEMATEL-NK approach. Comput Ind Eng 164:107833
2. Yuri R, Yashar R, Mariia D et al (2021) Approach to intellectualization of complete supply chain management processes using fuzzy expert systems. Intellectualization Logistics Supply Chain Manage 5:26–39
3. Xu Y, Gao Y, Liu R et al (2021) Robots built robots: nanorobots customized by intelligent robot. Cryst Growth Des 21(10):5508–5514
4. Chu Y, Wang J (2022) Application of measuring intelligent robot in building deformation monitoring. Procedia Comput Sci 208:206–210
5. Yang S, Wang S, Huang S (2023) ROS-based remote control of industrial robot joystick. Proc Inst Mech Eng Part C J Mech Eng Sci 237(1):160–169

6. Lima B, Amaral L, Nascimento-Jr G et al (2022) User-oriented natural human-robot control with thin-plate splines and LRCN. J Intell Rob Syst 104(3):50–59
7. Hasheminejad S (2022) A new protection method for the power transformers using Teager energy operator and a fluctuation identifier index. Electric Power Syst Res 213(Dec):1–9
8. Pei X, Zheng X, Wu J (2021) Intelligent bearing fault diagnosis based on Teager energy operator demodulation and multiscale compressed sensing deep autoencoder. Measurement 179(6):109452
9. Ahmad A, Mozelius P, Ahlin K (2021) Speech and language relearning for stroke patients-understanding user needs for technology enhancement. Proceedings of the eTELEMED 31–38
10. Yang L, Zeng J, Dong G (2023) Simulation of speech enhancement algorithm combined with high-efficiency channel attention mechanism. Comput Simul 40(03):258–262+535

Research on Security Technology of Power Dispatching Data Network Based on Deep Learning

Guoli Chen

Abstract Aiming at the problem that the traditional automatic backup system of power dispatching data lacks the scheduling processing of power data, which leads to the poor stability of its method, an automatic backup system of power dispatching data based on deep learning is proposed. The power system dispatching module uses the deep learning network to construct the bad data identification model, eliminate the bad data in the power system, and generate power dispatching data. Power dispatching data backup module adopts online data backup mode and combines with deep learning theory to perform unified backup, recovery, and backup media management of power dispatching data. The experimental results show that the system designed in this paper can meet the requirements of power dispatching data backup, and the system is stable, which can prove that the practical application effect of the system is better.

1 Introduction

Electric power, as the basis of national economic development, plays an important role in the development of national economy. In the past decade, with the rapid development of technology, the traditional electronic commerce has also brought about a new round of development. At present, intelligent grid technology is one of the important directions for the development and promotion of power grid both at home and worldwide [1]. It is a new type of power grid based on physical power grid, which combines advanced sensing and measuring technology, communication technology, computer technology, and control technology with power grid. It is a new type of power grid technology, which combines advanced sensing and measuring technology. To meet customers' energy demand, optimize the distribution system, ensure energy security and economic efficiency, optimize power supply system when facing environmental challenges, and provide users with reliable, economic, clean,

G. Chen (✉)
State Grid Hanzhong Electric Power Supply Company, Hanzhong, Shaanxi, China
e-mail: chenguoli@sn.sgcc.com

S. Yadav et al. (eds.), *Proceedings of 3rd International Conference on Artificial Intelligence, Robotics, and Communication*, Lecture Notes in Electrical Engineering 1172, https://doi.org/10.1007/978-981-97-2200-6_20

and interactive power supply and value-added services. With the development of urban power, the maximum of spatial distribution of urban power grid is constantly expanding. In order to ensure the stable operation of power grid, power grid should expand the scale of power grid to meet the maximum power supply. However, when the energy load is at a low level, the low utilization of the units can result in excessive waste of local resources [2, 3].

Power dispatch data is a product of power system dispatch automation. With the widespread application of power system dispatch automation, its carrying capacity is increasing day by day, and its role in the power system is gradually prominent. Therefore, the reliability and security of power dispatch data have become a hot topic of widespread concern in the power system. Human operations, virus invasion, hardware damage, natural disasters, and other reasons can all lead to the destruction and loss of power dispatch data. When there is damage or loss of power dispatch data, the primary concern is how to quickly recover the data and ensure the normal operation of the power system. For this reason, the power system needs to build an automatic backup system for power dispatch data, which can quickly recover data in case of data damage or loss, and ensure the normal operation of the power system. In the process of power system dispatching automation, because of external factors, it is easy to generate some bad data, which affects the performance of power system. In the process of automatic backup of power transmission data, the backup of bad data will still be in the backup environment, extending the backup time and data recovery time, and delaying the system recovery work. Therefore, it is necessary to check and remove the bad data during the power generation process. On this basis, an in-depth study based on automatic backup technology for data transmission system is made, and a non-destructive data model is established using the in-depth study method to eliminate the bad data. Based on the backup strategy, the backup of data transmission system is implemented, improving the stability of power system.

2 Methods

2.1 Overall Structure Design of Automatic Backup System for Power Dispatching Data

The automatic backup system for data transmission system is based on BECOS32x64 system and mainly consists of a set of electronic data acquisition module, a power supply module, a power supply data backup module, and an alarm system. The main function of data collection system is to store the operation data of the system. The main function of the power supply module is to use the power supply module as the data collection and to generate the power supply data from the bad data. The main function of the power supply data backup module is to back up the power supply data. The main task of monitoring and alarm system is to monitor and report all the malfunctions of the system [4].

2.2 Power System Operation Module Based on Deep Learning

(1) **Scheduling module**

The power system dispatching module dispatches the power system operation data collected by the data acquisition module, which mainly includes two parts: intelligent learning and real-time operation. Real-time operation part's main function is online dispatching of power system through remote terminal equipment. The main function of the intelligent learning part is to use the deep learning network for artificial intelligence learning and eliminate bad operation data, and its calculation formula is as follows:

$$K = \lambda \frac{(1 - P)^2}{x_k \cdot y_k} \tag{1}$$

where λ represents the power dispatching factor; P indicates bad operation data; and x_k and y_k represent two kinds of attributes of power data. The power system operation data and dispatching standard experience required by the intelligent learning part are provided by the real-time operation part, and the output results provide decision reference for the implementation of the operation part (that is, the implementation of power dispatching).

The intelligent learning part of power system dispatching module uses deep learning network to identify bad operation data, and the model includes two links: training and identification. The training link learns according to the historical dispatching data of power system and constructs the mathematical relationship between the data to be identified and the set type [5]. After processing the data to be identified, the identification link introduces it into the trained model, compares it with the model parameters, and identifies the bad operation data according to the similarity threshold set in advance. Deep learning network consists of input layer, feedforward layer, and decision layer. The main functions of input layer and feedforward layer are to receive the transformed power dispatching sample data and to simulate the deep processing of power dispatching sample data by human neural network. In essence, the identification of power operation data can be understood as a classification problem, so Softmax classifier is used as the decision-making layer of deep learning network, and finally bad power operation data is output to realize power system scheduling.

(2) **Data backup module**

Traditional power dispatching data backup mostly adopts host direct storage mode, and all power dispatching data backup is realized by using disk array. When a single point of failure occurs, it will lead to serious problems in the overall backup process [6]. Therefore, the power dispatching data backup module adopts the online data backup mode and performs unified backup, recovery, and backup medium management on the power dispatching data generated by the power system dispatching module based on deep learning. The calculation formula is as follows:

$$H = \alpha K - \frac{1}{2}\beta_m \sqrt{x_k \cdot y_k} \tag{2}$$

where a represents the backup feature vector of power data and β_m represents the storage management factor of power data. The PC assumes the function of backup client, adopts ES40 server as media server, and is equipped with RA3000 disk array for power dispatching data backup. The two servers back up the power dispatching data in the form of active and standby, that is, under normal backup conditions, one server backs up the data online and the other server maintains the standby state. If the online data backup server fails, the standby server automatically starts data backup to ensure the continuous operation of the power dispatching data backup system.

2.3 Scheduling Data Backup Strategy

In the automatic backup system of power dispatching data, the backup strategy is particularly important, which can be summarized into three main types: full backup, incremental backup, and differential backup. The main basis for choosing backup strategy is the criticality of power dispatching data and the urgency of data backup [7].

(1) **Full backup**

The full backup strategy is to back up all power dispatching data in days. The advantage of full backup strategy is intuitive. At the same time, when the power dispatching data is damaged or lost, the damaged and lost data can be recovered only by using the backup the day before the problem occurs. The disadvantage of the full backup strategy is that when the whole power system is fully backed up in days, there is a lot of duplicate backup data, which wastes a lot of backup space and reduces the economy of the power dispatching data backup system. Moreover, the amount of power dispatching data to be backed up under the full backup strategy is very large, which leads to a long backup time. For the power system with complex operation environment and high stable operation requirements, it is not scientific and economical to use full backup strategy.

(2) **Incremental backup**

Compared with the full backup strategy, the incremental backup strategy only backs up the power dispatching data of the previous day. The main advantage of incremental backup strategy is that there is no duplicate backup data in the backup process, which saves backup space and shortens backup time. However, the shortcomings of incremental backup strategy are also prominent. When using this strategy to back up power dispatching data, the data recovery process is extremely complicated, and the stability of data backup is poor. Due to the close connection between tapes of different connected disks, any disk with problems will cause the failure of the whole backup process [8].

(3) **Differential backup**

The backup data of differential backup strategy is the changed data added after the previous full backup. The manager makes a complete backup of power dispatching data in the first time period in a fixed time area. In the remaining time period, back up all the data that are different from the first time period. Compared with the first two backup strategies, the differential backup strategy can solve the defects of the first two backup strategies on the basis of integrating the main advantages of the first two backup strategies. When differential backup strategy is adopted, it is not necessary to completely back up all power dispatching data, which can save backup storage space and backup time. At the same time, differential backup strategy is simple in recovering data when power dispatching data is damaged or lost, and managers can recover data only by using the tapes of the last full backup and the previous backup. Therefore, the differential backup strategy is usually used for power dispatching data backup.

3 Experimental Analysis

In order to evaluate the performance of in-depth study based on automatic backup system for power dispatch information system designed in this article, the state-owned power system was used as the experimental object and its backup performance was tested in its control and control areas stored.

3.1 Backup Stability Test

Firstly, the backup stability of different systems is detected. The closer the output signal frequency is to 70 Hz, the smaller the fluctuation range is, indicating the better the stability of the system. The experimental results are shown in Fig. 1. As can be seen from Fig. 1, the output signal frequency of the traditional method is between 53 and 94 Hz, with a wide fluctuation range, while the output signal frequency of the system method in this paper is always close to 70 Hz, with a range from 65 to 73 Hz. The closer the output signal frequency of the system is to 70 Hz, the stronger the stability of the system. Through the above test results, it can be shown that this system is more stable and can meet the requirements of power dispatching data backup [9].

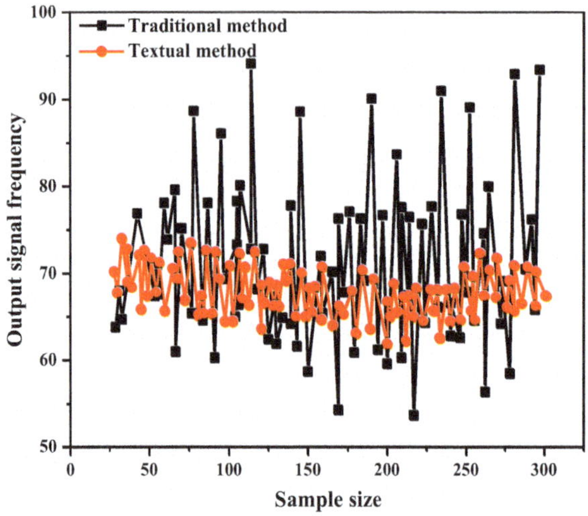

Fig. 1 Comparison diagram of backup stability

3.2 Backup Performance Test

Verify the stability of the backup operation of the system. For different equipments of data transmission, this system, OpenVMS based on backup technique and a backup strategy based on backup technique were used to backup data transmission, and the backup technique of different systems was compared. The results are given in Tables 1, 2 and 3. Among them, *Y* value represents the estimated performance of backup power scheduling data for different systems.

From the analysis of Tables 1, 2 and 3, it can be seen that the backup time of different sizes of power dispatching data backup sources is between 1.43 and 9.58 s, and the backup space occupancy rate is between 0.0003% and 0.0007%. The *Y* value shows a linear and gentle upward trend with the continuous increase of the size of power dispatching data backup sources. Under the condition of the same backup source size, the backup time and backup space occupancy rate of the backup system based on OpenVMS are higher than those of this system, and the *Y* value shows a significant fluctuation state. When the backup source is less than 100 MB, the backup time of the backup system based on incremental backup strategy is shorter than that

Table 1 Test results of system performance in this paper

Backup source size /MB	Backup time /s	Backup space occupancy 1%	*Y* value
1	1.43	0.0003	96.45
10	2.81	0.0003	96.48
100	4.06	0.0004	97.06
500	6.70	0.0005	97.17
1000	9.58	0.0007	97.62

Table 2 Performance test results of backup system based on OpenVMS

Backup source size /MB	Backup time /s	Backup space occupancy 1%	Y value
1	1.96	0.0006	74.12
10	4.02	0.0006	72.01
100	7.93	0.0008	75.24
500	11.22	0.0012	74.18
1000	17.68	0.0017	74.55

Table 3 Performance test results of backup system based on incremental backup strategy

Backup source size /MB	Backup time /s	Backup space occupancy 1%	Y value
1	1.31	0.0007	83.23
10	2.24	0.0008	80.08
100	3.87	0.0010	80.41
500	9.46	0.0016	79.23
1000	20.05	0.0026	77.34

of this system. When the backup source is more than 100 MB, the backup time shows a significant upward trend, which is higher than that of this system, while the backup space occupancy rate continues to be higher than that of this system, and the *Y* value shows a linear and gentle downward trend with the continuous increase of the backup source size of power dispatching data. Based on the above experimental results, the backup performance of this system is obviously better than that of the comparison system [10].

4 Conclusion

The backup of power transmission data is the guarantee for the stable operation of power system. On this basis, an automatic power dispatching data backup system based on deep study has been designed. The experimental results show that the real-time performance evaluation of the system in this paper has met the expected results, indicating that the system meets the requirements of power transmission data backup and has high stability. In the process of power system dispatching automation, because of external factors, it is easy to generate some bad data, which affects the performance of power system. In the process of automatic backup of power transmission data, the backup of bad data will still be in the backup environment, extending the backup time and data recovery time and delaying the recovery work of the system. Therefore, it is necessary to check and remove the bad data during the power generation process. On this basis, an in-depth study based on automatic backup technology for data transmission system is made, and a non-destructive data model is established using the in-depth study method to eliminate the bad data. Based on the backup strategy, the backup of the power dispatch information is accomplished, improving the stability of the power system.

References

1. Kai W, Rui Z, Lei S, Haibo L, Yi W, Jianhua P et al (2021) Research on intelligent technology of dispatching and control to ensure power supply based on multivariate information. J Phys Conf Ser 1846(1):012023 (9 pp)
2. Lu F, Xu L, Jiang Y, Luo P, Hu X, Liang J et al (2022) Smear character recognition method of side-end power meter based on PCA image enhancement. Nonlinear Eng 11(1):232–240
3. Chen W, Liqiang Z, Tianpeng Y, Tao J, Yijing J, Zhihao L (2021) Research on the state detection of the secondary panel of the switchgear based on the yolov5 network model. J Phys Conf Ser 1994(1):012030 (6 pp)
4. Wu C, Ma X, Kong X, Zhu H (2021) Research on insulator defect detection algorithm of transmission line based on centernet. PLoS ONE 16(7):e0255135
5. Tadano T, Ogata J, Shimada S, Kogaki T, Otake Y (2022) Inflow wind estimation based on deep learning for yaw misalignment correction. J Wind Energy JWEA 46(1):1–9
6. Zhao Y, Zhang J, Xiang S, Tang Y (2021) Research on intelligent analysis technology of network security risk based on big data. J Phys Conf Ser 1792(1):012036
7. Li Y, Liu R, Liu X, Li H, Sun Q (2021) Research on information security risk analysis and prevention technology of network communication based on cloud computing algorithm. J Phys Conf Ser 1982(1):012129
8. Bin DM, Yu T, Li X (2021) Research on load balancing dispatching method of power network based on cloud computing. J Phys Conf Ser 1852(2):022047
9. Yu T, Yin X, Yao M, Liu T (2021) Network security monitoring method based on deep learning. J Phys Conf Ser 1955(1):012040 (6 pp)
10. Mao J, Cao C, Wang LJ, Ye J, Zhong WJ (2021) Research on the security technology of federated learning privacy preserving. J Phys Conf Ser 1757(1):012192 (7 pp)

An Improved RRT Path Planning Method Incorporating Deep Reinforcement Learning for Space Multi-arm Robot

Shuhan Liu, Yunhua Wu, and Hao Zhang

Abstract Aiming at the complex internal structure of the satellite to be repaired in the on-orbit maintenance mission and the high requirements of the maintenance mission on the precision of robotic arm path planning, a multi-robotic arm path planning algorithm Deep Deterministic Policy Gradient RRT (DDPG-RRT) combined with deep reinforcement learning is proposed. Firstly, the initial environment is established. Then, on the basis of the traditional RRT algorithm, the idea of DDPG algorithm is introduced to set the dynamic step size to search the collision-free path between the start position and the target position of each robotic arm departing at the same time. Finally, a path optimization algorithm is introduced to smooth the planning path. The results show that compared with the traditional RRT and RRT-Connect obstacle avoidance algorithms, the average planning time, the total average path length, and the success rate of this method are reduced, which is feasible.

1 Introduction

The development of spacecraft in-orbit maintenance technology is essential to ensure stable spacecraft operation in orbit. Owing to the complexity of spacecraft systems and extreme operating environments, avoiding on-orbit failures is almost impossible. Even the failure of an individual component can lead to the in-orbit failure of billions of dollars of spacecraft [1]. Multi-armed space robots are able to cooperate and complete complex tasks with higher efficiency and play a crucial role in on-orbit service missions.

Multi-arm space robot path planning for on-orbit maintenance tasks, maintenance targets (such as satellites) internal structure is complex, the distance between the components is very close, so the maintenance of the robotic arm path accuracy requirements are very high, once the collision will cause serious consequences; due to the components around the target is not easy to dismantle, belongs to the narrow

S. Liu · Y. Wu (✉) · H. Zhang
Nanjing University of Aeronautics and Astronautics, Nanjing 210016, China
e-mail: yunhuawu@nuaa.edu.cn

S. Yadav et al. (eds.), *Proceedings of 3rd International Conference on Artificial Intelligence, Robotics, and Communication*, Lecture Notes in Electrical Engineering 1172, https://doi.org/10.1007/978-981-97-2200-6_21

space path planning problem; in order to improve the maintenance efficiency, multi-arm simultaneous maintenance, need to consider the collision between the multi-arm. However, traditional path planning algorithms such as fast search random tree algorithm does not complete the high-precision obstacle avoidance problem well [2], and collision is easy to occur between multiple arms; large amount of computation, offline planning over a long period of time, and no real-time.

At present, domestic and foreign scholars have a large number of research results in the field of robotic arm path planning. Common path planning methods include ant colony algorithm [3], A* algorithm [4], artificial potential field method [5], and so on. Ant colony algorithm is robust, widely used, easy to combine with other problems, but the convergence speed is slow, and the theoretical analysis is more difficult; A* algorithm is only applicable to two-dimensional space; the artificial potential field method is easy to fall into the local optimum. La Valle [6] firstly put forward RRT algorithm, which can be applied to multi-dimensional space, the principle of the algorithm is simple, and the applicability is strong, but it has a certain degree of randomness in the expansion of the direction, and the convergence speed is slow. Therefore, it also proposes RRT-Connect algorithm [7], and two search trees expand simultaneously to improve the convergence speed. Wang [8] proposed PS-RRT algorithm, which introduces a dynamic sub-area sampling mechanism to quickly fill the area near the parent node and plans a touch-free path for the robotic arm. Xu [9] applied the Bias-RRT algorithm to introduce a target bias strategy to set a certain probability P to define the target point as a random point, so that the search tree grows towards the target. Han [10] proposed Obi-RRT algorithm, which proposed three sampling spaces around these key points by pruning the paths to obtain key points and then sampling them. As the keypoints are constantly updated, this results in optimal or near-optimal paths. Wang [11] proposed NC-RRT algorithm, which makes use of obstacle information and restricts the sampling range to the circular area of variable size containing the target points, so as to avoid the planned paths from being too tortuous. To avoid the planned path being too tortuous. Jiang [12] combined the advantages of the RRT-Connect algorithm and the target-biased RRT algorithm to remove redundant nodes in the obstacle avoidance path using two-way deletion to take the shortest path.

These improved RRT algorithms still suffer from the following difficulties in implementing path planning: they do not complete the high-precision obstacle avoidance well and are prone to collision with obstacles; they are computationally intensive, have a long offline planning over time, and do not have real-time performance; and local bias sampling sacrifices exploring the randomness, which may lose the more optimal solution [13]. For these problems of RRT series algorithms, this paper proposes a fusion of deep reinforcement learning multi-mechanical arm path planning algorithm DDPG-RRT. The new node extension direction of this method is based on the random sampling point and the target point to participate in the decision, the degree of their influence on the direction of the two set the step size parameters p_1 and p_2 to control. Setting p_1 and p_2 as the action of DDPG algorithm, after determining the direction, the extension of new nodes is carried out according to the established step length starting from the nearest node to the random node and the target point

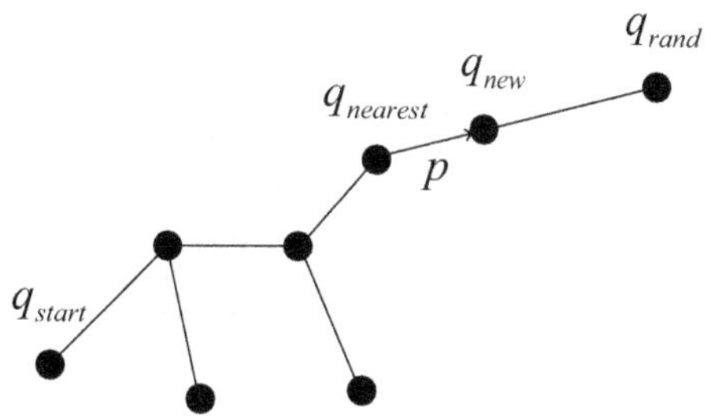

Fig. 1 Fundamentals of RRT algorithm

direction to get the new node of each robotic arm as the state quantity of DDPG algorithm. Finally, the intelligent body is given the corresponding reward value according to the two parameters of whether a collision occurs and the distance from the new node to the goal point. This reduces the time required for path planning, enhances the real-time nature of planning, and prevents falling into local optimum.

2 Basic Algorithm Principles

2.1 *Traditional RRT Algorithm*

The RRT algorithm is a search algorithm that takes q_{start} as the starting point and q_{goal} as the end point and forms a collision-free search tree between the starting point and the end point. q_{start} as the starting node, and then randomly generate a node q_{rand} in the configuration space, calculate the distance of all nodes to q_{rand} to get the smallest distance from q_{rand} point q_{nearest}; q_{nearest} will be along the direction of q_{rand} to expand the distance of a step p, to generate the new node q_{new}. If there is no collision with an obstacle between the connecting lines of q_{new} and q_{nearest}, the node is added to the search tree; if there is a collision, repeat the above process to regenerate q_{rand}. When the distance between q_{new} and q_{goal} is less than a preset threshold, the reverse tracing can be obtained without collision path; if the number of searches exceeds the upper limit, it means that the path planning fails. The basic principle of the RRT algorithm is shown in Fig. 1.

2.2 *DDPG Algorithm*

DDPG is a reinforcement learning algorithm based on the Actor-Critic framework. In this framework, the actor network is responsible for fitting a deterministic policy function in order to select the optimal action as the output given the current state. Whereas the Critic's task is to evaluate the merits of the actor's learning strategy, the Critic network is used to fit the Q-function (value function). Figure 2 depicts

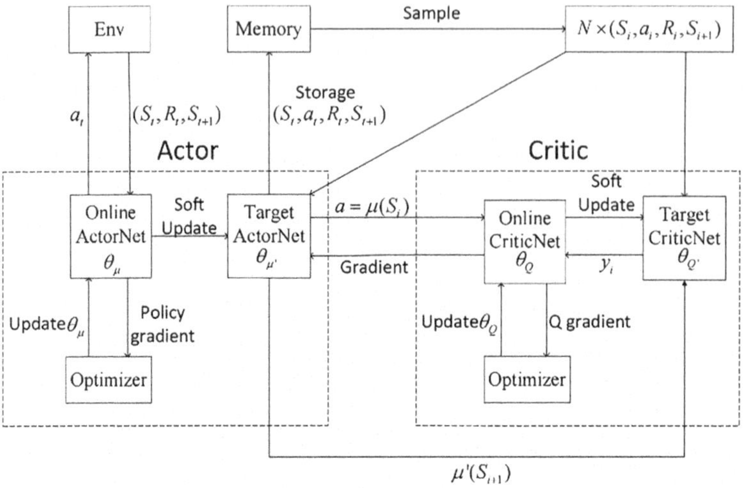

Fig. 2 DDPG algorithm network structure

the network framework of the DDPG algorithm, S_t, R_t, a_t are the state, reward, and action at time-step t, μ denotes the strategy, and θ is the parameter of the neural network.

The actor network consists of an online neural network and a target neural network, and both networks have the same initial values for the parameters θ_μ and θ'_μ. The parameter of the online policy network θ_μ is updated by calculating the policy gradient shown in (1).

$$\begin{aligned}\nabla_{\theta_\mu} J &\approx E_{S_t\sim\rho^\beta}\left[\nabla_{\theta_\mu} Q(s,a|\theta_Q)|_{S=S_t,a=\mu(S_t|\theta_\mu)}\right]\\ &= E_{S_t\sim\rho^\beta}\left[\nabla_a Q(s,a|\theta_Q)|_{S=S_t,a=\mu(S_t)}\nabla_{\theta_\mu}\mu(s|\theta_\mu)|_{S=S_t}\right].\end{aligned} \tag{1}$$

ρ^β is the action strategy used to collect data. A small incremental update of the target network is performed after each time-step to ensure the stability of the network:

$$\theta_{\mu'} \leftarrow \tau\theta_\mu + (1-\tau)\theta_{\mu'}. \tag{2}$$

The Critic network also consists of an online neural network and a target neural network, denoted as $Q(s, a|\theta Q)$ and $Q'(s, a|\theta Q')$, both of which have the same initial values of the parameter θQ and $\theta Q'$, and the parameter of the online Q-network θQ is updated by minimizing the loss function L shown in (3).

$$L = \frac{1}{N}\sum_i (y_i - Q(s_i, a_i|\theta_Q))^2. \tag{3}$$

$\theta Q'$ is updated in the same way as $\theta\mu'$:

$$\theta_{Q'} \leftarrow \tau\theta_Q + (1-\tau)\theta_{Q'}. \quad (4)$$

During the training process, when the number of samples in the memory bank reaches a set value, then the number of samples from it is taken as $N : \ N \times (S_i, a_i, R_i, S_{i+1})$.

3 DDPG-RRT Algorithm

The DDPG-RRT proposed in this paper introduces the deep reinforcement learning idea on the basis of the traditional RRT algorithm and sets the step size to the action of the DDPG algorithm when the new node expands, and the new node is used as the state quantity, so as to obtain the optimal node expansion scheme. Improve the real-time planning while preventing from falling into the local optimum.

3.1 Dynamic Step Size

The extension direction of the new node of each robotic arm is set based on the random sampling point and the target point, and the influence of the two on the direction of the step size Parameters p_1 and p_2 is set to control. The expansion method of the new node is shown in Fig. 3.

Set p_1 and p_2 as the actions of the DDPG algorithm, and after determining the direction, the new node extension is carried out in the direction of q_{rand} and q_{goal} in accordance with the established step size starting with q_{nearest}, and the new node of each robotic arm is obtained as the state quantity of the DDPG algorithm.

According to the joint angle q_{new} of each robotic arm, forward kinematics is used to detect whether collision occurs between robotic arms and whether collision occurs between robotic arms and obstacles. The robotic arm model is simplified as seven

Fig. 3 New node extension method of DDPG-RRT algorithm

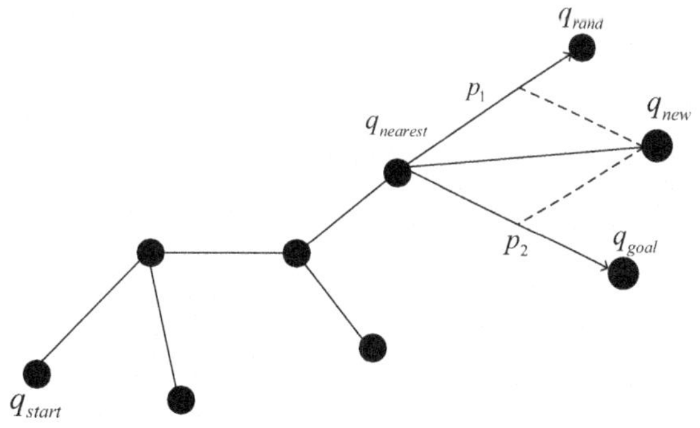

segments of cylinders with different radii, and the collision detection problem can be transformed into a mathematical problem to solve. If a collision occurs, a penalty value $R = -100$ is given, and the robotic arm re-extends the node; if no collision occurs, a reward value is given as (5):

$$R = -\sum_{i=1}^{n} \text{norm}(q_{\text{new}i} - q_{\text{goal}i,2}). \tag{5}$$

n is the number of robotic arms of the robot in the multi-arm space.

The training ends when the number of training rounds reaches the predefined maximum number of rounds, thus obtaining a collision-free path for multiple robotic arms in the planning space.

3.2 Path Optimization

The planning path is smoothed to optimize the result of path planning. It includes the following steps:

Firstly, connect q_{start} and q_{goal} with seven-dimensional joint vectors on the linear joint space of each robotic arm, and replace q_{goal} with the previous point if an obstacle is detected between the two points; after that, it repeatedly performs collision detection and backtracking forward from the original path point until there is no obstacle in between the two points from q_{start} to q_1. Record the first intermediate point q_1 after the smoothing path of each robotic arm; then take q_1 as the starting point and continue to connect q_{goal} directly. Then take q_1 as the starting point, continue to connect q_{goal}, and repeat the previous non-stop backtracking and collision detection method, which can ensure to find a path point q_2 furthest away from q_1, so that there is no collision between q_2 and q_1 on the direct vector path, and this q_2 is the second intermediate point after the smooth path; repeat the above process, and finally get the inter-office spatial path matrix of each robotic arm after the smooth path as path.

3.3 DDPG-RRT Algorithm Flow Description

Comprehensive above, DDPG-RRT algorithm applied to multi-mechanical arm path planning process is shown in Fig. 4.

Fig. 4 Flowchart of DDPG-RRT algorithm

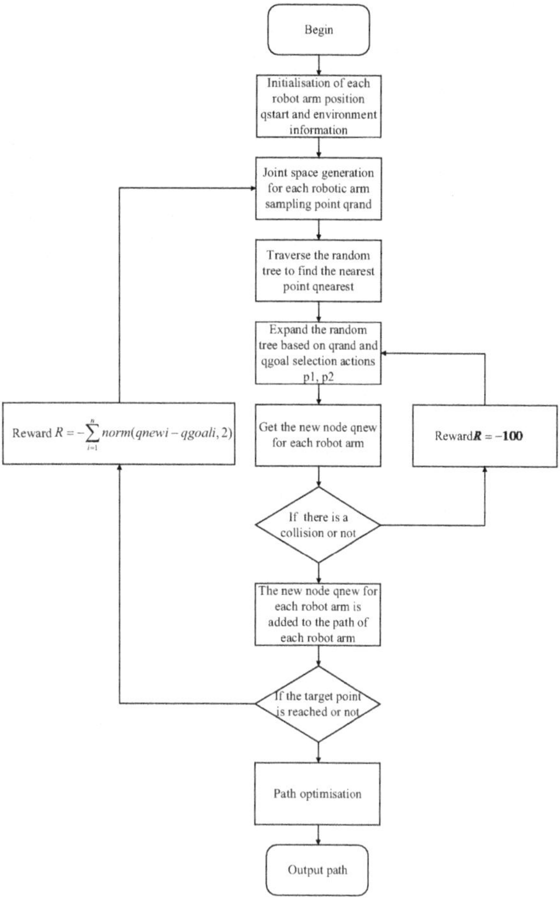

4 Simulation

4.1 Simulation Parameter Setting

In order to verify the effectiveness of the method, a simulation test platform is built based on Simscape. The initial parameters and environment information are as follows.

The robotic arm adopts a seven-degree-of-freedom redundant robotic arm, the kinematic model of each robotic arm is established by MDH modelling method, and the joint parameters are shown in Table 1.

The model of the satellite to be repaired is established as shown in Fig. 3. The four motion target point positions are the four screw screwing positions of the target parts to be repaired, which are marked in red in Fig. 5. The initial joint angles of the four seven-degree-of-freedom robotic arms are all set to [0,0,0,9*π/10,0,0,0]. The

Table 1 Parameters of seven-degree-of-freedom robotic arm

Joint	α_{i-1}/(rad)	a_{i-1}/(mm)	d_i/(mm)	θ_i/(rad)	offset/(rad)
1	0	0	121.94	θ_1	0
2	$\pi/2$	0	140.5	θ_2	$\pi/2$
3	$-\pi/2$	0	140.5	θ_3	0
4	π	408.04	0	θ_4	0
5	0	375.96	−229.82	θ_5	$\pi/2$
6	$-\pi/2$	0	−102.5	θ_6	$-\pi/2$
7	$-\pi/2$	0	−94	θ_7	0

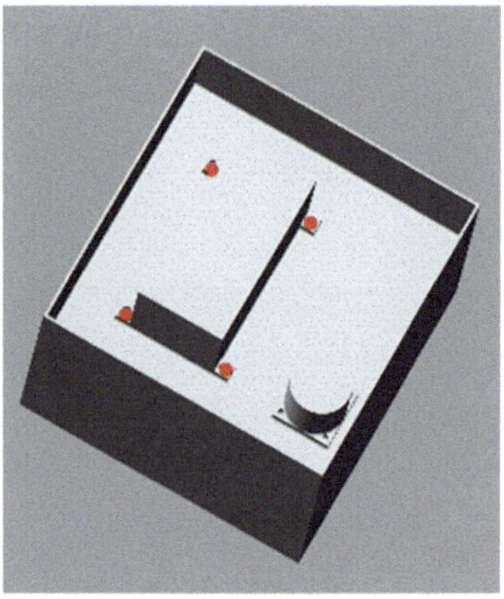

Fig. 5 Satellite model to be repaired

robot base is set up as a square with a side length of 1.2 m. The initial state of the environment is shown in Fig. 6.

In the DDPG algorithm, set the parameters as follows: the state quantity S_t is the joint angle of the new node expanded by each robotic arm, a total of 28-dimensional vector; the action a_t is the weight of the extension to the target point and the random point when the new node is obtained, a total of eight-dimensional vector; set the network structure of the Actor network and the Critic network, and both consist of two hidden layers with 200 neurons each; the hidden layers both use ReLU as the activation function, the output layer of Actor network uses tanh as the activation function, the initial value of the output action is set to 0.5/57.29, and the range of values is set to [0.1/57.29]; the output layer of the Critic network does not use the activation function; add random noise to adjust the action; set the learning rate $\alpha = 0.001$, the discount coefficient, and the update coefficient. Set the maximum training rounds to 10,000 with the maximum number of steps per round 1250. Take the DDPG-RRT algorithm step size as 0.2/57.29.

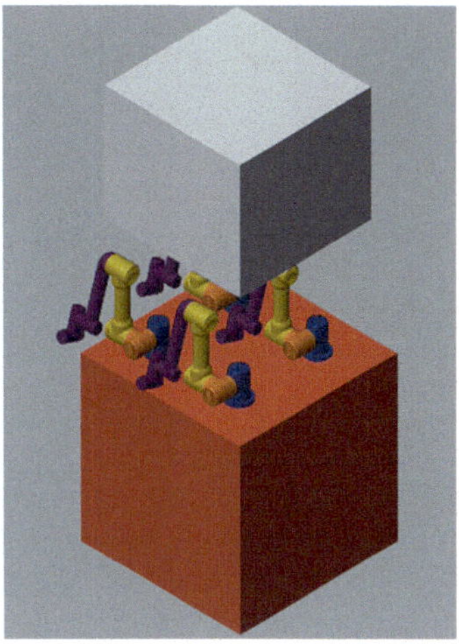

Fig. 6 Initial state of environment

4.2 Simulation Results

In order to verify the effectiveness of the DDPG-RRT algorithm, the RRT, RRT-Connect, and DDPG-RRT algorithms are subjected to multiple simulation experiments. A comparison of the obstacle avoidance effects of RRT and DDPG-RRT algorithms is shown in Figs. 7 and 8, where Fig. 7a–d shows the obstacle avoidance effects of the traditional RRT algorithm; and Fig. 8a–d shows the obstacle avoidance effects of the DDPG-RRT algorithm. Figure 9a–d shows the obstacle avoidance curves between each robotic arm (A1, A2, A3, A4) and obstacles when applying the DDPG-RRT algorithm. Figure 9e shows the obstacle avoidance curves between each robotic arm when applying the DDPG-RRT algorithm. It can be seen that when using the traditional RRT algorithm, there is a collision between each robotic arm and the obstacle, and even if the obstacle avoidance is completed, the distance error between the end of the robotic arm and the target point is also very large, with a low accuracy of obstacle avoidance, whereas when applying the DDPG-RRT, each robotic arm completes the task of obstacle avoidance very well and the distance error between the end of the robotic arm and the target point is very small, which means that the algorithm effectively solves the problem of high-precision obstacle avoidance between multiple robotic arms.

Finally, to demonstrate the generality of the algorithm, each algorithm was run 100 times in this scenario to calculate the average of planning time, total path length, and success rate. The comparison data of each algorithm metrics are compiled into Table 2. The results from Table 2 show that the average planning time, the average length of the total path, and the success rate of the DDPG-RRT algorithm are optimized compared with the traditional RRT and RRT-Connect obstacle avoidance algorithms. The average planning time is reduced by 40.4 and 14.1%, the total path average length is reduced by 16.4 and 9.5%, and the success rate is improved by 29.7%

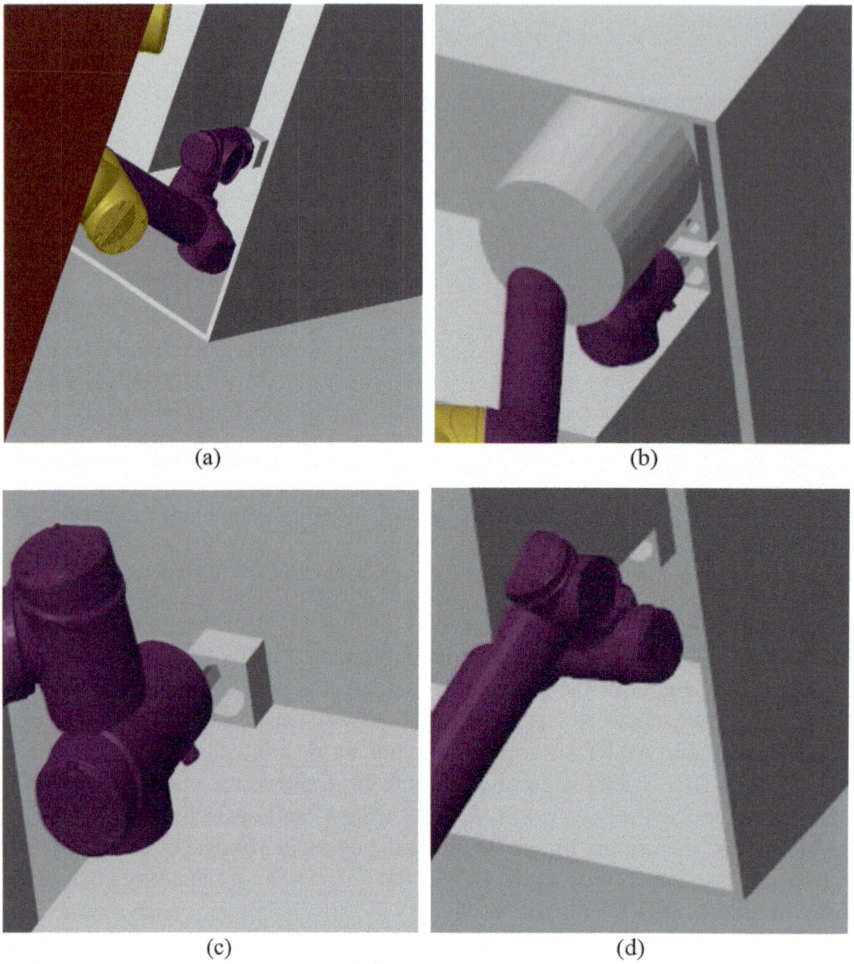

Fig. 7 Obstacle avoidance effect of RRT algorithm

and 17.1%, respectively, compared with the traditional RRT and RRT-Connect algorithms. Comprehensively, the DDPG-RRT algorithm reduces the average planning time, reduces the total path average length, and improves the success rate of obstacle avoidance, proving the effectiveness and superiority of the improved algorithm.

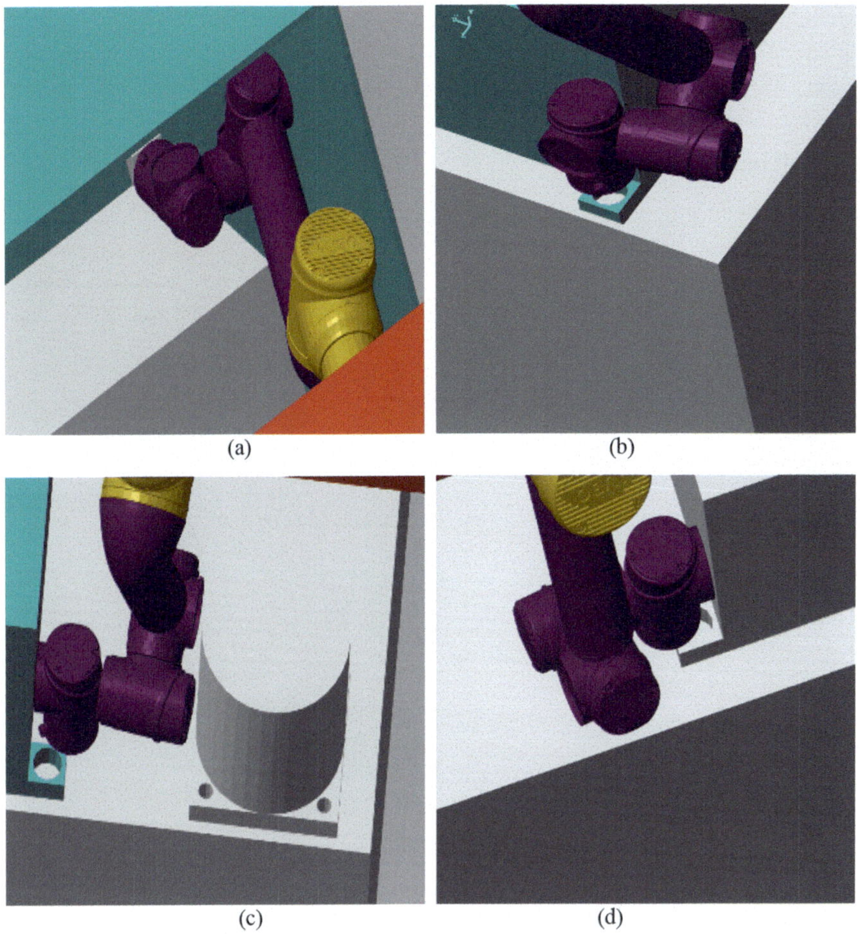

Fig. 8 Obstacle avoidance effect of DDPG-RRT algorithm

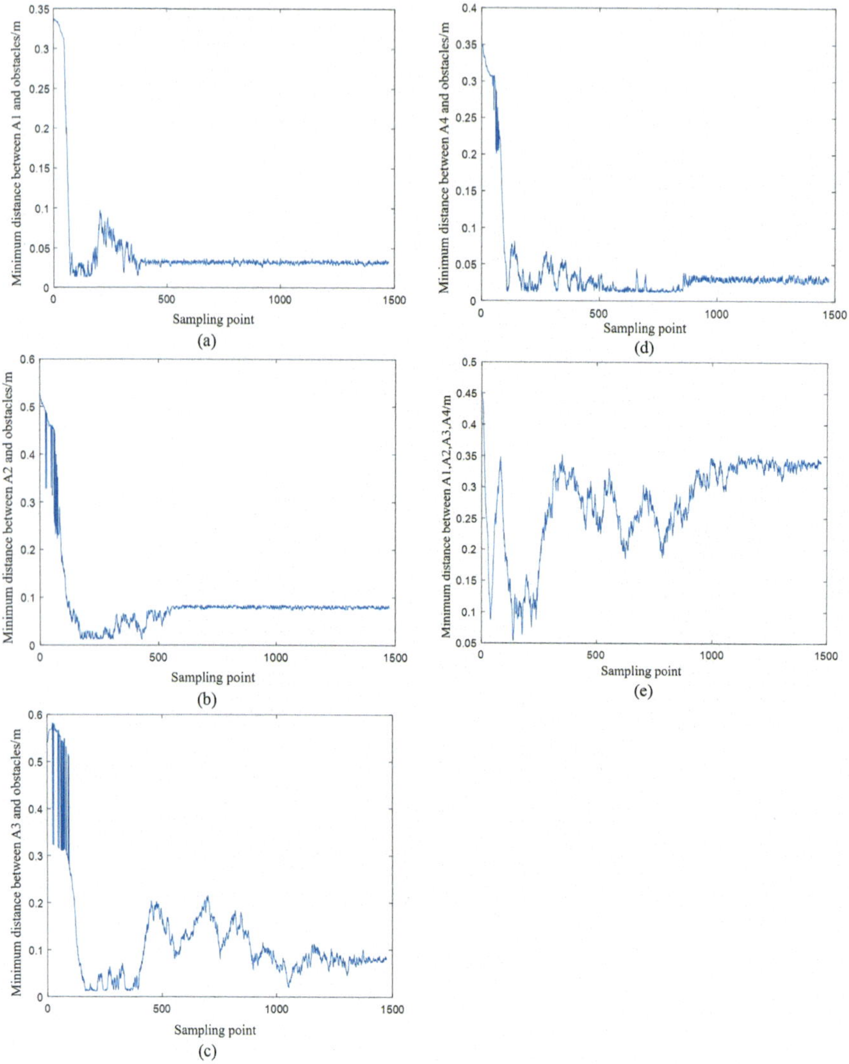

Fig. 9 Obstacle avoidance curves for A1, A2, A3, A4 of DDPG-RRT algorithm

Table 2 Comparison of performance indexes

Algorithms	Average planning time (s)	Average length of total path (m)	Success rate (%)
RRT	3056	4.392	74
RRT-Connect	2122	4.057	82
DDPG-RRT	1822.4	3.672	96

5 Conclusion

In this paper, a DDPG-RRT algorithm is proposed for seven-degree-of-freedom multi-robot arm path planning. Aiming at the problems of low obstacle avoidance accuracy and long offline planning over time of the traditional RRT algorithm, deep reinforcement learning idea is introduced, and when a new node expands, the step size is set to the action of the DDPG algorithm, and the new node is used as a state quantity, so as to obtain the optimal node expansion scheme, which improves the real-time planning and at the same time prevents it from falling into the local optimum. The improved algorithm reduces the average planning time, reduces the average length of the total path, and improves the success rate of obstacle avoidance. The effectiveness and superiority of the improved algorithm are proved through comparative simulation with RRT and RRT-Connect algorithms.

References

1. Cui N, Wang P, Guo J et al (2007) Overview of technological developments in space on-orbit services. Astronaut J 28:805–811
2. You H, He S, Liu H et al (2019) Nine-degree-of-freedom robotic arm path planning based on bi-RRT algorithm. Comput Simul 36:308–313
3. Liu H et al (2017) An improved ant colony algorithm for robot path planning. Soft Comput 21:5829–5839
4. Dolgov D, Thrun S, Montemerlo M et al (2010) Path planning for autonomous vehicles in unknown semi-structured environments. Int J Robot Res 29:485–501
5. Zhou H, Zhou S, Yu J et al (2020) Improved artificial potential field method applied for AUV path planning. Math Probl Eng 21(5):1–21
6. La Valle SM, Kuffner J (2000) Rapidly-exploring random trees: progress and prospects rapidly-exploring random trees. In: 4th International workshop on algorithmic foundations of robotics (Wellesley), pp 293–308
7. Kuffner J, La Valle SM (2000) RRT-connect: an efficient approach to single-query path planning. In: IEEE International conference on robotics and automation, pp 995–1001
8. Wang H, Xiang G, Zhu Y et al (2022) Robotic arm obstacle avoidance path planning based on PS-RRT algorithm. Comput Appl Res 40:42–46
9. Xu N, Chen X, Kong Q et al (2011) Motion planning for robot with nonholonomic constraints. Robot 33:666–672
10. Tan X, Liu G, Yi J (2023) Improved RRT algorithm for robotic arm path planning. Comput Integr Manuf Syst 1:1–18
11. Wang X, Han B, Chen Y et al (2020) Collision-free path planning method for robots based on an improved rapidly-exploring random tree algorithm. Appl Sci 10:131–138
12. Jiang H, Chao Y, Jou J (2023) Improving RRT algorithm for robotic arm path planning. Mech Des Manuf 1:21–27
13. Chen Q, Jiang H, Zheng Y (2019) A review of fast extended random tree algorithms for robot path planning. Comput Eng Appl 2:10–17

A Classification Denoising Algorithm Based on 3D Point Cloud Curvature

Haihao Zhao, Qi Ouyang, Xinglan Zhang, Jianyuan Zhang, and Xinyi Zhang

Abstract In the process of denoising and smoothing 3D point cloud data, it often faces the problems of poor results required for denoising and difficulty in maintaining the local feature details of the point cloud. These problems directly affect the accuracy and computational complexity of subsequent 3D model reconstruction. Therefore, this study proposes a curvature-based classification algorithm for denoising 3D point cloud data, where the point cloud data is classified into smooth and sharp regions based on a set curvature threshold. Statistical filtering combined with radius filtering and Gaussian filtering are used to achieve the processing of different regions, respectively, which effectively achieves fast denoising of the point cloud model. The experimental results show that the proposed method not only effectively removes the noisy points but also preserves the local features of the point cloud model data by applying this method and a single denoising method to the self-collected three-dimensional point cloud data.

1 Introduction

In the field of computer vision and computer graphics, 3D point cloud data is an important form of 3D scene representation, which is widely used in the fields of object reconstruction, environment modelling, and robot navigation. However, the acquired 3D point cloud data usually contains various noises, such as sensor errors, environmental disturbances, and sampling noises, which can seriously affect the quality of the point cloud data and the accuracy of the subsequent modelling.

At present, many scholars at home and abroad have done a lot of research in point cloud denoising. Wang et al. [1] proposed a point cloud denoising algorithm based on moving least squares surfaces to move the sampled points to their corresponding local surfaces to remove the noise. Cao et al. [2] proposed a bilateral filtering point cloud denoising algorithm based on feature selection, which solves the excessive

H. Zhao (✉) · Q. Ouyang · X. Zhang · J. Zhang · X. Zhang
Chongqing University of Technology, Chongqing 400044, China
e-mail: zhaozzz97@stu.cqut.edu.cn

S. Yadav et al. (eds.), *Proceedings of 3rd International Conference on Artificial Intelligence, Robotics, and Communication*, Lecture Notes in Electrical Engineering 1172, https://doi.org/10.1007/978-981-97-2200-6_22

smoothness phenomenon caused by the bilateral filtering algorithm while effectively maintaining the point cloud features, but the time consumption is high. Literature [3, 4] both propose a denoising algorithm based on the classification of point cloud noise scales, and the use of different denoising methods for different scales improves the computational efficiency, but the running time is too long due to too many parameter settings. Guo et al. [5] used a denoising method combining density k-mean clustering algorithm and improved bilateral filtering to improve the denoising efficiency of point clouds, but the denoising efficiency performance is average in small-scale point clouds. Wang et al. [6] proposed a fusion method of WLOP algorithm and bilateral filtering algorithm, which has better robustness, but is not applicable to massive point clouds. He et al. [7] combined voxel filtering and Gaussian filtering algorithms to improve the retention of point cloud features, but voxel filtering has too much impact on the number of point clouds and affects the characteristics of the point cloud.

For the above-mentioned problems of 3D point cloud denoising methods, this paper proposes a curvature-based classification algorithm. Geometric feature extraction is performed for the acquired 3D point cloud data, by using principal component analysis (PCA) to find the neighbourhood points containing N number of points for any point for normal vector estimation, redirection of the point cloud direction to ensure the correctness of curvature computation and the principal curvature information can better capture the changes of the surface and thus improve the accuracy of classification and set a curvature threshold for the classification display. The combination of statistical and radius filtering algorithms is used to denoise smooth regions smaller than the threshold, and Gaussian filtering algorithm is used to smooth and denoise sharp regions with large curvature changes larger than the threshold. This approach enhances the subsequent 3D point cloud modelling accuracy and visualisation effects.

2 3D Point Cloud Classification

In the method of this paper, principal component analysis (PCA) [8] is mainly used for the estimation of normal vectors of each point in the point cloud, which is based on the idea of finding the main directions in the data and projecting the data onto these main directions to obtain the eigenvectors of the data. For any point P, find a neighbourhood point containing N number of points as the main estimated direction. Suppose the coordinates of a point p_i in the point cloud are (x, y, z), calculate the centre of mass of each point as C_i.

$$C_i(x, y, z) = \frac{1}{n}\sum_{i=1}^{n} P_i \tag{1}$$

Compute the covariance matrix S within the neighbourhood points of the point cloud p_i:

$$S = \frac{1}{n}\sum_{i=1}^{n}(P_i - C)(P_i - C)^T \tag{2}$$

The eigenvalues of the covariance matrix S are calculated as $(\lambda_1, \lambda_2, \lambda_3)$, where the eigenvector $(\varepsilon_1, \varepsilon_2, \varepsilon_3)$ corresponding to the eigenvalues is the estimate of the normal vector.

$$S \cdot \varepsilon_i = \lambda_i \cdot \varepsilon_i \tag{3}$$

The point cloud normal vectors estimated by the PCA algorithm are dichotomous, and it is also necessary to redirect all the point cloud normal vectors n_i to the viewpoint p direction. The following equations are satisfied, and the normal vector directions are flipped if they are not satisfied.

$$n_i \cdot (p - p_i) > 0 \tag{4}$$

Figure 1 is the result after normal vector redirection.

In differential geometry, the principal curvature at a given point on a surface usually describes how a surface is oriented and how much it curves at that point. In three-dimensional Euclidean space, for any point p, consider the intersection of this point and the plane determined by a certain set of vectors with the surface on which the point p lies. This intersection is a plane curve that is the curvature of the point p. The points are classified according to a set curvature threshold dis. Points smaller than the threshold are point clouds of classified smooth regions, and points larger than the threshold are point clouds of sharp regions.

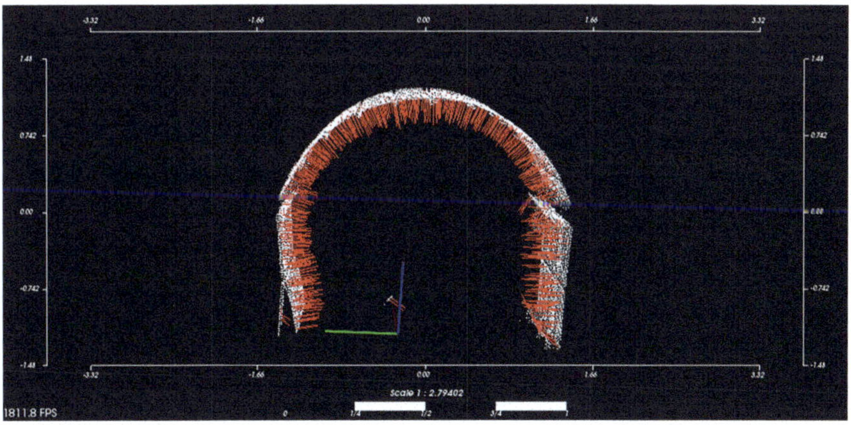

Fig. 1 Normal vector redirection results

3 Smoothing Denoising of 3D Point Cloud Data

3.1 Point Cloud Denoising of Smooth Areas

For the region that is less than the curvature threshold after classification, it means that the point cloud contains less feature information and the surface of the point cloud region is already relatively smooth, so we should focus on filtering the excess point cloud noise for this region. When statistical filtering is used in combination with radius filtering, it is for better denoising of point cloud data in smooth regions while retaining useful information. Statistical filtering is a method of denoising point cloud data based on its statistical properties.

Calculate the mean distance μ and variance σ in the neighbourhood of point p_i. Calculate the distance d_i from the neighbourhood points in the radius to p_i. If d_i satisfies the following equation then it is retained (k is a self-setting parameter) otherwise the point is considered to be an outlier and a noise point.

$$\mu - k\lambda < d_i < \mu + k\lambda \tag{5}$$

The core idea of radius filtering is a filtering algorithm that determines whether a point is an outlier based on the number of points in the neighbourhood of the point. For each point p_i, specify the radius size of the point's neighbourhood to form a spherical neighbourhood centred on the point p_i. Set a threshold M for the number of points in the neighbourhood of an outlier, if the number of points in the spherical neighbourhood is less than the threshold M, then the point p_i is identified as an outlier and is removed.

The idea when using statistical filtering in combination with radius filtering is to apply statistical filtering first to remove most of the noise while retaining more useful information. This method helps to eliminate obvious outliers. Then radius filtering is applied to further filter the remaining noisy points. Since radius filtering is based on local point density, radius filtering can be more effective in dealing with the remaining outlier points thus achieving effective denoising of smooth regions.

3.2 Point Cloud Denoising of Sharp Areas

For the region that is larger than the curvature threshold after classification and has more features, it indicates that the curvature of this region changes a lot, using the usual point cloud denoising method tends to lose too many obvious features in this part, in order to prevent the loss of point cloud features in this region caused by too much smoothing; Gaussian filtering is used for the smoothing and denoising operation for the sharp region. Gaussian filtering is a smoothing filtering technique used in image processing and point cloud processing. Gaussian filtering suppresses and smoothes out noise very well [9], its core idea is to blur the image or point cloud data and reduce the noise by applying a Gaussian function to perform a weighted average of each point and the points in its neighbourhood [10].

For any point cloud $p_i(x_i, y_i, z_i)$ calculate the standard deviation σ to define its neighbourhood size. For point p_i calculate the Euclidean distance d_{ij} between point p_j. Then the weights w_{ij} of u neighbouring points in the neighbourhood are calculated from the Gaussian distribution function and summed to obtain a sum of weights as W_i:

$$\begin{cases} d_{ij} = \sqrt{(x_i - x_j)^2 + (y_i - y_j)^2 + (z_i - z_j)^2} \\ w_{ij} = e^{-\frac{d_{ij}^2}{2\sigma^2}} \\ W_i = \sum_{j=1}^{u} w_{ij} \end{cases} \tag{6}$$

For point p_i, the Gaussian filter calculates the new coordinate value of the weighted average of the points in the neighbourhood of the point as $\left(x_i', y_i', z_i'\right)$.

$$\begin{cases} x_i' = \frac{1}{W_i} \sum_{j=1}^{u} w_{ij} x_j \\ y_i' = \frac{1}{W_i} \sum_{j=1}^{u} w_{ij} y_j \\ z_i' = \frac{1}{W_i} \sum_{j=1}^{u} w_{ij} z_j \end{cases} \tag{7}$$

The above steps are repeated for each point in the 3D point cloud to calculate new coordinate values and adjust them. The new smoothed and denoised point cloud is obtained when all points are weighted and averaged. By using Gaussian algorithm, the closer points in the point cloud have higher weights in the average and the further points have lower weights, thus achieving the smoothing effect.

4 Experimental Preparation and Analysis of Results

4.1 Experimental Preparation

In this experiment, the LR-16FIS-C1 model LIDAR was used to collect the original point cloud in the experimental tunnel, based on the ROS development platform under the Ubuntu 18.04.6 LTS operating system, with the processor AMD Ryzen 7 5800h, the graphic card NVIDIA GeForce RTX 3060 and 16.0 GB of RAM. The PCL-based C++ code was used for programming, and after classifying and visualising the 3D point cloud data by setting a threshold value, it can be seen that the point cloud of sharp areas containing more feature information is obviously marked by red. The experimental results with the original pictures are shown in Fig. 2.

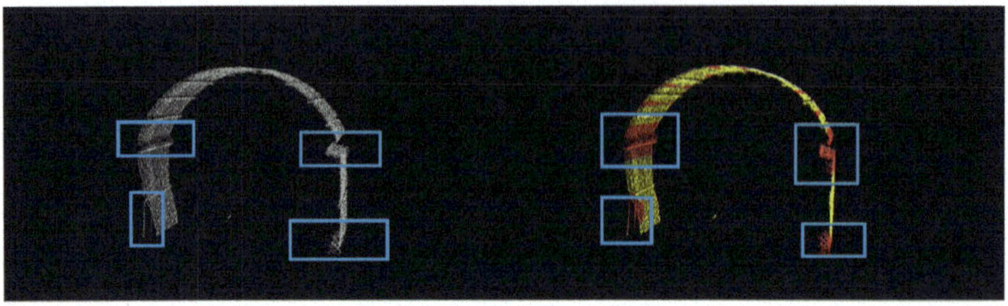

Fig. 2 Comparison between original and classified point cloud

4.2 Analysis of Experimental Results

Parameterisation and visualisation of statistical filtering reveals that too small a parameter results in the formation of large voids in the 3D point cloud and fails to filter out noise points, while too large a parameter results in the formation of small voids and fails to filter out the point cloud. The specific parameters used and visualised are shown in Table 1 and Fig. 3.

Therefore, after adding the radius filter to the original statistical filter and re-adjusting the parameters and visualising the results, it is found that the combination of the statistical filter and the radius filter can well filter the point cloud in the smooth region without too much reduction of the overall number of point clouds, which maintains the overall characteristics of the point cloud well. The specific parameters used and visualised are shown in Table 2 and Fig. 4.

Table 1 Statistical filter denoising parameter settings

Initial number of point clouds	*K*	Std.	Number of filtered point clouds
13,954	10	1.0	11,816
13,954	10	2.0	13,660

Fig. 3 Statistical filtering denoising effect

Table 2 Statistical filtering combined with radius filtering denoising parameter settings

Initial number of point clouds	*K*	Std.	*R*	MinN	Number of filtered point clouds
13,954	15	1.0	0.1	20	9768
13,954	20	2.0	0.2	20	13,755

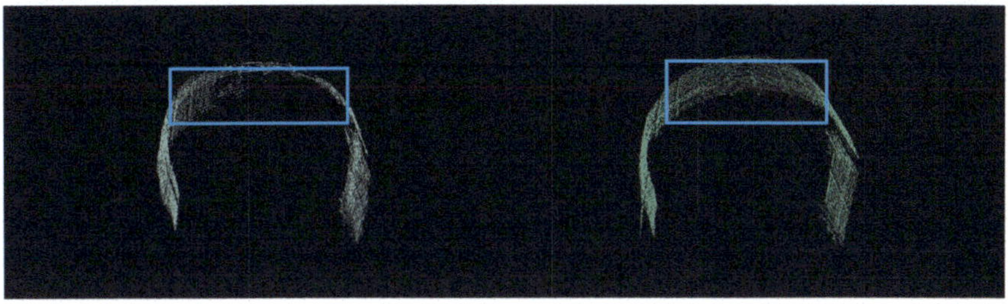

Fig. 4 Statistical filtering combined with radius filtering denoising effect graph

Gaussian filtering is performed for the classified sharp regions, and after visualising the effect before and after processing and the local features, it is found that the processed point cloud is shifted while maintaining the local features and appears smoother, and the specific visualisation effect is shown in Fig. 5.

In order to better demonstrate the feasibility and visualisation effect of the algorithm in this paper, it is found that using the method in this paper not only filters the point cloud well, but also preserves the local details, which ensures the effect of three-dimensional reconstruction of the point cloud model, through the three-dimensional

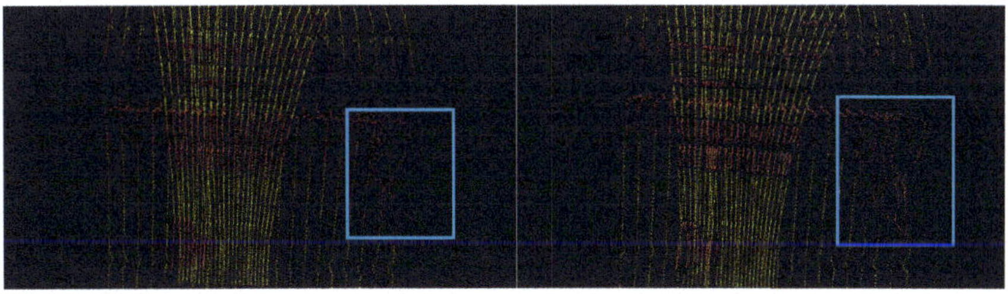

Fig. 5 Before and after Gaussian filtering

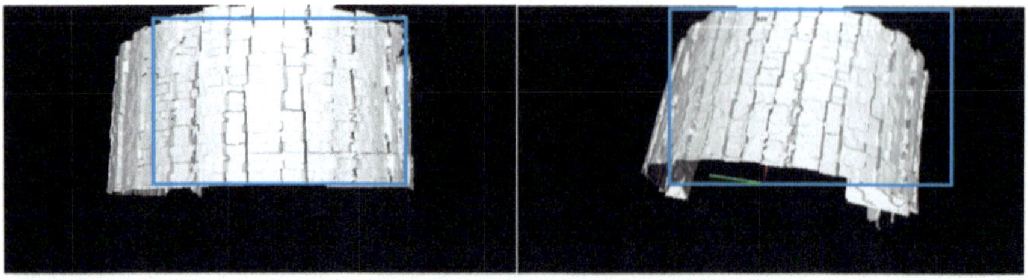

Fig. 6 Comparison of 3D reconstruction before and after processing by this method

surface modelling of the point cloud data before and after Gaussian filtering and visualisation. The specific 3D surface reconstruction is shown in Fig. 6.

5 Conclusion

By comparing the processing images of point cloud noise, it is found that effective removal of point cloud noise cannot be achieved by only a single statistical filter, radius filter or Gaussian filter. In addition, the point cloud noise with large variations in curvature characteristics may also cause excessive smoothing problems. However, the noise classification proposed in this paper not only effectively removes the point cloud noise, but also ensures the local characteristics of the 3D point cloud. It effectively prevents excessive smoothing in the denoising process of 3D point cloud to the extent that the features are lost in the reconstruction of the 3D model in the later stage and ensures the accuracy and visualisation effect of the 3D reconstruction work in the later stage.

References

1. Wang X, Li Z, Ma Y, et al (2011) Robust denoising of unorganized point clouds. In: Proceedings of 2011 international conference on intelligent computing and integrated systems, pp 268–270
2. Shuang C, Jianping Y, et al (2013) Bilateral filtering point cloud denoising algorithm based on feature selection. J Southeast Univ (Natural Sci Ed) 43:351–354
3. Hua Y, Jiankeng P et al (2015) Bilateral filtering point cloud denoising algorithm based on noise classification. Comput Appl 35:2305–2310
4. Pengfei L, Haie W et al (2016) Noise classification denoising algorithm for point cloud models. Comput Eng Appl 52:188–192
5. Jin G, Xiaoning C, et al (2016) Adaptive denoising algorithm for point clouds using density k-means and improved bilateral filtering. Sens Microsyst 35:147–149+153
6. Xiaohui W, Lushen W, et al (2019) Denoising scattered point cloud data based on normal vector distance classification. J Jilin Univ (Eng Ed) 50:278–288
7. XinYu H, Lijuan Q (2022) Research on combining improved voxel filtering algorithm with Gaussian filtering algorithm based on. Inf Record Mater 23:229–231

8. Zhe L, Tian Z (2019) An improved PCA-based alignment algorithm for ICP point clouds. J Natural Sci Heilongjiang Univ 36:473–478+505
9. Deng G, Cahill LW (1993) An adaptive Gaussian filter for noise reduction and edge detection. IEEE Conf Rec Nucl Sci Symp Med Imaging Conf 3:1615–1619
10. Uğur E, Gökrem L, Enginoğlu S (2018) Different applied median filter in salt and pepper noise. Comput Electr Eng 70:789–798

Comparative Analysis of the Effect of KPCA in PSO-GRU Combination Model

Renli Gu, Yanqiao Chen, and Peng Wang

Abstract Considering the issues of strong coupling, complex fault mode and nonlinear relationship between variables in marine condensate feed water system, the kernel principal component analysis (KPCA) is suggested as it offers significant benefits in addressing nonlinear problems. Combined with gated recurrent unit (GRU), which can improve the advantages of gradient explosion phenomenon in dealing with complex timing problems. The KPCA-PSO-GRU combined model is constructed for fault classification prediction. Samples were constructed from typical faults in the operation of the condensate feed water system, and the main features of monitoring parameters were extracted by kernel principal component analysis (KPCA) to reduce the dimensions of the original data, which were then input into the PSO-GRU model, and the original data without KPCA were directly input into the GRU model for comparative analysis. The results show that KPCA has greater advantages in extracting features from nonlinear data, and the KPCA-PSO-GRU combined model has higher accuracy in predicting fault classification, which can shorten the time for fault prediction and diagnosis of condensate feed water system, and has certain guidance.

1 Introduction

Condensing water supply system is one of the most complicated systems in marine power system. The system consists of many equipment, complicated pipeline connection and strong coupling relationship. Once failure occurs, it has a direct impact on the stability and safety of the entire power system operation [1, 2]. And, the monitoring parameters of the system are numerous, relying only on human monitoring will

R. Gu (✉) · Y. Chen
Naval University of Engineering, Wuhan 430033, China
e-mail: rlgu2021@163.com

P. Wang
Troops 91184, Wuhan, China

S. Yadav et al. (eds.), *Proceedings of 3rd International Conference on Artificial Intelligence, Robotics, and Communication*, Lecture Notes in Electrical Engineering 1172, https://doi.org/10.1007/978-981-97-2200-6_23

inevitably have flaws. Hence, it is imperative to devise intelligent algorithm-based fault feature extraction and diagnostic techniques.

In view of the characteristics of condensate feed water system, such as multiple monitoring parameters, complex coupling relationship between parameters and timing of parameters, it is necessary to carry out research on feature extraction methods to retain the main features of the original monitoring parameters and achieve data dimensionality reduction. Based on the outstanding performance of gated recurrent unit (GRU) in processing time series [3], this study employs kernel principal component analysis (KPCA) to extract the principal component of original monitoring data, thus reducing the dimension of input model data, and then uses particle swarm optimization (PSO) to optimize the parameters of gated recurrent unit (GRU) model and finally the KPCA-PSO-GRU fault prediction model is constructed.

2 Model Principles

2.1 Kernel Principal Component Analysis (KPCA)

In the age of big data, larger amounts of data contain more information but also entail a greater presence of redundant information. Consequently, this leads to increased computational requirements and model complexity. Kernel and principal component analysis (KPCA) is a method that uses kernel function to map the original data to a high-dimensional space [4], and applies principal component analysis (PCA) to map the data from this high-dimensional space to another low-dimensional space to achieve data feature extraction. Thus, this paper employs kernel principal component analysis (KPCA) to extract primary features from the original data, eliminating redundant information between the data, reducing the data dimension of the input model, and thus improving the model efficiency and prediction accuracy [5].

2.2 Particle Swarm Optimization (PSO)

Particle swarm optimization (PSO) is a group search-based optimization method that draws inspiration from the foraging behavior of birds. In this method, the group works together to find the optimal destination by sharing information. Consequently, the algorithm mimics the foraging behavior of birds and leverages information sharing among each particle in the group to discover the most optimal destination. The method is widely utilized in the optimization of complex nonlinear problems due to its simple structure, limited adjustment parameters, convenient operation, and resistance to falling into the local optimal solution.

The particle updates its speed and position according to the following formula:

$$v_{i+1} = wv_i + c_1r_1(p\text{best}_i - x_i) + c_2r_2(g\text{best}_i - x_i) \tag{1}$$

$$x_{i+1} = x_i + v_{i+1} \tag{2}$$

2.3 Gated Recurrent Unit (GRU)

Gated recurrent unit (GRU) was proposed by Cho et al. [6, 7] in 2014. Based on RNN, the concept of two gates, reset gate and update gate, was introduced to modify the calculation method of hidden states in the recurrent neural network, better capture the dependency relationship in the timing series, and improve the phenomena such as gradient disappearance or gradient explosion. The operation principle of GRU structural unit is as follows:

$$z_t = \sigma\left(W_z \cdot \left[h_{t-1}, x_t\right]\right) \tag{3}$$

$$r_t = \sigma\left(W_r \cdot \left[h_{t-1}, x_t\right]\right) \tag{4}$$

$$\widetilde{h_t} = \tanh\left(W \cdot \left[r_t * h_{t-1}, x_t\right]\right) \tag{5}$$

$$h_t = z_t \cdot \widetilde{h_t} + (1 - z_t) \cdot h_{t-1} \tag{6}$$

2.4 KPCA and PSO Optimized GRU Model Construction

The structure of GRU model consists of three parts: input layer, hidden layer, and output layer. Among them, the selection of hidden layer parameters will have a great impact on the training effect of GRU model, so it is necessary to optimize this part of parameters. Particle swarm optimization (PSO) can optimize the parameters of GRU neural network due to its high convergence performance, few parameters set and easy implementation [8, 9]. Therefore, this paper utilizes PSO to optimize the three hyperparameters of GRU: the number of hidden units, the maximum number of iterations, and the initial learning rate. The specific process is as follows (Fig. 1).

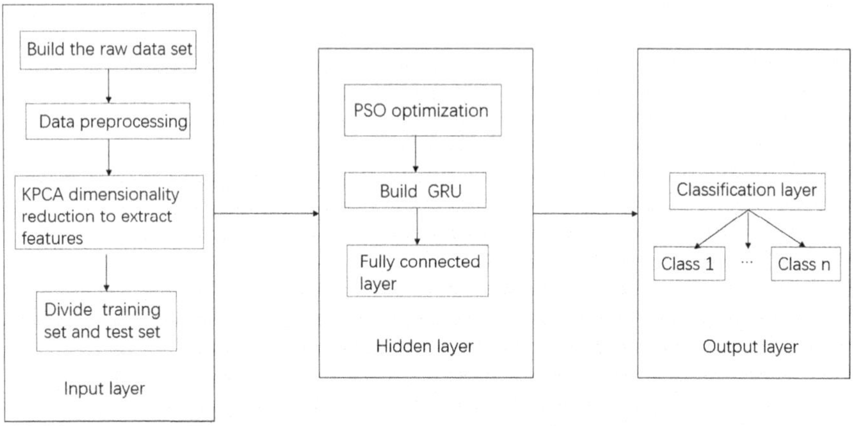

Fig. 1 Flowchart of the model

3 Example Analysis

Taking the ship power system as the research object, the data generated when the system fails is obtained by the fault simulator, and several typical faults are selected for analysis. A data was recorded every 1 s, a total of 675 s was recorded, and 29 parameters were detected for each sample (Tables 1 and 2).

The 29 parameters selected are as follows:

Kernel principal component analysis (KPCA) is used to analyze the original sample. First, the sample is standardized to eliminate the impact of different dimensions on data analysis. The standardized method adopted here is *z*-score.

Then the polynomial kernel function is selected to map the normalized data to the high-dimensional space, and then the feature extraction is carried out. The extracted results are shown in Fig. 2. Two principal elements are extracted, the contribution rate of the first principal element is 78%, the contribution rate of the second principal element is 21%, and the cumulative contribution rate of the two principal elements is over 98%, reducing the data dimension from 29 to 2.

Table 1 Information of fault sample

Fault type	Sample size	No.
30% internal leakage of water storage regulation valve	300	1
50% internal leakage of water storage regulation valve	90	2
70% internal leakage of water storage regulation valve	90	3
30% throttle valve stuck	40	4
50% obstruction of ion filter and other pipelines	140	5
70% recirculation valve stuck	15	6

Table 2 Fault parameter

No.	Parameter	No.	Parameter	No.	Parameter
1	Speed of main turbine unit	11	Feed water temperature of deaerator	21	Outlet pressure of no. 1 feedwater unit
2	Condensing water temperature of main turbine unit	12	Water level of deaerator	22	Outlet pressure of no. 2 feedwater unit
3	Condenser vacuum	13	Main pressure of feedwater	23	Flow rate of condenser throttle valve
4	Condenser water level	14	Opening of no. 5 boiler feed water regulating valve	24	Flow rate of recirculation valve
5	Inlet pressure of condensate pump	15	Opening of no. 6 boiler feed water regulating valve	25	Opening of feed water bypass load relief valve
6	Onlet pressure of condensate pump	16	Pressure difference before and after feed water regulating valve of no. 5 boiler	26	Opening of deaerator return valve
7	Opening of condenser throttle valve	17	Pressure difference before and after feed water regulating valve of no. 6 boiler	27	Opening of emergency water supply valve
8	Opening of recirculation valve	18	Water level of no. 5 boiler	28	Opening of the water supply valve
9	Condensing pressure before deaerator	19	Water level of no. 6 boiler	29	Speed of no. 2 feed water unit
10	Pressure of deaerator	20	Speed of no. 1 feed water unit		

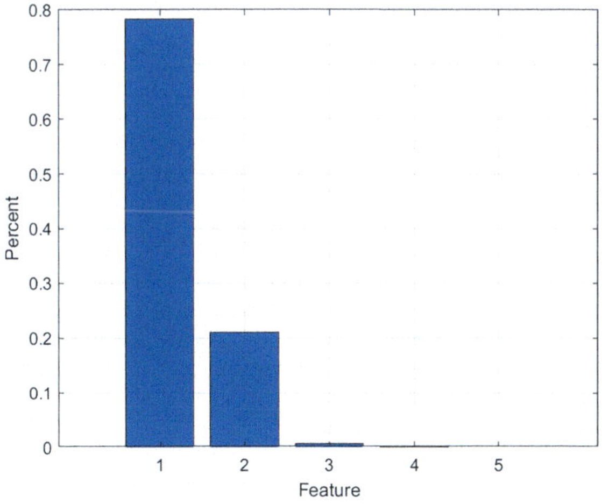

Fig. 2 Extract results of KPCA

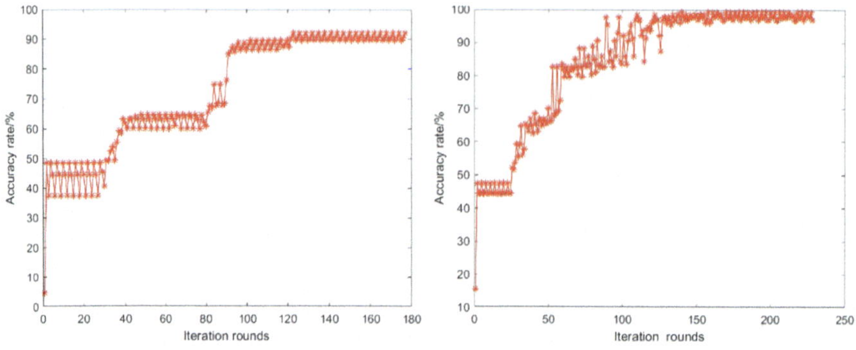

Fig. 3 Model result

Split the sample with a 70%training sample and a 30% test sample. The GRU parameters optimized by PSO were obtained by using the training set as 116 hidden units, 182 maximum iterations, and 0.007 initial learning rate, respectively. The accuracy curve obtained by KPCA-PSO-GRU model were used. Compared with raw data directly input into the PSO-GRU model, the results are shown in Fig. 3.

The data extracted by KPCA takes 795 s to enter into the PSO-GRU model, achieving an accuracy of over 90%. Without feature extraction of the original data, the accuracy of direct input into the PSO-GRU model can reach more than 95%, but it takes more than 1000 s. It indicated that the prediction effect of KPCA is not as good as that of the original data input model, but the time of model processing is greatly shortened and precious time is saved for fault prediction and diagnosis.

4 Conclusion

(1) To address the problem of excessive parameters monitored by complex systems, the kernel principal component analysis (KPCA) is adopted to extract data features, which can reduce the dimension of data and save the time of data processing on the basis of retaining data information.
(2) Input the data processed by kernel principal component analysis (KPCA) into the GRU model, and predict it with an accuracy of over 90%, which verifies the accuracy of fault prediction of KPCA-PSO-GRU model, and greatly saves the time of fault prediction and diagnosis compared with the direct processing of raw data.

References

1. Yuanlong Y, Jinxiang W, Zidu Z et al (2019) Response characteristics of ship's steam power system under switching conditions of turbine feedwater unit. Chin Ship Res 14(02):150–155. https://doi.org/10.19693/j.issn.1673-3185.01248
2. Haibo Q, Jiashan J, He N (2018) Research on cavitation causes of Marine condensate pump based on potential path analysis. Marine Eng 40(10):64–68. https://doi.org/10.13788/j.cnki.cbgc.2018.10.064.
3. Shufang Z, Xiaoyu D (2018) Research on speech recognition based on improved LSTM deep neural network. J Zhengzhou Univ (Eng Technol) 39(05):63–67. https://doi.org/10.13705/j.issn.1671-6833.2018.02.004
4. Jiang Q, Yan X (2018) Parallel PCA-KPCA for nonlinear process monitoring. Control Eng Practice 80:17–25. https://doi.org/10.1016/j.conengprac.2018.07.012
5. Li J, Chen Y, Xiang H, Cai Z (2020) Remaining useful life prediction for aircraft engine based on LSTM-DBN. Syst Eng Electron 42(7):1637–1644. https://doi.org/10.3969/j.issn.1001-506X.2020.07.28
6. Gong S (2011) Personalized recommendation system based on association rules mining and collaborative filtering. In: International symposium on quantum, nano and micro technologies
7. Yuyan W, Bolin L, Chen P et al (2021) Recursive neural network research review. J Jishou Univ (Natural Sci Ed) 1:41–48. https://doi.org/10.13438/j.carolcarrollnkiJDZK.2021.01.007
8. Chao W (2021) Research on short-term power prediction of photovoltaic power station based on PSO-GRU. Ind Control Comput 10:101–102, 105. https://doi.org/10.3969/j.issn.1001-182X.2021.10.039
9. Yang MD (2020) Temperature prediction based on improved PSO-LSTM neural network. Mod Inf Technol 4(4):110–112. https://doi.org/10.19850/j.carolcarrollnki.2096-4706.2020.04.031

Comparison and Hyperparameter Analysis of Four Reinforcement Learning Algorithms in the Lunar Lander Environment

Yuyang Lu

Abstract Reinforcement learning trains an intelligent agent to take actions in an environment in order to maximize the cumulative reward. Provided the OpenAI Gym environment LunarLander-v2, both policy-based and value-based methods are implemented to train the lander model in order to achieve an excellent landing, considering factors such as the moving speed, time consumed, and landing position. The performances were compared and further optimized. In this paper, the work compares Deep Q-Network (DQN), Double Deep Q-Network (DDQN), Policy Gradient (PG), and Advantage Actor Critic (A2C) to train the model. By adjusting the values of key hyperparameters, several ways are identified to improve the efficiency and effectiveness of the reinforcement learning process. Ultimately, DQN emerged as the most effective option. This work employed a neural network with 128 replay samples and 256 hidden layer units to the agent and selected a greedy policy with ε-decay set to 0.996 and a discount value of 0.99 in the Q-function to train the agent, which displayed the best training performance.

1 Introduction

The exploration of the moon has always been one of the most important parts of the aerospace industry over the last 100 years. The lunar lander, which can collect important data and samples from the lunar surface, plays a crucial role in the research of lunar structure and properties [1]. There have been many historical moments of successful landing, with China's landing on the far side of the moon and the US manned moon landing being particularly impressive. As a result, this work is interested in the problem of teaching an agent how to independently make a successful landing on a dynamic terrain environment. The intelligent lander has to take actions (fire left/right orientation engines, fire main engine, and engine off) to ensure landing

Y. Lu (✉)
Faculty of Electronic and Information Engineering, Xi'an Jiaotong University, Xianning West Road No.28, Xi'an, Shaanxi 710049, China
e-mail: luyuyang0918@stu.xjtu.edu.cn

S. Yadav et al. (eds.), *Proceedings of 3rd International Conference on Artificial Intelligence, Robotics, and Communication*, Lecture Notes in Electrical Engineering 1172, https://doi.org/10.1007/978-981-97-2200-6_24

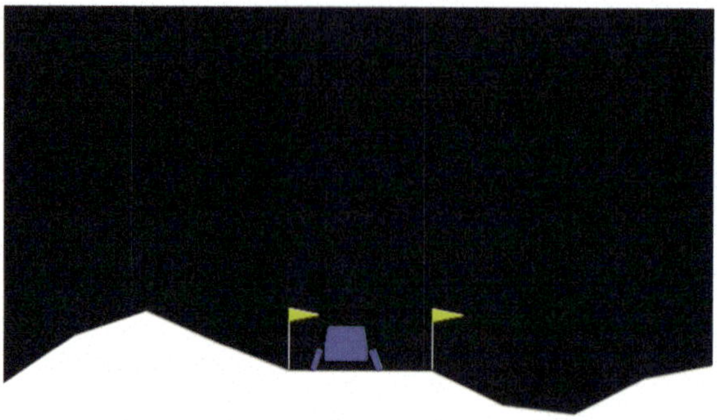

Fig. 1 Good performance of landing reaching over 200 rewards [2]

in the required position gently [2]. This work introduces Deep Q-Network (DQN), Double Deep Q-Network (DDQN), Policy Gradient (PG), and Advantage Actor Critic (A2C) to tackle this problem and compare the efficiency between these four methodologies [3]. Then, the appropriate values of hyperparameters are explored by choosing the best options, from which the lunar lander problem could be clearly demonstrated, and it may have practical applications in the future.

This work introduces the Lunar Lander environment from OpenAI Gym to model the lander and moon surface environment [2]. In this environment, a random ground surface is generated for each episode. The environment creates a situation where the lander should land at a specific position interval under low-gravity conditions. A representative scenario is demonstrated in Fig. 1. The lander model in this environment can adopt both discrete and continuous actions. The optimal solution is to fire the engine at full throttle or turn it off in Pontryagin's maximum principle. For this reason, a discrete action space is chosen in this environment, which includes firing the left or right orientation engine, main engine, or shutting off the engine. Finally, the reward considers two main behaviors: (1) safely landing on the ground and (2) moving slower and consuming less time. Therefore, the goal is to train the agent to guide the lander to the landing pad softly and consume fuel as efficiently as possible.

2 Method

There are four models leveraged in this work, including DQN, DDQN, PG, and A2C. By analyzing results of the four methods, the characters of each method are summarized and work out which method fit this environment best. In this way, deeper insights into the lunar lander problem are achieved.

2.1 DQN

Q-Learning. Q-Learning is grounded in the Q-function concept. The Q-function of a policy π, $Q^{\pi}(s,a)$, evaluates the predicted return or discounted sum of rewards obtained from a single state s by adopting action a and then trailing policy π. $Q^{*}(s,a)$ represents the optimal Q-function starting from observation s, acting a, and taking the optimal policy. The optimal Q-function can be expressed by the Bellman optimality equation:

$$Q^{*}(s,a) = \mathbb{E}\left[r + \gamma \max_{a'} Q^{*}\left(s', a'\right)\right] \tag{1}$$

This equation divides the maximum return from state s and action a into the immediate rewards r and the discounted Q-function return (discounted by γ) evaluated by choosing to follow policy which can always get the best rewards since action a until reaching the last episode. The present Q-function is determined by both immediate rewards r and all the possible next states s'.

Q-Learning is a process to update the Q-value iteratively using the Bellman optimality equation:

$$Q_{i+1}(s,a) \leftarrow \mathbb{E}\left[r + \gamma \max_{a'} Q_{i}\left(s', a'\right)\right] \tag{2}$$

and after enough episodes, the value can converge to the optimal Q-function, i.e., $Q_i \rightarrow Q^{*}$ as $i \rightarrow \infty$.

In general, agents only know possible states and actions. They must experience every state and actions at least once to know the rewards, and multiple times to estimate the probabilities. And rewards are not immediately observable, neither are transportation probabilities. As a result, Q-table is introduced to replace the Q-values for each state s and action a. Then, the Q-Learning algorithm could be achieved [4, 5].

$$Q(s_t, a_t) \leftarrow (1 - a_t)Q(s_t, a_t) + a_t\left(r_t + \gamma \max_{a_{t+1}} Q(s_{t+1}, a_{t+1})\right) \tag{3}$$

Deep Q-Learning. In most problems, it is impractical to simply use a value table which contains all the condition s and a to represent the Q-function. Instead, a neural network with parameters θ is trained to assess the Q-values, i.e. $Q(s,a;\theta) \approx Q^{*}(s,a)$. This can be done by trying to minimize the computing loss at each step i:

$$L_i(\theta_i) = \mathbb{E}_{s,a,r,s' \sim \rho(\cdot)}\left[(y_i - Q(s,a;\theta_i))^2\right], \tag{4}$$

where

$$y_i = r + \gamma \max_{a'} Q\left(s', a'; \theta_{i-1}\right) \quad (5)$$

Here, y_i represents the temporal difference (TD) target, and $y_i - Q$ is regarded as TD error. ρ is the combination of different behavior, which includes transitions $\left\{s, a, r, s'\right\}$ collected from the environment.

The Target Network is introduced in which parameters of the network is taking few iterations as an updating period instead of one iteration. For this reason, the parameters in the previous iteration θ_{i-1} are fixed and not updated.

Different from using the policy $a = arg\ \max_a Q(s, a; \theta)$ in learning process, Q-Learning uses another behavior policy called ε-greedy policy to select the action in the environment. This policy sets the agent $1 - \varepsilon$ probability to take the ideal action with the best Q-values and ε probability to randomly explore the environment which can ensure better covering the state-action space [6].

Experience Replay. If using merely the last transition $\left\{s, a, r, s'\right\}$ to compute loss, DQN will reduce to standard Q-Learning. As a result, DQN loss should be computed following gradient descent to avoid calculating the full expectation in the DQN loss.

To deal with the challenge that successive samples are highly correlated, this work introduced experience replay to update the network in a more stable way. During data collection, the transitions are added to the replay buffer. A mini-batch of replay samples are used to compute loss as well as its gradient during training, instead of using the latest transition. In this way, not only better data efficiency can be achieved as the replay buffer reuse transition in many updates, but DQN can also obtain better stability with uncorrelated transitions in a batch [7].

Training Details. This research focuses on hyperparameters: discount parameter γ, speed of decay in ε and parameters in neural network (batch size and units in hidden layers) that may affect the training efficiency. As for evaluation of agent training, three standards are taken into consideration: speed of iteration, maximum training reward, and training stability. The influence of each hyperparameters is evaluated in all these three standards.

A three-layer neural network is leveraged, shown in Fig. 2, with 128/256/512 neurons in the hidden layers. RELU activation is chosen for hidden layers and LINEAR activation for output layer. The number of hidden neurons is changed to monitor variations of outcomes. The learning rate in the experiment is 0.0005.

As for experience replay, samples are reserved in 64/128/256 that will be passed through to the network at one time. With 128 samples to replay randomly, increase in training rewards is expected to be promoted but also spend more time in each iteration.

In ε-greedy policy, exploration is initially chosen as little information is known about the environment ($\varepsilon = 1$). Then after each step, the value of ε decays by multiplying a constant d (always ranging from 0.99 to 1) until it reaches 0.01.

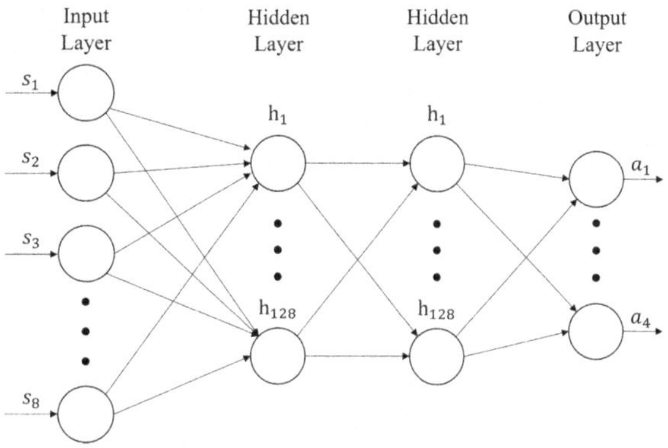

Fig. 2 Architecture of neural network in the deep learning model. *Figure Credits* Original

2.2 DDQN

Consider the Q-value:

$$Q_\pi(s, a) = \mathbb{E}_\pi[r + \gamma \max_{a'} Q_\pi(s', a') | S_t = s, A_t = a] \tag{6}$$

When using the maximum estimate of the maximum value, it takes the maximum overestimated values implicitly. Learning bias is thus introduced due to the systematic overestimation. Sometimes this overestimation in action-values can result in unstable training performance and low-quality policy. To tackle this problem, DDQN tries to decouple action selection and value calculation. The Q-Network determines the action a^* and the Target Network further determines the quality $Q(s', a^*)$, which is selected for target computation, which is

$$Q_\pi(s_t, a_t) = r_t + \gamma Q\left(s_{t+1}, arg \max_{a'} Q'(s_t, a_t)\right) \tag{7}$$

the Q-value of Deep Q-Network update followed by the estimated Q-value from Target Network and the parameters of Target Network update based on the parameters of Deep Q-Network per several iterations [8].

2.3 PG

Policy Gradient is a policy-based method, different from value-based method which estimate the expected value of a particular state or action, it aims to directly learn the policy to define the agent's behavior. Policy gradient introduces gradient ascent to solve the maximization problem, where the stepping is kept through the parameters using the following update rule:

$$\theta_{t+1} = \theta_t + \alpha \nabla J(\theta_t). \tag{8}$$

According to the Policy Gradient Theorem, the gradient of the expected reward is

$$\nabla \mathbb{E}_{\pi_\theta}[r(\tau)] = \mathbb{E}_{\pi_\theta}\left[r(\tau) \nabla \log \pi_\theta(\tau)\right] \tag{9}$$

$$\pi_\theta(\tau) = P(s_0) \prod_{t=1}^{T} \pi_\theta(a_t|s_t) p(s_{t+1}, r_{t+1}|s_t, a_t). \tag{10}$$

Equivalently, taking the log on both sides, the output changes to

$$\log \pi_\theta(\tau) = \log P(s_0) + \sum_{t=1}^{T} \log \pi_\theta(a_t|s_t) + \sum_{t=1}^{T} \log p(s_{t+1}, r_{t+1}|s_t, a_t), \tag{11}$$

$$\nabla \mathbb{E}_{\pi_\theta}[r(\tau)] = \mathbb{E}_{\pi_\theta}\left[r(\tau) \sum_{t=1}^{T} \nabla \log \pi_\theta(a_t|s_t)\right] \tag{12}$$

As a result, $\nabla J(\theta)$ could be easily represented and maximized the "expected" reward [9].

2.4 A2c

Actor Critic is a methodology which combines value-based method and policy-based method. When adding a baseline to reduce the variance, allowing return values estimation according to subsequent state values in the policy gradient, the methodology turns into Actor Critic. In Actor Critic, Actor is the policy, which inputs the state and outputs the best action, learning the policy to control the behavior of the agent, Critic is the learnable value function V_w. In A2C, instead of critic to learn V_w, the work makes them to learn the advantage values [10].

$$A(s, a) = Q(s, a) - V(s) \tag{13}$$

3 Result

Here are the results of four different methods. Value-based methods are compared, including DQN and DDQN in one graph and policy-based method PG and A2C in another one.

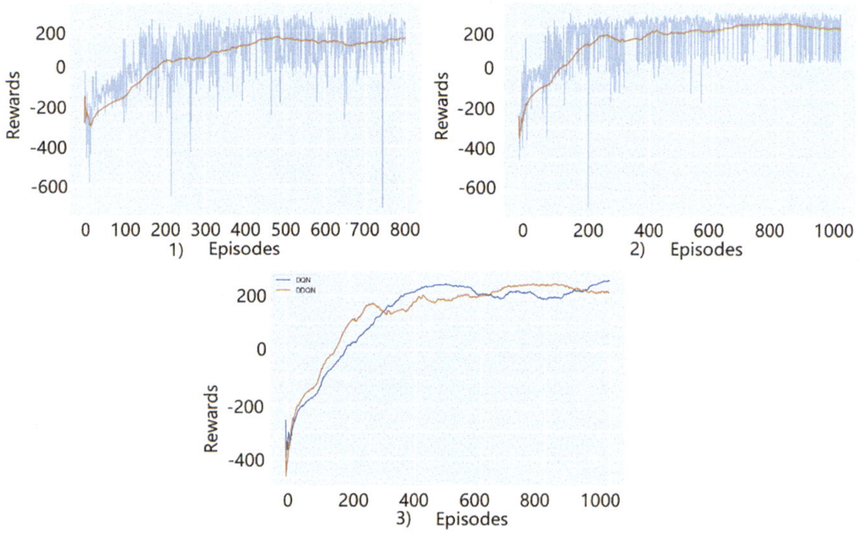

Fig. 3 Performance comparison between DQN and DDQN. *Figure Credits* Original

3.1 Training Rewards

It could be found that both DQN and DDQN has a favorable training result while DQN reaches around 180 rewards in 800 episodes (Fig. 3(1)) and DDQN reaches approximately 210 rewards in 1000 episodes (Fig. 3(2)). Also, it could be found that DDQN appears much stable as more extremely low scores occur in Fig. 3(1). From Fig. 3(3) two methods are compared, as DQN may get a higher reward and spend less time, DDQN appears to be the better choice for its steadiness.

Viewing policy-based methods A2C and PG, they have faster iteration speed but in the meanwhile tend to be unstable, even reaching −1500 reward in an episode. Comparing training rewards between A2C (Fig. 4(1)) and PG (Fig. 4(2)), if could be found that the variance of A2C training results is smaller than that of PG. A2C performs more stable than PG and can fit many environments as its training performances are always good in several attempts. Moreover, after enough training episodes (about 3000), A2C gets an ideal increase in training rewards, while the reward in PG decreases less than 0 (Fig. 4(3)).

3.2 Hyperparameter Research

As demonstrated in Figs. 3(3) and 4(3), value-based methods perform more stable and appear more suitable in this environment. More deeply, choosing different value of some parameter is considered to select the best training agent.

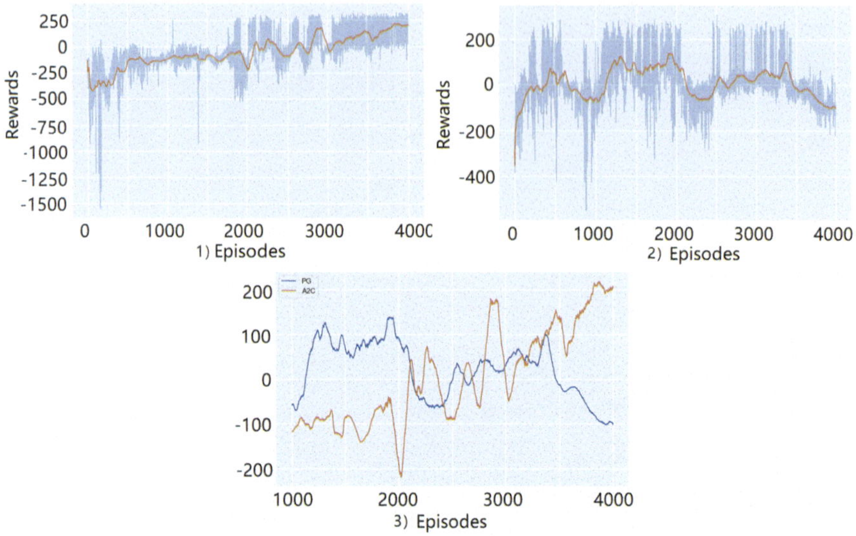

Fig. 4 Performance comparison between A2C and PG. *Figure Credits* Original

From Fig. 5, the agent with 128 samples to experience replay performs best. As agent with 256 samples to replay is expected to get better rewards, its reward decreases from 200 to 100 in the last 200 episodes. The agent with batch size in 64 appears not that stable.

From Table 1, different number of neurons are chosen in the hidden layers. The agent with 128 units in a hidden layer needs more training episodes to reach 200

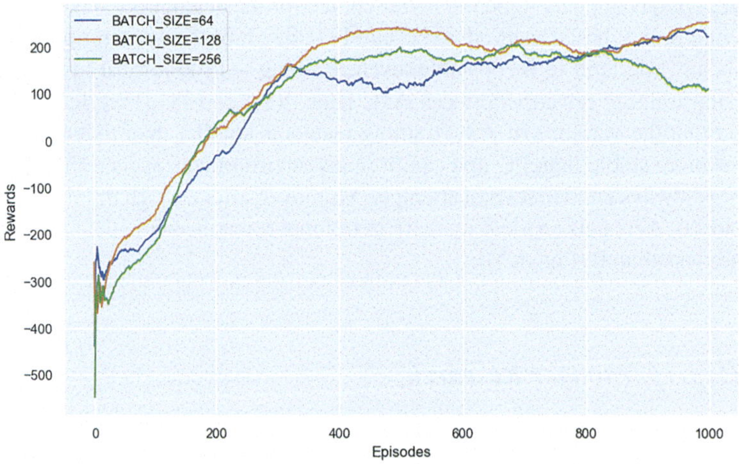

Fig. 5 DQN with different number of replay samples. *Figure Credits* Original

Table 1 Q-Network in different hidden layer units

Agent	Episodes (when reaching 200 reward)	Iteration speed
DQN 256 units	593	2.03 s/it
DQN 128 units	778	2.14 s/it
DQN 512 units	569	2.93 s/it

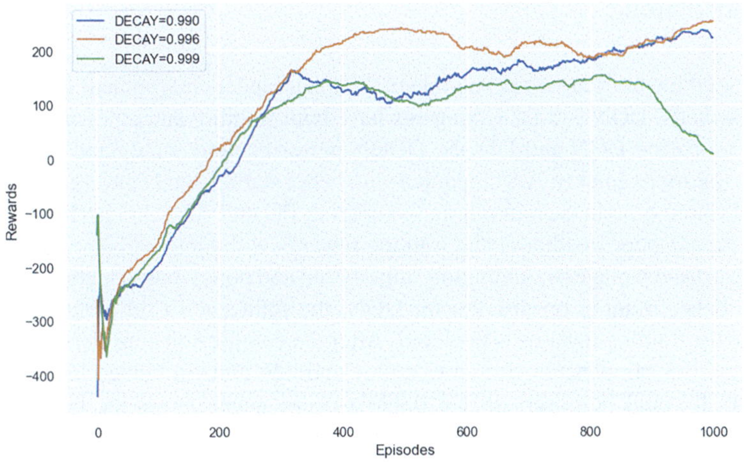

Fig. 6 DQN with different ε-decay. *Figure Credits* Original

reward (778), while agents with 256 and 512 units spend less than 600 episodes to accomplish. The agent with 256 units has a more efficient iteration.

Taking both convergence and iteration speed into account, 128 samples are selected to experience replay and 256 units in hidden layers to conduct future research. From Fig. 6, it could be found that with larger decay (DECAY = 0.999), the training rewards decrease respectively for the remaining possibility in exploration bringing stochasticity. The training rewards with smaller decay (DECAY = 0.990), however, rise smoothly but have slower convergence speed compared to that of decay in 0.996. As a result, choosing 0.996 as decay constant is more ideal in this experiment.

In Table 2, the impact of different discount value γ is evaluated on training quality. Although each agent can get maximum reward larger than 270, the speed of convergence differs in different γ. The agent with γ value of 0.99 reaches 200 rewards in only 593 episodes faster than 0.90 and 0.80. As a result, the lunar lander problem focuses more on the future rewards than present.

Table 2 DQN in different γ

Agent	Episodes (when reaching 200 reward)	Maximum reward
DQN = 0.99	593	296
DQN = 0.90	782	285
DQN = 0.80	> 1000	276

4 Conclusion

Value-based methods are more suitable in the lunar lander environment than policy-based methods. DQN and DDQN agents have both stability and efficiency in iterations. Comparing DQN and DDQN, DDQN performs more stably and its training rewards rise more quickly. Accompanied with better stability and convergence speed, it spends more time in each iteration. Comparing PG and A2C, PG performs better at first but has poor steadiness and training rewards with large variance. A2C optimizes this disadvantage by combining value-based and policy-based methods getting a much better training reward. As for DQN, the influence of different parameter value on the training rewards is explored. After the comparison considering training efficiency, 128 samples are chosen to replay and 256 units in hidden layers for Q-Network. It could be observed that 0.996 ε-decay agent shows the best training rewards, as the environment needs moderate exploration as well as exploitation. The lunar lander problem emphasizes more on future return as the discount value in 0.99 performs better than 0.90 and 0.80.

References

1. Furfaro R, Bloise I, Orlandelli M, Di Lizia P, Topputo F, Linares R (2018) Deep learning for autonomous lunar landing. Adv Astronaut Sci 167:3285–3306
2. Brockman G, Cheung V, Pettersson L, Schneider J, Schulman J, Tang J, Zaremba W (2016) Openai gym. arXiv preprint arXiv:1606.01540
3. Levine S, Kumar A, Tucker G, Fu J (2020) Offline reinforcement learning: tutorial, review, and perspectives on open problems. arXiv preprint arXiv:2005.01643
4. Jang B, Kim M, Harerimana G, Kim JW (2019) Q-learning algorithms: a comprehensive classification and applications. IEEE access 7:133653–133667
5. Watkins CJ, Dayan P (1992) Q-learning. Machine learning 8:279–292
6. Fan J, Wang Z, Xie Y, Yang Z (2020) A theoretical analysis of deep Q-learning. In: Learning for dynamics and control. PMLR, online, pp 486–489
7. Lin LJ (1992) Reinforcement learning for robots using neural networks. Carnegie Mellon University
8. Gu J, Fang Y, Sheng Z, Wen P (2020) Double deep Q-network with a dual-agent for traffic signal control. Appl Sci 10(5):1622
9. Kakade SM (2001) A natural policy gradient. In: Advances in neural information processing systems, vol 14. The MIT Press, Vancouver, pp 1–8
10. Xin X, Karatzoglou A, Arapakis I, Jose JM (2022) Supervised advantage actor-critic for recommender systems. In: Proceedings of the fifteenth ACM international conference on web search and data mining. Association for Computing Machinery, New York, pp 1186–1196

Active Learning Based on Prototype Vector Feature Adaptive Fusion

Xiaohao Fang, Wenjie Cheng, and Xia Ji

Abstract From successfully embedding the KNN classifier into a network using prototype vectors, to exploring various active learning methods in deep learning to find uncertain and representative samples for active learning, feature fusion in the learned feature space has been continuously investigated. In this paper, we aim to further combine prototype vectors with feature fusion and propose an active learning based on prototype vector feature adaptive fusion. Firstly, we explore the features of prototype vectors in the NCE-Net model, where they represent common features for each class. Secondly, we construct a prototype vector matrix based on the feature space. Then, we fuse the prototype vector matrix with the feature matrix of all unlabeled samples and add the samples with inconsistent pseudo-labels before and after feature fusion to the candidate set. Finally, we cluster the candidate samples to consider the representativeness of the selected samples in active learning. Experimental results on classic datasets for image classification show that the proposed method achieves higher accuracy compared to traditional active learning methods.

1 Introduction

With the rapid development of artificial intelligence, deep learning has become one of the most influential technologies in the field. The success of deep learning models heavily relies on a large amount of labeled data for training. However, annotating large datasets is challenging and costly, posing a significant challenge to the development of deep learning [1]. To address this issue, an increasing number of active learning approaches [2–6] have been applied to deep learning. Wan et al. [3] combined the prototype vectors proposed by Yang et al. in prototype convolutional learning (CPL) [7] with a nearest-neighbor classifier to design a nearest-neighbor classifier embedding network. By utilizing prototype vectors, they solved the problem of high

X. Fang · W. Cheng · X. Ji (✉)
The School of Computer Science and Technology, Anhui University, Hefei, China
e-mail: jixia1983@163.com

S. Yadav et al. (eds.), *Proceedings of 3rd International Conference on Artificial Intelligence, Robotics, and Communication*, Lecture Notes in Electrical Engineering 1172, https://doi.org/10.1007/978-981-97-2200-6_25

computational complexity when embedding a KNN classifier, thus avoiding expensive computation costs. This method employed two metrics, rejection confidence and fuzzy confidence, based on model output to guide the sample selection in active learning, but it did not further exploit the features of the samples. However, recent work by ALFA-Mix [8] introduced an active learning (AL) method based on interpolation between labeled and unlabeled samples. This method first calculates the mean of feature vectors for each labeled sample in each class, constructing an anchor point set based on these mean feature vectors. Then, a certain strategy is used to fuse the features of the constructed anchor point set with the unlabeled sample features to select samples for active learning. This approach utilizes the current feature space information learned by the model, but it also relies on the softmax classifier, raising concerns about its generalization performance.

To address the aforementioned issues, we introduce prototype vectors to represent the current state of labeled samples for feature fusion. We extend the feature fusion method in [8] ALFA-Mix from softmax classification to distance-based classification, thereby enriching the application of feature fusion methods across different types of classifiers. In this paper, we propose a simple and efficient AL solution, referred to as the prototype vector feature adaptive fusion active learning method. Firstly, we construct a prototype vector matrix between unlabeled samples and all prototype vectors in the feature space learned by NCE-Net. Then, we perform feature fusion between the prototype vector matrix and the feature matrix of all unlabeled samples. Finally, we add the unlabeled samples whose pseudo-labels change before and after feature fusion to the candidate subset and select final labeled samples using the K-means method to avoid excessive overlap of labeled samples in the feature space.

2 Related Work

In deep learning, convolutional prototype learning (CPL) [7] proposed a framework for prototype learning. During the classification phase, CPL classifies objects through prototype matching, which involves finding the nearest prototype based on Euclidean distance and assigning the class of that prototype to a specific object. Inspired by the CPL approach [3, 7] and others addressed the generalization extension issue of discriminative classifiers in networks and proposed using a soft nearest-neighbor classifier instead of a softmax classifier. They predicted the class label of a test image based on its soft distance to labeled images. However, this method requires computing the distances between all unlabeled sample features and all labeled image features, leading to high computational complexity, especially in high-dimensional feature spaces, making it difficult to implement and use. In order to avoid the computational complexity, N prototype vectors are learned from labeled images, and a nearest-neighbor classifier is embedded into the network through the introduction of prototype vectors. The ALFA-Mix method proposed by [8] utilizes the representations learned by the current network to compute the mean of labeled samples of the same class, constructing an anchor point set. Linear interpolation is performed between the anchor point set and unlabeled instances, and active learning sample selection is based on the change in pseudo-labels before and after interpolation. The hyperparameters, initial weights and biases of the MLP, and other parameters in the closed-form interpolation scheme proposed in this method are quite sensitive, which also affects the practicality of the model. Based on the limitations of the above-mentioned methods, we introduce prototype vectors and utilize them for feature fusion, which enables a more accurate exploration of highly uncertain samples at different class boundaries. Furthermore, our method is based on the feature space, which also facilitates easy integration with clustering methods such as K-means, making it conducive to combining various clustering methods to incentivize representative indicators for active learning [9].

3 Our Proposed Method

Figure 1 illustrates the workflow of active learning based on our prototype vector feature adaptive fusion method. Overall, it consists of three steps. Firstly, we construct a prototype vector matrix using the distance relationship between current unlabeled samples and prototype vectors (Sect. 3.1). Then, we perform feature fusion between the constructed prototype vector matrix and the intermediate features of unlabeled samples (Sect. 3.2). Finally, we apply K-means algorithm to the candidate subset I obtained from feature fusion to obtain the final labeled samples.

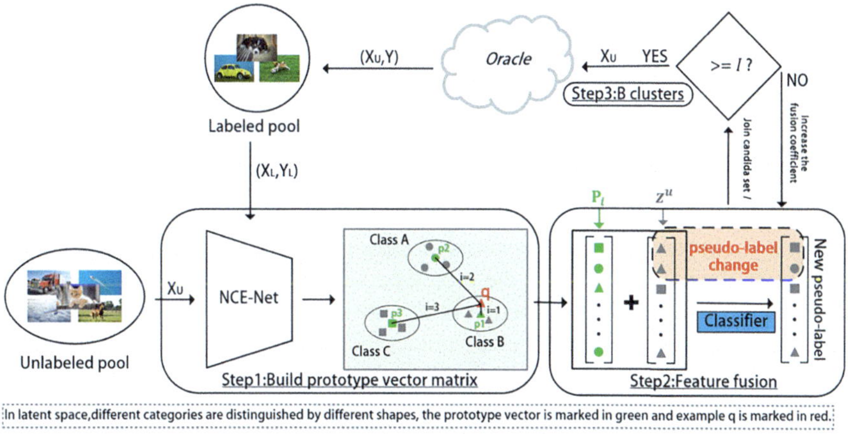

Fig. 1 Schematic diagram of the process for active learning using our proposed method

3.1 Constructing a Prototype Vector Matrix

In convolutional neural networks, prototype vectors obtained through prototype learning via convolution can effectively represent the common features of samples within a category. For different categories, we maintain and learn a prototype vector for each category, denoted as p_c ($c = 1, \ldots, C$, where C represents the number of categories in the dataset), and based on the classic ResNet-18 network backbone, p_c is set as a 1×512 dimensional vector. The classification likelihood based on prototype vectors in NCE-Net is defined as:

$$p_c(f_x) = \delta(-d(f_x, p_c)). \tag{1}$$

Among them, p_c corresponds to the prototype vector of the c-th category, which has the same dimension as the features extracted by the neural network. p_c is randomly initialized and during training, p_c is learned together with the feature extractor by jointly updating the network parameters through gradient descent. f_x represents the feature representation extracted by the neural network. In Eq. (1), $d(\cdot)$ is defined as the Euclidean distance or cosine distance. Taking the Euclidean distance as an example, the activation function $\delta(\cdot)$ adopts a Gaussian kernel function. The likelihood of image x for category c is defined as:

$$p_c^{Euc}(f_x) = \exp\left(\frac{-f_x - p_c^2}{2R_c^2}\right), \tag{2}$$

where R_c is a learnable parameter representing the rejection radius for class c. If $d(\cdot)$ is measured using the cosine distance, the activation function $\delta(\cdot)$ adopts the Sigmoid function. The definition of the classification likelihood is then:

$$p_c^{\mathrm{Cos}}(f_x) = \frac{1}{1 + \exp(-\cos(f_x, p_c) \cdot R_c)}. \tag{3}$$

For better understanding, Fig. 2 further illustrates the detailed process of constructing $\boldsymbol{P}_1$ and $\boldsymbol{P}_2$. . When the dataset consists of three categories, the unlabeled samples p and q are shown in relation to the prototype vectors (denoted by solid markers) in the feature space. The construction of $\boldsymbol{P}_1$ involves finding the prototype vector that is closest to the current unlabeled sample (indicated by the red lines connecting the prototype vectors to p and q in Fig. 2), while the construction of $\boldsymbol{P}_2$ involves finding the prototype vector that is the second closest to the current unlabeled sample (indicated by the green lines connecting the prototype vectors to p and q in Fig. 2). Finally, we save the prototype vector matrices constructed under different distance metrics on the right side of Fig. 2.

Specifically, when the model classifier is a distance classifier, the predicted probability output of unlabeled samples also reflects the distance relationship between themselves and the prototype vectors learned for different categories in the feature space. To avoid redundant distance calculations in the feature space, we sort the

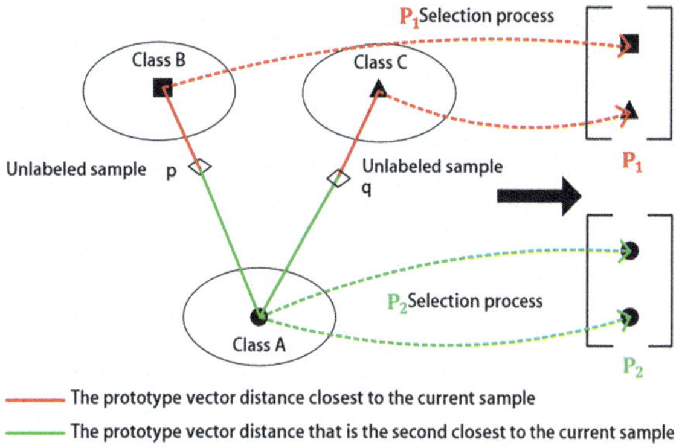

Fig. 2 Specific details of the construction of the prototype vector matrix

model's probability predictions output $p(f_x)$ in descending order and denote the sorted output as $\hat{Y}$. Here, $\hat{Y}$[:,i] represents the predicted probability value of the class to which the prototype vector, closest to the current unlabeled sample in the feature space, belongs. We can utilize the sorted predictions $\hat{Y}$[:,i] to find the corresponding prototype vectors based on their distances and construct the corresponding $\boldsymbol{P}_i$ matrix. This approach avoids the need for distance calculations in high-dimensional feature spaces. Therefore, in our method, we first extract the sorted predicted outputs $\hat{\mathrm{Y}}$ and then iterate over the indices of $\hat{Y}$[:,i] to extract the corresponding prototype vectors in the feature space and combine them to obtain the corresponding $\boldsymbol{P}_i$.

3.2 *Feature Fusion Based on the Prototype Vector Matrix*

We perform feature fusion by combining the prototype vector matrix $\boldsymbol{P}_i$ with the current representation of unlabeled samples $\boldsymbol{Z}^u$. By incorporating the changes in pseudo-labels before and after feature fusion, we add the unlabeled samples to the candidate subset I, allowing our method to more accurately explore high-uncertainty samples at different class boundaries. By setting a reasonable number of candidates in the subset I, we can select the final budgeted samples that possess both uncertainty and representativeness using clustering methods. Specifically, we use linear interpolation between the prototype vector matrix $\boldsymbol{P}_i$ and the current representation $\boldsymbol{Z}^u$ of all unlabeled instances in the model, defined as follows:

$$\tilde{\boldsymbol{Z}}\boldsymbol{\alpha} = \alpha \mathbf{P}_i + (1 - \boldsymbol{\alpha})\boldsymbol{Z}^u, \tag{4}$$

where $\boldsymbol{Z}^u$ is a large matrix composed of the current model's learned representations of all unlabeled samples. $\boldsymbol{\alpha} \in [0, 1)$ is the interpolation fusion coefficient, which is initialized to a very small value and can be adaptively incremented based on the different quantities of the candidate subset I. $\boldsymbol{P}_i$ is the set of prototype vectors shown in Fig. 2 that satisfy the distance relationship condition for each unlabeled sample corresponding to $\boldsymbol{Z}^u$.$\tilde{\boldsymbol{Z}}\boldsymbol{\alpha}$ represents the new representation obtained by linearly interpolating the representations of all unlabeled samples with the prototype vector matrix.

Considering that different intervals of the classification function may have varying sensitivities to distance, in order to make samples that are far from the prototype vectors more easily discovered, we introduce a simple interpolation coefficient based on pseudo-labels, called the distance sensitivity coefficient, defined as follows:

$$\boldsymbol{dc} = 1 - \hat{\boldsymbol{Y}}[:, 0], \tag{5}$$

where $\hat{\boldsymbol{Y}}$ represents the predicted probability output for each unlabeled sample sorted in descending order for all classes by the current model.$\hat{\boldsymbol{Y}}[:, 0]$ represents the maximum predicted probability value (i.e., pseudo-label) for each unlabeled sample.

4 Experimental Analysis

4.1 Image Classification Experiments and Results Analysis

Dataset: For the classification task, we selected CIFAR-10 and CIFAR-100 as they are classic image recognition datasets that simulate the varying number of images per class in real-world scenarios. Moreover, recent works have evaluated their performance. CIFAR-10 and CIFAR-100 both consist of 60,000 $32 \times 32 \times 3$ images, with 50,000 images for training and 10,000 images for testing. CIFAR-10 has ten classes with 6000 images per class, while CIFAR-100 has 100 classes with 600 images per class. We applied random cropping, random horizontal flipping for data augmentation, as well as mean and standard deviation vector normalization.

Comparison methods: We used the publicly available code provided by the authors for these comparison methods and followed their recommended parameter settings. Additionally, we included a random selection method as a baseline. For NCE-Net, which is based on a distance classifier, we used the suffix (-cos) to indicate the cosine distance and the suffix (-euc) to indicate the Euclidean distance. During evaluation, these methods were assessed using the same target model.

Training details: Our settings were as follows, implemented using PyTorch and executed on a T4-16G cloud platform. On CIFAR-10, we initially initialized the labeled set with 1000 randomly selected images. For each active learning cycle, we chose 1000 images from the unlabeled set and moved them to the labeled set until the labeled set reached 10,000. The classification model was trained for 200 epochs in each cycle, with a learning rate of 0.1 for the first 160 epochs and 0.01 for the remaining epochs. The batch size was set to 128, and momentum and weight decay were set to 0.9 and 0.0005, respectively. For CIFAR-100, we initialized the labeled set with 5000 randomly selected images and selected 2500 images in each cycle until the labeled set reached 20,000. The classification model was trained for 200 epochs in each cycle, with a learning rate of 0.9 for the first 120 epochs, reduced to 0.09 at 120 epochs, and further reduced to 0.009 at 160 epochs. The batch size was set to 128, and momentum and weight decay were set to 0.9 and 0.00009, respectively.

Performance measurement: We evaluated the performance of these methods in image classification by measuring the average accuracy over 5 experiments. In each experiment, all methods started with the same initial labeled pool, and the target model used for evaluating accuracy was a classic 18-layer residual network (ResNet-18).

Table 1 shows the performance gaps between our method and the random baseline, SRAAL, TA-VALL, LL4AL, and NCE methods on the CIFAR-10 dataset. The top two accuracy rankings for each round of active learning are highlighted in bold. Firstly, when using the cosine function as the distance classifier, our method improved the accuracy by 0.868% compared to the second-ranked method, and by 0.206% when using the Euclidean distance, achieving the highest accuracy. Secondly, our method consistently outperformed other methods in terms of accuracy in each round of active learning tasks, before reaching the maximum budgeted sample count. This indicates that our method utilizes prototype vectors for feature fusion, enabling the selection of samples with higher information content that are far from the current labeled samples, i.e., samples that are easily confused at category boundaries, thus improving the effectiveness of active learning.

Figure 3 shows the performance of our method compared to state-of-the-art approaches on the CIFAR-100 dataset. It can be observed that our method achieves the highest accuracy even with an increasing number of dataset categories, outperforming SRAAL, TA-VAAL, and LL4AL significantly. Compared to the NCE-cos method, our approach improves accuracy by 0.324% when the number of samples reaches 20,000. This indicates that our method maintains its advantage in handling multi-class datasets.

4.2 Sample Selection Experiments and Result Analysis

In Fig. 4, we visualize the distribution of budgeted samples selected by three active learning methods in a certain round. In Fig. 4c, the samples selected by our proposed method are distributed at the farthest boundaries of each class. However, in Fig. 4a, b, the active learning methods select a small number of labeled samples that are located close to the class centers, which result in budget waste. Therefore, compared to the other methods, our approach can select boundary samples that are far away from the current class centers. This is possibly because prototype vectors can effectively represent the common characteristics of each class, and when unlabeled samples far from the prototype vectors undergo feature fusion, they are more likely to undergo label changes than those closer in distance. As a result, our method can easily capture boundary samples as shown in the visualization diagram.

Table 1 Performance comparison of our method with state-of-the-art methods in active learning on the CIFAR-10 dataset

Method	1000	2000	3000	4000	5000	6000	7000	8000	9000	10,000
NCE-Random-euc	55.870	66.318	73.848	78.782	81.426	83.922	85.122	86.182	87.116	87.796
SRAAL (2020CVPR)	53.694	65.166	72.452	80.404	83.152	85.028	86.282	87.490	88.466	89.190
TA-VAAL (2021CVPR)	46.816	58.094	71.282	78.576	81.361	83.852	86.276	87.746	89.090	89.662
LL4AL (2019CVPR)	46.470	59.976	71.046	80.104	82.632	86.026	87.510	88.656	89.590	89.844
NCE-cos (AAAI21)	54.814	66.990	76.080	81.242	83.486	86.320	87.680	88.932	89.602	90.230
NCE-euc (AAAI21)	56.386	67.314	75.712	81.110	83.760	**86.490**	87.884	89.122	**90.120**	90.976
Ours-cos	54.856	**67.498**	**76.636**	**81.954**	**84.252**	86.482	**88.114**	**89.272**	90.094	**91.098**
Ours-euc	55.898	**67.466**	**76.346**	**81.496**	**84.114**	**86.562**	**88.080**	**89.220**	**90.412**	**91.182**

Bold faced values indicate better results

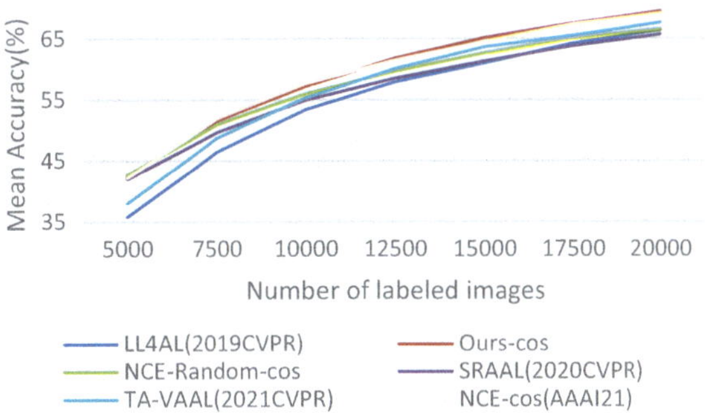

Fig. 3 Performance comparison of our method with state-of-the-art methods in active learning on the CIFAR-100 dataset

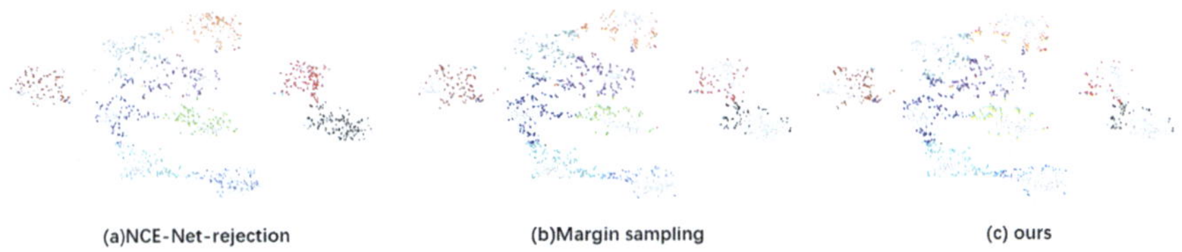

Fig. 4 Visualizing the sample selection behavior of three active learning methods in a certain active learning cycle in the feature space using t-SNE. The gray color represents the feature distribution of all unlabeled samples in this active learning cycle, while the light color represents the budgeted samples labeled by different methods in this active learning cycle

5 Conclusion

In this paper, we propose an active learning based on prototype vector feature adaptive fusion. This method fully exploits highly uncertain samples at different class boundaries in the feature space and further combines clustering methods to consider representative samples for final sample selection. Experiments in image classification demonstrate that our model outperforms previous state-of-the-art methods. With the increasing demand for labeling budget in each round, our method has greater advantages. In the future, we will also consider learning multiple prototypes for each category, allowing our method to perform better while dealing with more complex tasks.

References

1. Yasufumi K et al (2022) Non-deep active learning for deep neural networks. Sensors 22(14):5244
2. Donggeun Y, In So K (2019) Learning loss for active learning. In: Proceedings of the IEEE/CVF conference on computer vision and pattern recognition. pp 93–102
3. Fang W et al. (2021) Nearest neighbor classifier embedded network for active learning. In: Proceedings of the AAAI conference on artificial intelligence, vol. 35.11, pp 10041–10048

4. Beichen Z et al (2020) State-relabeling adversarial active learning. In: Proceedings of the IEEE/CVF conference on computer vision and pattern recognition, pp 8756–8765
5. Kwanyoung K et al (2021)) Task-aware variational adversarial active learning. In: Proceedings of the IEEE/CVF conference on computer vision and pattern recognition, pp 8166–8175
6. Ozan S, Silvio S (2017) Active learning for convolutional neural networks: a core-set approach". In: arXiv preprint arXiv:1708.00489
7. Hong-Ming Y et al (2018) Robust classification with convolutional prototype learning. In: Proceedings of the IEEE conference on computer vision and pattern recognition, pp 3474–3482
8. Amin P et al (2022) Active learning by feature mixing. In: Proceedings of the IEEE/CVF conference on computer vision and pattern recognition, pp 12237–12246
9. Burr S (2009) Active learning literature survey

Dual Neural Network Approach for Dual-Energy Imaging and Material Decomposition

Ye Shen, Junru Ren, Yizhong Wang, Xinrui Zhang, Shaoyu Wang, Ailong Cai, Ningning Liang, Lei Li, and Bin Yan

Abstract Dual-energy computed tomography (DECT) is a promising technology which uses X-ray beams with different energies to obtain the material attenuation information of the scanned object. However, in contrast to single-energy CT (SECT), DECT imaging suffers from high hardware costs, complex systems and increased radiation doses. In order to obtain DECT images easily, in this paper, we propose a dual neural network structure that achieves spectral mapping and material decomposition. Specifically, we developed the mapping U-Net and the material decomposition U-Net, respectively. Finally, we verified the feasibility of the proposed method on the clinical cranial cavity. Numerical results show that the proposed method can obtain high-quality DECT images from SECT images and realize accurate basis material decomposition.

1 Introduction

Computed tomography (CT) is a promising technology which can provide cross-sectional and volumetric images of internal structures. CT has been widely used in medical imaging [1], industrial nondestructive testing [2] and security inspection [3]. Compared with single-energy computed tomography (SECT), dual-energy computed tomography (DECT) can effectively reduce the artifacts caused by beam hardening effect and provide accurate quantitative information based on transmission data under different spectral. Thanks to its advantages, many advances have been made in the clinical application of DECT in recent years, such as the diagnosis of acute bowel ischemia [4] virtual monochromatic imaging [5] and pulmonary functional imaging [6].

Despite the wide applications of DECT, there are still some issues that need to be solved. For one thing, DECT suffers from the complexity of system and processing

Y. Shen · J. Ren · Y. Wang · X. Zhang · S. Wang · A. Cai · N. Liang · L. Li (✉) · B. Yan
Department of Henan Key Laboratory of Imaging and Intelligent Processing, Information Engineering University, Zhengzhou 450000, China
e-mail: leehotline@163.com

S. Yadav et al. (eds.), *Proceedings of 3rd International Conference on Artificial Intelligence, Robotics, and Communication*, Lecture Notes in Electrical Engineering 1172, https://doi.org/10.1007/978-981-97-2200-6_26

steps. For another, in comparison to SECT, DECT produces high- and low-energy spectral which means increased radiation doses. Last but not least, the cost of DECT scanners is usually more expensive than SECT scanners, which makes it difficult for people in many underdeveloped areas to access and use the technology.

To overcome these problems, the efforts on reducing the requirement of dual-energy data have received increasing attention. In terms of the compelling inferential capability of deep learning, it is a common tool to make up for data missing in these works [7–9]. For example, Lyu et al. [10] proposed a convolution network to realize DECT imaging using a low-energy image and a single-view high-energy projection. Further, in 2021, Wang et al. [11] used a fully convolutional neural network to obtain the high-energy CT images from low-energy CT images. Subsequently, Liu et al. [12] designed comparative experiments indicating that deep learning has the potential to generate brain DECT images from brain SECT images.

Due to the CT images obtained at different energies having high similarity [10], inspired by the above approach, we aim to further use only single-spectrum to approximate the energy-dependent information in this paper. Specifically, a U-Net model is developed to map single-spectrum images to dual-spectrum images at pre-specified energy levels. Then achieve material decomposition through a mirrored dual-channel U-Net. This based on deep learning technology DECT estimation method is used to obtain material decomposition images that are close to the images produced by DECT using two spectral projections.

2 Methods

2.1 Material Decomposition for Mapping-Based Dual-Energy CT

Figure 1 shows the process of generating high- and low-energy from single-energy maps and performing material decomposition. Material decomposition by DECT requires the attenuation of information of high- and low-energy, and the difference between the two energies should be as large as possible. Therefore, we choose middle-energy images as the input of model. Firstly, mapping U-Net obtains high- and low-energy images by learning the features of middle-energy images. Secondly, a mirrored dual-channel U-Net (DC-U-Net) is applied to perform material decomposition tasks. The network details are described in detail below.

2.2 Network Details

Mapping U-Net. It is well known that U-Net has extraordinary performance in medical image processing [13]. Therefore, we developed a mapping U-Net to identity

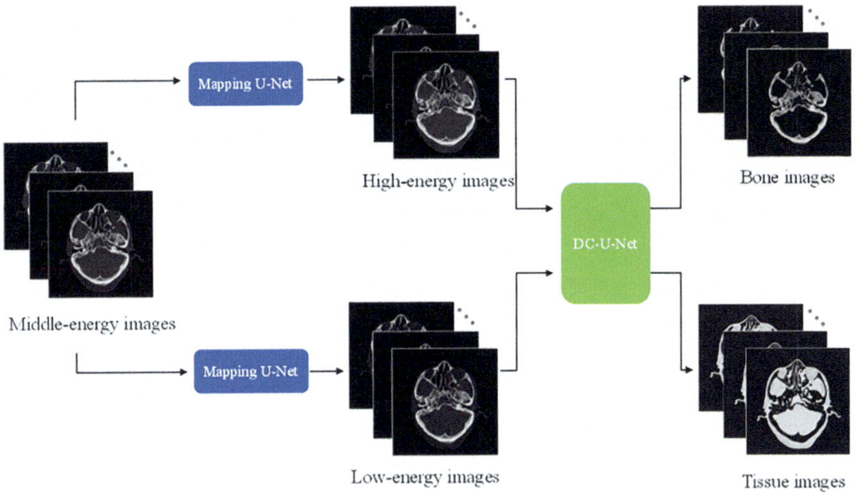

Fig. 1 Workflow of material decomposition for mapping-based dual-energy CT

map single-spectrum images into another spectrum images at pre-specified energy level. The architecture of proposed U-Net is shown in Fig. 2. The network can be divided into two parts: encoder and decoder. In the encoder section, there are 12 convolutional layers and 5 pooling layers. The convolution layer adopts 3 × 3 convolution kernel, followed by a batch norm (BN) and rectified linear unit (ReLU). The strides for convolution lays are all 1. Every two convolution layers form a group, and the number of convolution kernels in each group is 32, 64, 128, 256, 512 and 1024, respectively. Meanwhile, to reduce the size of the feature map and prevent overfitting, the pooling layers adopt 2 × 2 maximum pooling module with the strides of 2. Decoder is slightly different from encoder, it consists of five connected modules, and the information of each connected module is jointly determined by the encoder convolution layer and the upsampling layer. Specially, in decoder, the number of convolutions is in reverse order and decreases in turn. Convolution plus sampling is used for each convolution layer. Additionally, upsampling uses an interpolation template with stride size 2, ensuring a match with the pooling process information.

Material Decomposition U-Net. As shown in Fig. 3, the mirrored dual-channel U-Net (DC-U-Net) has a similar structure to mapping U-Net which we mentioned in Fig. 2. Actually, DC-U-Net is exactly composed of two symmetric U-Nets. The main part of the network is composed of two symmetrical U-Net networks, each of which contains compression and expansion paths. The compression path is used to extract the features of bone and tissue images, and the expansion path is used to recover the details of the base material image. In addition, the information of the two paths can interact with each other, resulting in better noise reduction. There is information transmission between the two paths inside mapping U-Net, and the adjacent U-Net paths also need information transmission in DC-U-Net. Therefore, the connection of the corresponding layer is added between the two U-Net extension paths. In the

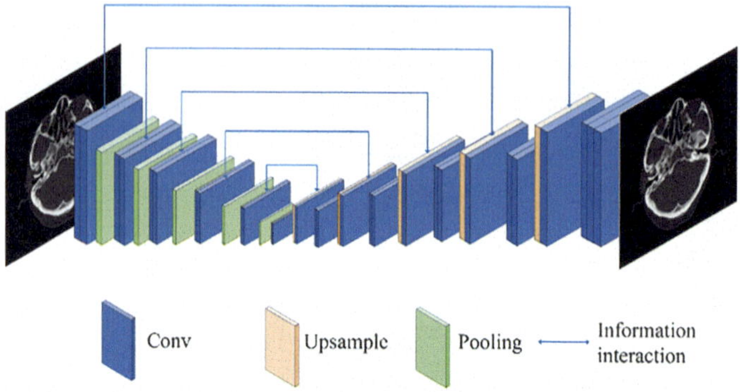

Fig. 2 Architecture of Mapping U-Net. Input SECT images at a certain energy. After the encoder and decoder, another spectrum images at pre-specified energy level are obtained

process of realizing the decomposition of base material images, this operation plays a key role in the learning of the hidden information in the two energy spectrum images and the extraction and reconstruction of the base material features.

Loss Function. The choice of loss function is very important in the construction of network model. In this paper, we consider mean absolute error and marginal error to evaluate the training effect of the model both mapping U-Net and DC-U-Net.

The L_{mean} loss is the mean absolute error of different energy images, which can be expressed as

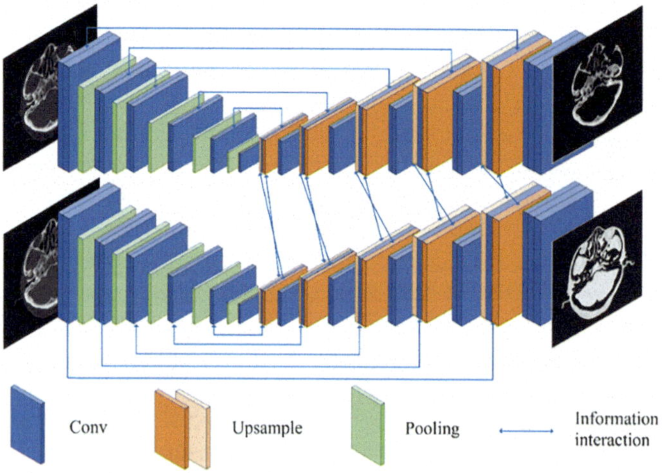

Fig. 3 Architecture of mirrored dual-channel U-Net (DC-U-Net). The mapped high- and low-energy images as the input into DC-U-Net; after the network, the decomposed bone and tissue images are obtained

$$L_{mean} = \frac{|V(X) - Y|}{n}, \tag{1}$$

where X denotes the input image; $V(X)$ denotes the output of the network which is the predicted image; Y represents the ground truth, and n is the number of image pixel. The mean absolute error measures the network's detection results for the overall structure and internal patterns of images. When organizing images, some edges are very similar, so edge error is also required to improve the network's detection of image edges and other details. The definition of edge error is as follows:

$$L_{edge} = ||grad_x X| - |grad_x Y||^2 + \left||grad_y X\right| - \left|grad_y Y\right||^2, \tag{2}$$

where $grad_x$ and $grad_x$ denote the respective gradient values along different image directions. Thus, the mixing loss can be expressed as

$$L_{mix} = L_{mean} + \varphi L_{edge}, \tag{3}$$

where φ is the weight of different loss. By adjusting the value of φ, a balance between preserving overall image information and edge information can be achieved. Through minimizing the L_{mix} loss, mapping U-Net and DC-U-Net can generate higher-quality mapping energy images and material images.

3 Experiments

In this paper, a simulation experiment based on the data from 1615 cranial cavity slice images of 6 patients verified the effectiveness of the proposed method. The size of each image is 256 × 256 pixels. The distance between the x-ray source to the detector was set to be 300 mm, and the distance between the object to the detector was set to be 600 mm. SpekCalc software was applied to generate low-, middle- and high-energy data at 80, 110 and 140 kVp polychromatic spectrum, respectively. The dataset was split into training and testing sets. Moreover, in the image mapping phase, 1430 single-energy images at 110 kVp and 1430 single-energy images at 80 kVp formed training dataset I, while 1430 single-energy images at 110 kVp and 1430 single-energy images at 140 kVp formed training dataset II. Single-energy CT images at 110 kVp were input to the mapping U-Net I and II to estimate single-energy CT images at 80 and 140 kVp, respectively. The remaining 185 data were used as the network testing dataset. In material decomposition, dual-energy CT images at 80 and 140 kVp as the input of DC-U-Net, and bone and tissue images were used as labels.

All training and testing procedures were performed in Python and TensorFlow on a PC computer with an NVIDIA Tian V GPU of 12 GB memory. Figure 4 shows the convergence curve of two networks. As can be seen from the figure, both network models converge gradually with the increase of iteration rounds.

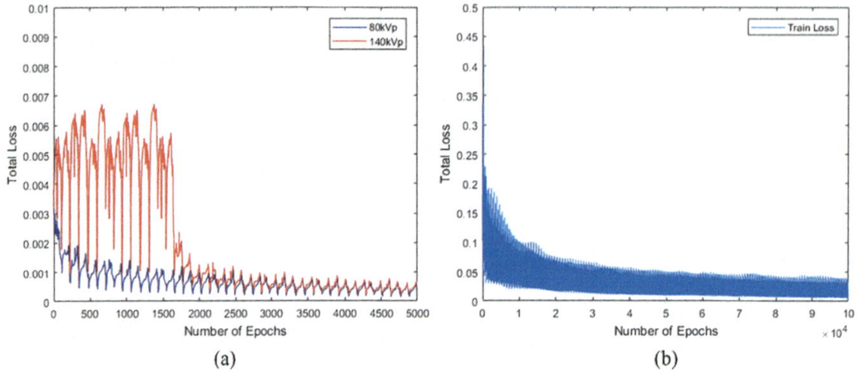

Fig. 4 Convergence curves of the mapping U-Net and DC-U-Net during the training process. The total loss of mapping U-Net is shown in figure (**a**) and the total loss of DC-U-Net is shown in figure (**b**)

4 Results

4.1 DECT Imaging

For SECT to realize DECT Imaging, we first use mapping U-Net to learn the features of dual-energy CT at pre-specified energy levels. Figure 5 shows the high- and low-energy images generated by middle-energy. We can find that the mapping U-Net can generate almost the same images and leave just a few errors.

We calculated peak signal-to-noise ratio (PSNR) and structural similarity (SSIM) for 185 mapping DECT images produced by our method in Table 1. We can see from these data that the mapping U-Net delivered high-quality DECT images from middle-energy CT images.

4.2 Basis Material Decomposition

Through the above analysis, we have obtained high-quality DECT mapping images, which are almost the same. Then the estimated DECT images are input to DC-U-Net network to perform material decomposition. In this part, decomposition by directly inverting (DI) the material attenuation coefficient is introduced as contrast method. Figure 6 shows the result of decomposition; regions interests (ROI) are marked with red rectangle. We can find that the performance of the proposed method in detail preserving and global structure is better than DI method, especially in ROIs. Moreover, the proposed method is closer to the true value in the numerical value of the material visually.

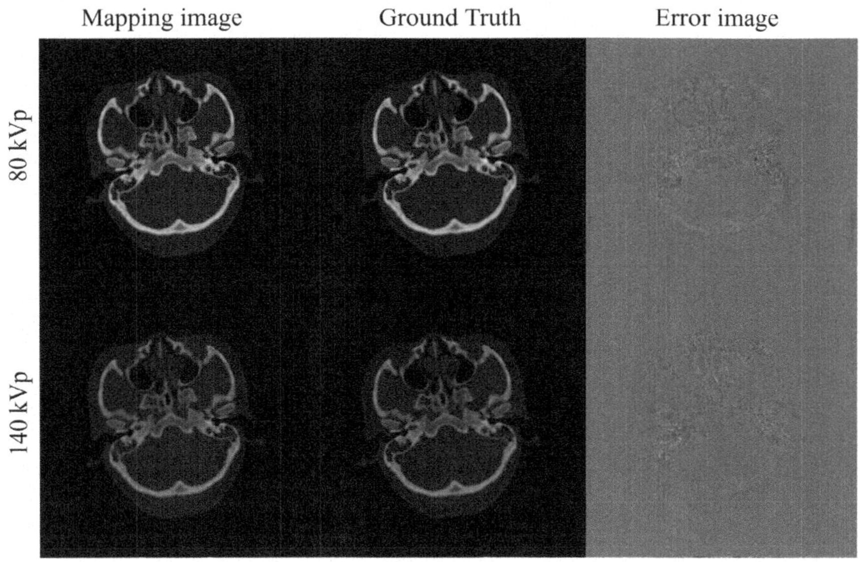

Fig. 5 Result of mapping images, ground truth and error images. The displayed window of the first two column is [0 0.08], and the last one is [−0.01 0.01]

Table 1 Quantitative of mapping DECT images

Energy value	Index	Mapping DECT images
80 kVp	PSNR	62.1150
	SSIM	0.9994
140 kVp	PSNR	66.6960
	SSIM	0.9996

Table 2 lists the quantitative of material decomposition by DI and the posed method. PSNR and SSIM are introduced to test the quality of material decomposition, as can be seen from the table; composed with DI, the PSNR and SSIM have significantly improved, which proves the effectiveness of the proposed method. Furthermore, it can be seen that the SSIM of the proposed method in tissue material has improved a lot. Generally, the quantitative results further demonstrate the superiority of our methods, which aligns with the results presented in the figure.

5 Conclusions

DECT has the advantage of being able to differentiate different materials even if their X-ray attenuation coefficients are in the same range. However, DECT imaging in clinical applications suffers from high costs, complex process and increased doses. In this

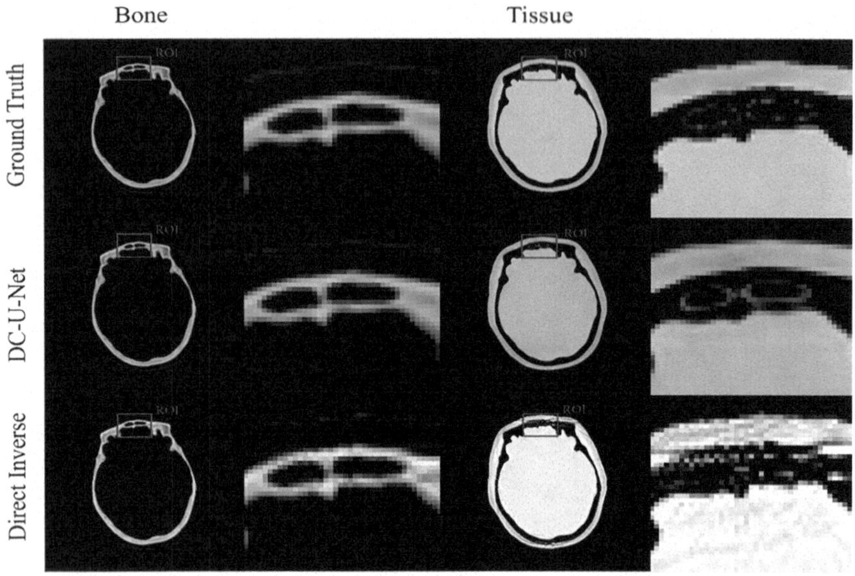

Fig. 6 Result of different methods. Top to bottom rows denote ground truth, the material decomposition result of DC-U-Net and DI, respectively. And from left to right, represent decomposed bone, ROIs of decomposed bone, decomposed tissue and ROIs of decomposed tissue. All the displayed grayscale is [0 1.2]

Table 2 Quantitative of material decomposition by different methods

Material	Index	DI	DC-U-Net
Bone	PSNR	20.2908	29.4789
	SSIM	0.7170	0. 8272
Tissue	PSNR	11.7872	19.8803
	SSIM	0.7943	0.8876

paper, we developed a mapping U-Net network to convert single-spectrum images to approximate DECT counterparts and used another network to obtain high-precision basis material images. Finally, we have verified the feasibility of the proposed method on the clinical cranial cavity and the quantitative of results shows the excellent performance.

In future work, we will introduce the proposed method to unsupervised network frameworks and explore the possibility in reducing training data.

Acknowledgements This work was supported by the National Key Research and Development Project of China (Grant No. 2020YFC1522002). This work was also supported by the National Natural Science Foundation of China (Grant No. 62101596, Grant No. 62201616).

References

1. Ghayvat H, Awais M, Bashir AK et al (2023) AI-enabled radiologist in the loop: novel AI-based framework to augment radiologist performance for COVID-19 chest CT medical image annotation and classification from pneumonia. Neural Comput Appl 35(20):14591–14609. https://doi.org/10.1007/s00521-022-07055-1
2. Ramírez IS, Márquez FPG, Papaelias M (2023) Review on additive manufacturing and non-destructive testing. J Manuf Syst 66:260–286. https://doi.org/10.1016/j.jmsy.2022.12.005
3. Hai C, He J, Li B et al (2023) Dual-domain metal trace inpainting network for metal artifact reduction in baggage CT images. Measurement 207:112420. https://doi.org/10.1016/j.measurement.2022.112420
4. Xu JJ, Ulriksen PS, Jawad S et al (2023) Iodine density mapping for the diagnosis of acute bowel ischemia using fast kV-switching dual-energy CT. Abdom Radiol 1–8.https://doi.org/10.1007/s00261-023-04097-4
5. Kraft J, Lutyj J, Grabenbauer F et al (2023) (2023) Assessment of dual-energy computed tomography derived virtual monochromatic imaging for target volume delineation of brain metastases. Radiother Oncol 187:109840. https://doi.org/10.1016/j.radonc.2023.109840
6. Ozawa Y, Ohno Y, Nagata H et al (2023) Advances for pulmonary functional imaging: dual-energy computed tomography for pulmonary functional imaging. Diagnostics 13(13):2295. https://doi.org/10.3390/diagnostics13132295
7. Cong WX, Xi Y, Fitzgerald P et al (2020) Virtual monoenergetic CT imaging via deep learning. Patterns 1(8). https://doi.org/10.1016/j.patter.2020.100128
8. Zhang YK, Hu DL, Yan ZH et al (2023) TIME-Net: transformer-integrated multi-encoder network for limited-angle artifact removal in dual-energy CBCT. Med Image Anal 83:102650. https://doi.org/10.1016/j.media.2022.102650
9. Li YS, Tie X, Li K et al (2023) A quality-checked and physics-constrained deep learning method to estimate material basis images from single-kV contrast-enhanced chest CT scans. Med Phys 50(6):3368–3388. https://doi.org/10.1002/mp.16352
10. Lyu TL, Zhao W, Zhu YS et al (2021) Estimating dual-energy CT imaging from single-energy CT data with material decomposition convolutional neural network. Med Image Anal 70:102001. https://hal.science/hal-03190848
11. Wang YZ, Yu XH, Feng ZW et al (2021) Image-domain material decomposition for single-energy CT images using cascaded network. In: 2021 international conference on artificial intelligence and electromechanical automation (AIEA), IEEE, pp 273–276. https://doi.org/10.1109/AIEA53260.2021.00064.
12. Liu CK, Liu CCY et al (2021) Generation of brain dual-energy CT from single-energy CT using deep learning. J Digit Imaging 34(1):149–161. https://doi.org/10.1007/s10278-020-00414-1
13. Siddique N, Paheding S, Elkin CP et al (2021) (2021) U-Net and its variants for medical image segmentation: a review of theory and applications. IEEE Access 9:82031–82057. https://doi.org/10.1109/ACCESS.2021.3086020

Fixed-Time Control for Cooperative Target Encirclement and Tracking by Multi-UAV

Hongjin Liao, Xin Chen, Jiangbo Jia, Guangyuan Zhu, and Weizhen Wang

Abstract In the research, a new fixed-time control method is proposed for multiple unmanned aerial vehicles (multi-UAV) to address cooperative target encirclement and tracking. Initially, only the position information about the target is used in this method. Subsequently, based on an interconnected system and a fixed-time lemma, a guidance law is proposed to control all UAVs to move along a circle centered on the target and the distance between every pair of UAVs remains consistent. The fixed-time stability of the system is rigorously mathematically proved, based on which the bounds on the convergence time can be accurately derived. Finally, a simulation demonstrates the feasibility of the designed strategy.

1 Introduction

Currently, multi-UAV cooperation is widely utilized in many areas, such as disaster rescue [1], target reconnaissance [2], and target tracking [3, 4]. Among them, target tracking is the basic mission of UAVs in military operations and has important research significance. Now, there are two main ways for multi-UAV cooperative target tracking, fixed structure formation tracking [5] and cooperative encirclement tracking [6]. When executing tracking tasks, the method of encirclement tracking is often used for fixed-wing UAVs, which means the UAVs keep a fixed distance with the target while moving around it in a circling manner. This tracking method not only ensures a minimum cruising speed for fixed-wing UAVs [7], but also reduces the risk of exposure. Therefore, many scholars are beginning to research on the application of UAVs in cooperative target encirclement and tracking.

In some studies [8], the UAV on a mission can directly acquire all the state information about the target. However, since the target is usually non-cooperative, most of its state information is not accessible. In [6], it is considered that only

H. Liao · X. Chen · J. Jia (✉) · G. Zhu · W. Wang
College of Automation Engineering, Nanjing University of Aeronautics and Astronautics, Jiangjun Ave.211106, Nanjing, China
e-mail: jiajiangbo@nuaa.edu.cn

S. Yadav et al. (eds.), *Proceedings of 3rd International Conference on Artificial Intelligence, Robotics, and Communication*, Lecture Notes in Electrical Engineering 1172, https://doi.org/10.1007/978-981-97-2200-6_27

the position information about the target can be detected, then based on distance measurement, a distributed target tracking guidance law is proposed.

For the problem of target tracking, the system convergence speed plays a critical role. However, traditional asymptotically stable control methods only ensure that the system converges to equilibrium points in infinite time. To enhance the convergence rate of the system, in [9], a finite-time control strategy is introduced, which means that the system will converge in a finite time. Compared to infinite-time control, finite-time control provides more rapid convergence. Subsequently, in [10], a finite-time control law is implemented by combining finite-time theory with backstepping control to address the tracking target problem for a set of helicopter formations. However, due to the differences in tracking methods, the aforementioned methods do not take into account the condition of target encirclement. Subsequently, in [11], finite-time robust guidance laws for encirclement and tracking are designed, but the problem of multi-UAV cooperation is not considered. In addition, although finite time enhances the convergence rate of the system, its convergence time relies on the initial states of the system, which is not applicable in some cases. Therefore, some researches introduce fixed-time control into the multi-agent cooperative control problem. Unfortunately, few works consider the problem of fixed-time target encirclement and tracking.

In this study, the problem of multi-UAV cooperation in encircling and tracking of the target is considered, and a novel fixed-time control strategy is proposed to solve this problem. The main novelties of this paper are listed as follows:

(1) Unlike most works that only consider a single UAV, multi-UAV cooperation is considered in this work, which means multi-UAV can be uniformly distributed on a circle centered on the target while executing tracking tasks.
(2) Compared to other methods [8] additionally need the velocity information about the target, only the position information about the target is required in this method.
(3) A fixed-time UAV guidance law is proposed to implement the surrounding and tracking mission. In contrast to most existing methods such as [3, 11], the guidance law ensures the fixed-time convergence of the multi-UAV system.

The paper is proceeded by the following organization. In Sect. 2, some lemmas and a model for the problem are introduced. Then the design of the guidance law is carried out in Sect. 3. A simulation is carried out in Sect. 4. Finally, the conclusion is given in Sect. 5.

2 Preliminaries

In this section, several technical lemmas will be introduced and the kinematic model of UAVs will be given. Throughout the paper, $sig^a(x) = |x|^a \text{sgn}(x)$, where $\text{sgn}(x) = x/|x|$ for $x > 0$, and $\text{sgn}(0) = 0$. $1_N = [1, 1, ..., 1]^T \in R^N$.

2.1 Technical Lemma

Consider a nonlinear interconnected system:

$$\dot{x}_i = f_i(t, x_i) + d_i(t, x), i = 1, 2, ..., m, \tag{1}$$

where $x_i \in R^{n_i}$, $n_1 + ... + n_m = n$, $x = [x_1^T, ..., x_m^T]$. Functions f_i and d_i are continuous in their arguments and satisfy $f_i(t, 0) = 0$, $d_i(t, 0) = 0$, for any $i \in m$. The system (1) could be considered as a disturbance of a normal system $\dot{x}_i = f_i(t, x_i)$.

Lemma 1 [12] If terms [C1]-[C2] are satisfied:

[C1] The Lyapunov function $V_i(t, x_i)$ of system $\dot{x}_i = f_i(t, x_i)$ satisfies:

$$c_1\|x_i\|^2 \le V_i(t, x_i) \le c_2\|x_i\|^2 \tag{2}$$

$$\frac{\partial V_i(t, x_i)}{\partial t} + \frac{\partial V_i(t, x_i)}{\partial x_i} f_i(t, x_i) \le -\alpha_i \phi_i^2(x_i) \tag{3}$$

$$\left\| \frac{\partial V_i(t, x_i)}{\partial x_i} \right\| \le \beta_i \phi_i(x_i) \tag{4}$$

$$\|d_i(t, x)\| \le \sum_{i=1}^{m} \gamma_{ij} \phi_j(x_j). \tag{5}$$

[C2] α_i and β_i are positive constants, and γ_{ij} is nonnegative constants, $\phi_i(x_i)$ is a class K function.

Then there is a positive Lyapunov function $V(t, x) = \sum_{i=1}^{m} V_i(t, x_i)$, whose derivative along (1) satisfies:

$$\dot{V}(t, x) \le -\frac{1}{2}\phi^T (S + S^T)\phi, \tag{6}$$

where $\phi = [\phi_1, ..., \phi_N]^T$, S is a $N \times N$ matrix, and elements of S are defined by:

$$s_{ij} = \alpha_i - \beta_i \gamma_{ij}, i = j \; s_{ij} = -\beta_i \gamma_{ij}, i \ne j. \tag{7}$$

Lemma 2 [13] Consider a system $\dot{x}(t) = f(t, x)$. It is fixed-time stable if there exists a positive function $V(x) \in R$, whose derivative along $\dot{x}(t) = f(t, x)$ satisfies:

$$\dot{V}(x) \le -\alpha_0 V^{r_1}(x) - \beta_0 V^{r_2}(x), \tag{8}$$

where α_0 and β_0 are positive constants, and $0 < r_1 < 1$ and $r_2 > 1$. Additionally, the convergence time T is bounded by:

$$T \le \frac{1}{\alpha_0(1-r_1)} + \frac{1}{\beta_0(r_2-1)}. \tag{9}$$

2.2 UAV Kinematic Model

It's assumed the attitude loop of the UAV is designed, and UAVs track the target at a certain cruising altitude. The kinematic model of $UAV_i(i = 1, 2, , N)$ is simplified as:

$$\begin{cases} \dot{x}_i = v_i cos\theta_i \\ \dot{y}_i = v_i sin\theta_i \ , \\ \dot{\theta}_i = \omega_i \end{cases} \tag{10}$$

where $[x_i, y_i]^T \in R^2$ represents the $2D$ position of UAV_i, $\theta_i \in [0, 2\pi]$ indicates the heading angle, $\omega_i \in R$ and $v_i \in R$ are angle velocity and linear velocity, respectively. $[x_t, y_t]^T \in R^2$ is defined as the position of the target, and the distance between $\mathrm{UAV_i}$ and the target is defined as: $\rho_i = \sqrt{(x_i - x_t)^2 + (y_i - y_t)^2} \in R^+$. Meanwhile, the velocities v_i and ω_i of UAVs should satisfy the physical constraints:

$$v_{min} \le v_i(t) \le v_{max}, \ -\omega_{max} \le \omega_i(t) \le \omega_{max}, \tag{11}$$

where v_{min}, v_{max}, and ω_{max} are known positive constants. Normally, the target's speed is much less than the UAV. Figure 1 depicts the mission of target surrounding and tracking for multi-UAV, where ρ_d denotes the anticipated distance between UAVs and the target. $\theta_{i,i-1}$ denotes the phase angle between UAV_i and UAV_{i-1}. χ_i is the bearing angle, according to Fig. 1, one has $\chi_i = \theta_i - \sigma_i$, where σ_i denotes the angle between the UAV-to-target line and the forward orientation of the x-axis. In order to facilitate the analysis, the model of (10) is converted into the polar coordinate form:

$$\begin{cases} \dot{\rho}_i = -v_i cos\chi_i \\ \dot{\chi}_i = \omega_i - \dfrac{v_i}{\rho_i} sin\chi_i \\ \dot{\theta}_{i,i-1} = \omega_i - \omega_{i-1} \end{cases} \tag{12}$$

where $\chi_i = \frac{\pi}{2}$ and $\chi_i = \frac{3\pi}{2}$ denote the clockwise and counterclockwise motion of the UAV, respectively. For convenience, only the clockwise motion is considered in this paper. The purpose of this work is to design a encirclement and tracking guidance law, so that each UAV can be uniformly distributed on a circle centered on the target with radius ρ_d and move in a circular motion in a fixed time T, which means:

Fig. 1 Target surrounding and tracking

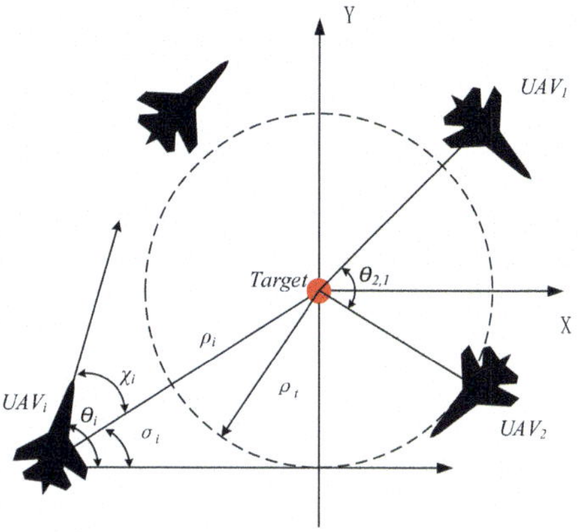

$$\lim_{t\to T}\rho_i=\rho_t,\ \lim_{t\to T}\chi_i=\frac{\pi}{2},\ \lim_{t\to T}\theta_{i,i-1}=\frac{2\pi}{N}. \tag{13}$$

3 Main Result

3.1 Fixed-Time Guidance Law Design

Next, based on Lemma 1, the following fixed-time guidance law is proposed:

$$\begin{cases}\omega_i=-k_1\dfrac{q_1(e_{\rho_i})+q_1(e_{\chi_i})}{e_{\chi_i}cos(\chi_i)}+\dfrac{v_ie_{\rho_i}}{e_{\chi_i}}+\dfrac{v_{i-1}e_{\chi_i}}{\rho_{i-1}}+\dfrac{v_i}{\rho_i}\\ v_i=\dfrac{\rho_i}{\rho_{i-1}}v_{i-1}-k_2\rho_iq_2(e_{\overline{\theta}_i})\end{cases}, \tag{14}$$

where $e_{\chi_i}=sin(\chi_i)-1$, $e_{\rho_i}=\rho_i-\rho_d$, $e_{\overline{\theta}_i}=\theta_i-\theta_{i-1}-\frac{2\pi}{N}$. $q_1(x_i)=|x_i|^{r_1+1}+|x_i|^{r_2+1}$, $q_2(x_i)=sig^{r_1}(x_i)+sig^{r_2}(x_i)$, r_1, r_2 are positive constants, satisfying:$0<r_1<1$, $r_2>1$. k_1, k_2 will be determined later.

For computational convenience, the following replacement is used:

$$\eta_{i,1}=\frac{q_1(e_{\rho_i})+q_1(e_{\chi_i})}{e_{\chi_i}cos(\chi_i)},\ \eta_{i,2}=\frac{v_ie_{\rho_i}}{e_{\chi_i}},\ \eta_{i,3}=\frac{v_{i-1}e_{\chi_i}}{\rho_{i-1}},\ \eta_{i,4}=q_2(e_{\overline{\theta}_i}). \tag{15}$$

Then the guidance law (14) can be transformed as:

$$\begin{cases} \omega_i = -k_1\eta_{i,1} + \eta_{i,2} + \eta_{i,3} + \dfrac{v_i}{\rho_i} \\ v_i = \dfrac{\rho_i}{\rho_{i-1}} v_{i-1} - k_2\rho_i\eta_{i,4} \end{cases}. \tag{16}$$

Substituting the guidance law (16) into the system model (12) yields a nonlinear composite system as follows:

$$\begin{cases} \dot{\rho}_i = -v_i cos(\chi_i) \\ \dot{\chi}_i = -k_1\eta_{i,1} + \eta_{i,2} + k_2 e_{\chi_i}\eta_{i,4} \\ \dot{\theta}_{i,i-1} = -k_2\eta_{i,4} + k_1(\eta_{i-1,1} - \eta_{i,1}) + \eta_{i,2} - \eta_{i-1,2} + \eta_{i,3} - \eta_{i-1,3} \end{cases}. \tag{17}$$

Decomposition of the system into two interconnected subsystems:

$$\dot{x}_{i,1} = f_{i,1}(x_{i,1}) + d_{i,1}(x_{i,2}), \dot{x}_{i,2} = f_{i,2}(x_{i,2}) + d_{i,2}(x_{i,1}) \tag{18}$$

where $x_{i,1} = [\rho_i, \chi_i]^T$, $x_{i,2} = \overline{\theta}_i = \theta_i - \theta_{i-1}$, with

$$f_{i,1}(x_{i,1}) = \begin{bmatrix} -v_i cos(\chi_i) \\ -k_1\eta_{i,1} + \eta_{i,2} \end{bmatrix}, d_{i,1}(x_{i,2}) = \begin{bmatrix} 0 \\ k_2 e_{\chi_i}\eta_{i,4} \end{bmatrix} \tag{19}$$

and

$$f_{i,2}(x_{i,2}) = -k_2\eta_{i,4},\ d_{i,2}(x_{i,1}) = k_1(\eta_{i-1,1} - \eta_{i,1}) + \eta_{i,2} - \eta_{i-1,2} + \eta_{i,3} - \eta_{i-1,3}. \tag{20}$$

Rewriting (18) in the form of a vector, the error dynamics of the whole multi-UAV system can be obtained as:

$$\dot{x}_1 = f_1(x_1) + d_1(x_2),\ \dot{x}_2 = f_2(x_2) + d_2(x_1) \tag{21}$$

where $x_1 = [x_{1,1}, x_{2,1}, ..., x_{N,1}]^T$, $x_2 = [x_{1,2}, x_{2,2}, ..., x_{N,2}]^T$, with

$$\begin{cases} f_1(x_1) = \sum_{i=1}^{N} f_{i,1}(x_{i,1}) \\ d_1(x_2) = \sum_{i=1}^{N} d_{i,1}(x_{i,2}) \end{cases}, \begin{cases} f_2(x_2) = \sum_{i=1}^{N} f_{i,2}(x_{i,2}) \\ d_2(x_1) = \sum_{i=1}^{N} d_{i,2}(x_{i,1}) \end{cases}. \tag{22}$$

Theorem 1 Under the guidance law (14), for any initial state, there exists a fixed time T, such that $\lim_{t\to T} x_1 = 1N \otimes [\rho_d, \frac{\pi}{2}]^T$, $\lim_{t\to T} x_2 = 1N \otimes \frac{2\pi}{N}$, which means:

$$\lim_{t \to T} \rho_i = \rho_d, \lim_{t \to T} \chi_i = \frac{\pi}{2}, \lim_{t \to T} \theta_{i,i-1} = \frac{2\pi}{N}. \tag{23}$$

Proof: Lemma 1 will be used for the proof, organized in four steps. $Step(1)$ and $Step(2)$ will analyze the stability of the isolated nominal subsystems $\dot{x}_1 = f_1(x_1)$ and $\dot{x}_2 = f_2(x_2)$, respectively. Then $Step(3)$ and $Step(4)$ will consider the effect of the interconnection term $d_1(x_2)$ and $d_2(x_1)$. Finally, combined with Lemma 2, the fixed-time convergence of the system will be proved.

$Step(1)$: For the nominal subsystem $\dot{x}_1 = f_1(x_1)$, consider the following Lyapunov function:

$$V_{i,1} = e_{\rho_i}^2 + e_{\chi_i}^2, \; V_1 = \sum_{i=1}^{N} V_{i,1} = \sum_{i=1}^{N} \left(e_{\rho_i}^2 + e_{\chi_i}^2\right). \tag{24}$$

Deriving the time derivative of V_1 along $\dot{x}_1 = f_1(x_1)$ yields:

$$\begin{aligned} \frac{\partial V_1}{\partial t} + \frac{\partial V_1}{\partial x_1} f_1(x_1, t) &= \sum_{i=1}^{N} \left(\frac{\partial V_{i,1}}{\partial t} + \frac{\partial V_{i,1}}{\partial x_{i,1}} f_{i,1}\left(x_{i,1}, t\right) \right) \\ &\leq - \sum_{i=1}^{N} 2k_1 \left(\begin{bmatrix} |e_{\chi_i}|^{r_1+1} + |e_{\chi_i}|^{r_2+1} \\ +|e_{\rho_i}|^{r_1+1} + |e_{\rho_i}|^{r_2+1} \end{bmatrix} \right) \\ &\leq - \sum_{i=1}^{N} 2k_1 \left(|V_{i,1}|^{(r_1+1)/2} + |V_{i,1}|^{(r_2+1)/2} \right). \end{aligned} \tag{25}$$

From (25), one can set $\phi_1^2(x_1) = \sum_{i=1}^{N} \left(|V_{i,1}|^{(r_1+1)/2} + |V_{i,1}|^{(r_2+1)/2} \right)$ and $\alpha_1 = 2k_1$. Furthermore, one can obtain:

$$\left\| \frac{\partial V_1}{\partial x_1} \right\| = \left\| \sum_{i=1}^{N} \frac{\partial V_{i,1}}{\partial x_{i,1}} \right\| = \sqrt{\left(\sum_{i=1}^{N} 2e_{\rho_i} \right)^2 + \left(\sum_{i=1}^{N} 2e_{\chi_i} \right)^2} \leq 4\sqrt{N}\phi_1(x_1). \tag{26}$$

Thus, one has $\beta_1 = 4\sqrt{N}$.

$Step(2)$: For the nominal subsystem $\dot{x}_2 = f_2(x_2)$, select the Lyapunov function as:

$$V_{i,2} = e_{\bar{\theta}_i}^2, \; V_2 = \sum_{i=1}^{N} V_{i,2} = \sum_{i=1}^{N} e_{\bar{\theta}_i}^2. \tag{27}$$

The time derivative of V_2 along $\dot{x}_2 = f_2(x_2)$ yields:

$$
\begin{aligned}
\frac{\partial V_2}{\partial t} + \frac{\partial V_2}{\partial x_2} f_2(x_2, t) &= \sum_{i=1}^{N} \left(\frac{\partial V_{i,2}}{\partial t} + \frac{\partial V_{i,2}}{\partial x_{i,2}} f_{i,2}(x_{i,2}, t) \right) \\
&= \sum_{i=1}^{N} -2k_2 \left(|e_{\overline{\theta}_i}|^{r_1+1} + |e_{\overline{\theta}_i}|^{r_2+1} \right) \\
&= -2k_2 \sum_{i=1}^{N} \left(V_{i,2}^{(r_1+1)/2} + V_{i,2}^{(r_2+1)/2} \right).
\end{aligned} \tag{28}
$$

Then one can set $\phi_2^2(x_2) = \sum_{i=1}^{N} \left(V_{i,2}^{(r_1+1)/2} + V_{i,2}^{(r_2+1)/2} \right)$, $\alpha_2 = 2k_2$. In addition, it can be obtained that:

$$
\left\| \frac{\partial V_2}{\partial x_2} \right\| = \left\| \sum_{i=1}^{N} \frac{\partial V_{i,2}}{\partial x_{i,2}} \right\| = \sqrt{\left(\sum_{i=1}^{N} 2e_{\overline{\theta}_i} \right)^2} \leq 4\sqrt{N}\phi_2(x_2). \tag{29}
$$

From (29), one has $\beta_2 = 4\sqrt{N}$.

Step(3): For the disturbance term $d_1(x_2)$ of the system $\dot{x}_1 = f_1(x_1) + d_1(x_2)$:

$$
\|d_1(x_2)\| = \left\| \sum_{i=1}^{N} d_{i,1}(x_{i,2}) \right\| = \sqrt{\left(\sum_{i=1}^{N} k_2 e_{\chi_i} \eta_{i,4} \right)^2} \leq \sqrt{N} k_2 k_3 \phi_1(x_1), \tag{30}
$$

where $k_3 = max\{q_2(e_{\overline{\theta}_i})\}$. So one has

$$
\gamma_{11} = \sqrt{N} k_2 k_3, \ \gamma_{12} = 0. \tag{31}
$$

Step(4): For the disturbance term $d_2(x_1)$ of the system $\dot{x}_2 = f_2(x_2) + d_2(x_1)$, one has:

$$
\|d_2(x_1)\| = \left\| \sum_{i=1}^{N} d_{i,2}(x_{i,1}) \right\| = \sum_{i=1}^{N} \left(\begin{bmatrix} k_1(\eta_{i-1,1} - \eta_{i,1}) + \eta_{i,2} \\ -\eta_{i-1,2} + \eta_{i,3} - \eta_{i-1,3} \end{bmatrix} \right) = 0. \tag{32}
$$

Then it can be obtained that:

$$
\gamma_{21} = \gamma_{22} = 0. \tag{33}
$$

In summary, it follows that $\alpha_1 = -2k_1$, $\alpha_2 = -2k_2$, $\beta_1 = \beta_2 = 4\sqrt{N}$, $\gamma_{11} = \sqrt{N} k_2 k_3$ and $\gamma_{12} = \gamma_{21} = \gamma_{22} = 0$. Now, select the following Lyapunov function:

$$
V_3 = V_1 + V_2 = \sum_{i=1}^{N} \left(V_{i,1} + V_{i,2} \right). \tag{34}
$$

According to Lemma 1, the derivative of (45) along the trajectory of the composite system (21) is satisfied in the following form:

$$\dot{V}_3 \le -\frac{1}{2}\phi^T(S+S^T)\phi, \tag{35}$$

where $\phi = [\phi_1, \phi_2]^T$, with

$$S = \begin{bmatrix} \alpha_1 - \beta_1\gamma_{11} & -\beta_1\gamma_{12} \\ -\beta_2\gamma_{21} & \alpha_2 - \beta_2\gamma_{22} \end{bmatrix} = \begin{bmatrix} 2k_1 - 4k_2k_3N & 0 \\ 0 & 2k_2 \end{bmatrix}. \tag{36}$$

From (36), it can be seen that if set $k_2 > 0$ and $k_1 > 2k_2k_3N$, the matrix $(S+S^T)$ is positive definite. Then it can be deduced that:

$$\begin{aligned} \dot{V}_3 &\le -\frac{1}{2}\lambda_{min}(S+S^T)(\phi_1^2+\phi_2^2) \le -k_4\sum_{i=1}^{N}\begin{pmatrix} V_{i,1}^{(r_1+1)/2} + V_{i,1}^{(r_2+1)/2} \\ +V_{i,2}^{(r_1+1)/2} + V_{i,2}^{(r_2+1)/2} \end{pmatrix} \\ &\le -k_4(V_3^{c_1} + V_3^{c_2}), \end{aligned} \tag{37}$$

where $k_4 = \frac{1}{2}\lambda_{min}(S+S^T)$, $c_1 = \frac{r_1+1}{2}$, $c_2 = \frac{r_2+1}{2}$. Because $0 < r_1 < 1$, $r_2 > 1$, one has $0 < c_1 < 1$, $c_2 > 1$. According to Lemma 2, system (21) is fixed-time stable. And it is calculated that the convergence time T is bounded by:

$$T \le \frac{1}{k_4(1-c_1)} + \frac{1}{k_4(c_2-1)}. \tag{38}$$

Thus, $\lim_{t\to T}\tilde{\rho}_i = \rho_d$, $\lim_{t\to T}\tilde{\chi}_i = \frac{\pi}{2}$, $\lim_{t\to T}\tilde{\theta}_{i,i-1} = \frac{2\pi}{N}$. The proof is completed.

4 Simulation

Next, a simulation will be presented to demonstrate the proposed algorithm. Consider four UAVs and a moving target, the communication graph of them is shown in Fig. 1. And the motion of the target is $v_t = 0.2(t-10)(\sin(0.15(t-10)\pi)+1)$, $\omega_t = 0.05\sin(0.05(t-10)\pi)$, where T is the simulation duration. Besides, it is assumed that the control inputs v_i and ω_i of UAVs satisfy the following constraints:

$$v_i(t) \in \left[3 - 1.8\sqrt{2}\right], \omega_i(t) \in [-0.8, 0.8] \tag{39}$$

The starting information of the target and UAVs is presented in Table 1. Then, parameters are set as follows: $r_1 = 0.8$, $r_2 = 1.2$, $k_1 = 0.02$, $k_2 = 0.15$. One set $\rho_d = 8$. The simulation duration is set as $500s$ with the step size is $0.02s$.

Table 1 Starting information of UAVs and the target

Name	Target	UAV_1	UAV_2	UAV_3	UAV_4
Position	(0, 0)	(−20, 15)	(−20, −20)	(15, −20)	(30, 5)
Heading angle	0	π	$-5\pi/6$	$\pi/3$	$-\pi/3$

Figure 2 depicts the trajectories of four UAVs and the target, which shows that the target encirclement and tracking mission is successfully completed within 15 s. Figure 3 shows the control inputs of UAVs. Figures 4 and 5 illustrate the changes of the system state variables, that is, distance between UAVs and the target, bearing angle of each UAV, and phase angle between each pair UAVs, from which one can see that they all converge to the equilibrium points within 20 s and remain stable, which shows that the proposed control method is effective to solve the problem.

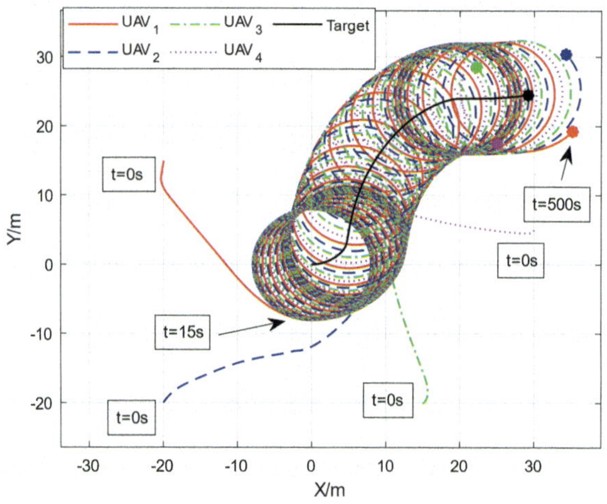

Fig. 2 Trajectories of four UAVs and the target

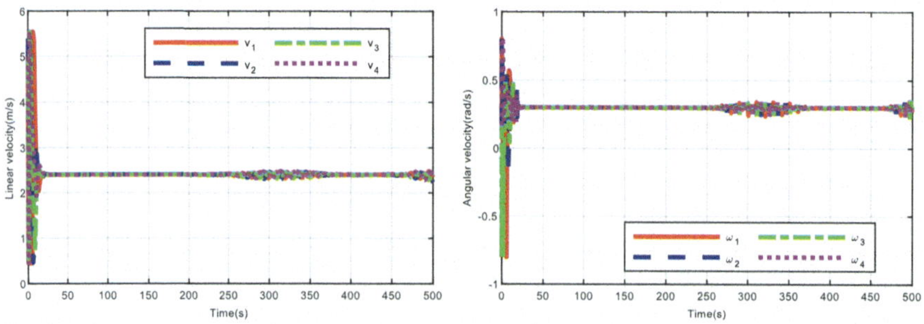

Fig. 3 Control inputs ω_i and v_i of UAVs

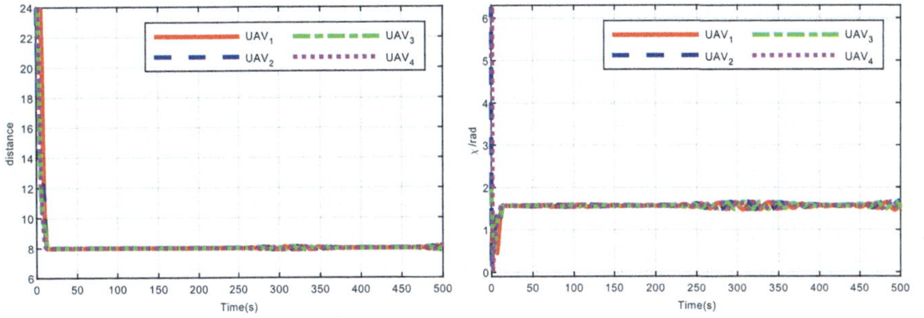

Fig. 4 Distance ρ_i and bearing angle χ_i of UAVs

Fig. 5 Phase angle between each pair UAVs

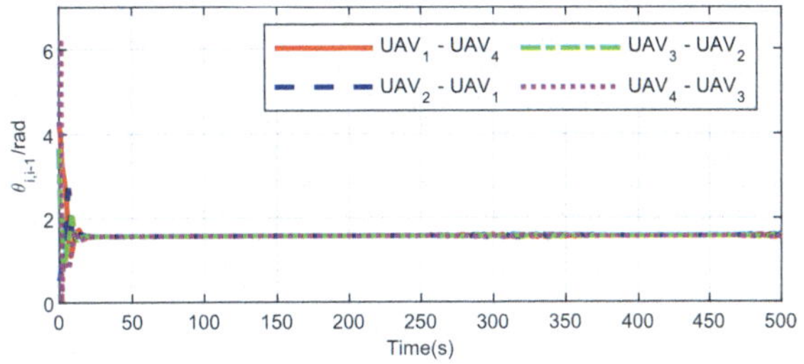

5 Conclusion

This study researches on fixed-time cooperative control for multi-UAV to implement target encirclement and tracking. A fixed-time guidance law is designed for multi-UAV to enable them to track a target-centered circle and maintain a fixed distance from each other. The convergence time is determined only by the control parameters. Finally, a simulation is presented to demonstrate that the control strategy is effective. To be more practical, future work will consider the application of fixed-time control to the case of multi-target tracking by multi-UAV and internal collision avoidance and obstacle avoidance for multi-UAV will be considered.

References

1. Liu X, Ansari N (2018) Resource allocation in UAV-assisted M2M communications for disaster rescue. IEEE Wireless Commun Lett 8:580–583. https://doi.org/10.1109/LWC.2018.2880467
2. Shames I, Fidan B, Anderson BD (2008) Close target reconnaissance using autonomous UAV formations. In: 2008 47th IEEE conference on decision and control. Cancun. pp 1729–1734. https://doi.org/10.1109/CDC.2008.4739440
3. Dong F, You K, Song S (2020) Target encirclement with any smooth pattern using range-based measurements. Automatica 116:108932. https://doi.org/10.1016/j.automatica.2020.108932

4. Doostmohammadian M, Taghieh A, Zarrabi H (2021) Distributed estimation approach for tracking a mobile target via formation of UAVs. IEEE Trans Autom Sci Eng 19:3765–3776. https://doi.org/10.1109/TASE.2021.3135834
5. Jia J, Chen X, Wang W, Wu K, Xie M (2023) Distributed observer-based finite-time control of moving target tracking for UAV formation. ISA Trans 140:1–17. https://doi.org/10.1016/j.isatra.2023.06.017
6. Jia J, Chen X, Wang W, Zhang M (2023) Distributed control of target cooperative encirclement and tracking using range-based measurements. Asian J Control 25:4595–4608. https://doi.org/10.1002/asjc.3124
7. Zhang M, Tian P, Chen X, Wang X (2017) Ground target tracking guidance law for fixed-wing unmanned aerial vehicle: a search and capture approach. J Dyn Syst Meas Contr 139:104503. https://doi.org/10.1115/1.4036563
8. Bai A, Luo Y, Zhang H, Li Z (2022) L2-gain robust trajectory tracking control for quadrotor UAV with unknown disturbance. Asian J Control 24:3043–3055. https://doi.org/10.1002/asjc.2711
9. Xiao F, Wang L, Chen J, Gao Y (2009) Finite-time formation control for multi-agent systems. Automatica 45:2605–2611. https://doi.org/10.1016/j.automatica.2009.07.012
10. Du H, Zhu W, Wen G, Wu D (2017) Finite-time formation control for a group of quadrotor aircraft. Aerosp Sci Technol 69:609–616. https://doi.org/10.1016/j.ast.2017.07.012
11. Kokolakis N-MT, Koussoulas NT (2021) Robust standoff target tracking with finite-time phase separation under unknown wind. J Guid Control Dyn 44(6):1183–1198
12. Fiedler M, Pták V (1962) On matrices with non-positive off-diagonal elements and positive principal minors. Czechoslovak Mathematic J 12:382–400. http://dml.cz/dmlcz/100526
13. Polyakov A (2011) Nonlinear feedback design for fixed-time stabilization of linear control systems. IEEE Trans Autom Control 57:2106–2110. https://doi.org/10.1109/TAC.2011.2179869

Motion Performance Analysis and Application of Quadruped Robot for Safety Production Supervision

Jiazhuo Xuan, Huafeng Luo, Xiaoming Huang, Ye Sheng, Hui Zhou, and Jun Yu

Abstract Aiming at the special application environment of substation safety production supervision and the requirements of the carrying inspection instrument on the motion smoothness, the motion performance of the quadruped robot, and the method to improve the motion smoothness are analyzed. Firstly, the overall design of the quadruped robot for safety production supervision is carried out, and the kinematic analysis of the single leg of the quadruped robot is carried out to derive the motion angle of inverse kinematics. Then, the virtual prototype model is constructed based on Adams software. And driving the joint connection of the drive wheels, the simulation analysis is carried out before and after the horizontal constraint of the quadruped robot body, respectively. The curves of the robot's center-of-mass displacement and velocity, the contact force, the joint torque, the drive torque curve, and the curve of the power consumption are obtained, which proves the idea that keeping the body of the quadruped robot horizontal can help to improve the quadruped robot kinematics performance in the inspecting process. Finally, the developed prototype is successfully applied in the substation safety supervision.

1 Introduction

With the continuous deepening of smart grids and the rapid development of robotics technology, various types of robots are widely used in safety production supervision in substations, transmission lines, transmission towers, and other areas. Robots can accurately identify and locate power equipment through autonomous navigation

J. Xuan (✉) · H. Luo · X. Huang
State Grid Zhejiang Electric Power Co.Ltd., Electric Power Research Institute, Hangzhou 310014, China
e-mail: xjz01@163.com

Y. Sheng · H. Zhou · J. Yu
State Grid Zhejiang Electric Power Co.Ltd., Hangzhou 310007, China

S. Yadav et al. (eds.), *Proceedings of 3rd International Conference on Artificial Intelligence, Robotics, and Communication*, Lecture Notes in Electrical Engineering 1172, https://doi.org/10.1007/978-981-97-2200-6_28

and perception technology. Improved the efficiency and stability of safety production supervision, gradually replacing manual safety production supervision of power equipment [1–3].

From the perspective of mobility, intelligent safety production supervision robots can be divided into wheeled robots, tracked robots, and rail-mounted robots, which are used to adapt to different application scenarios [4]. In the safety production supervision of substations, mobile robots are equipped with various sensors such as infrared thermal imagers, visible light CCD cameras, LiDAR, etc. Which can be operated independently or remotely, safety supervision of outdoor high-voltage equipment can be carried out in unmanned or few manned substations to timely detect unsafe hazards in the operation of substation equipment. Most of these robots use four-wheel mobile robots [5–9]. They perform well on flat and inclined surfaces, but have relatively poor abilities in obstacle crossing, ditch crossing, and climbing stairs, making them more suitable for walking on fixed flat lines in outdoor substations. Due to the height difference between the installation position of the equipment in the substation and the road surface. In response to the shortcomings of the aforementioned robots, reference [10] constructed a three-dimensional safety production supervision system for substations with wheeled robots, track robots, and fixed point cameras as the core, but invisibly increased the cost of safety supervision. Therefore, in this situation, it is necessary to consider using other types of robots to complete safety supervision tasks. A feasible option is a legged robot, which uses multiple independently controllable legs to move and can adapt to various terrains, including uneven roads, steps, and large height differences. These robots are usually equipped with stable attitude control and intelligent navigation systems, enabling them to safely traverse complex terrain and observe the operation status of substation equipment up close, achieving substation safety production supervision [11–14]. The footed robot Jueying Mini Lite developed by Hangzhou Yunshen Technology Company has flexible running and jumping skills and a comprehensive integration of intelligent technology and sports ability. It is equipped with multiple sensors such as LiDAR, depth camera, wide-angle camera, and ultrasound to achieve intelligent safety production supervision in substations [15, 16].

The walking methods of wheeled and tracked inspection robots have their advantages and disadvantages, but they cannot meet the safety production supervision needs of substations. Although legged robots can meet the inspection needs of substations, there is still room for improvement. This article presents an effective method to improve the motion performance of quadruped robots. By keeping the quadruped robot's body level, the robot's grip ability is increased, wear between joints is reduced, acceleration performance is improved, power consumption is reduced, and the efficiency of intelligent safety production supervision in substations can be effectively improved.

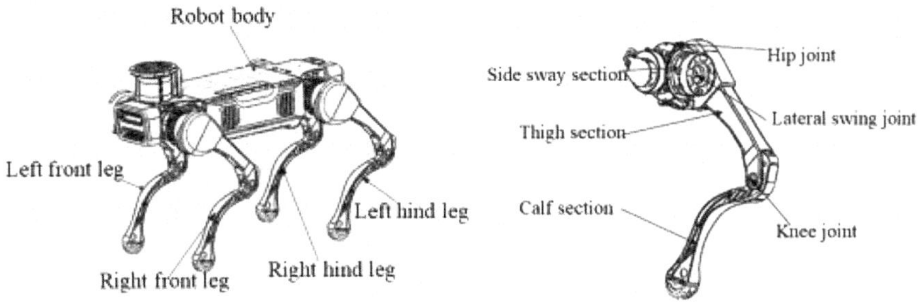

Fig. 1 Overall and leg structure of the robot

2 Structural Design of Safety Production Supervision Robot for Foot-Style Substation

The structure of the foot-type substation safety production supervision robot mainly consists of the robot body and four symmetrical leg units, the robot body can be equipped with various safety production supervision equipment required by the substation, each leg unit consists of three parts, namely the side sway section, thigh section, and calf section, the three parts of the robot body and each leg unit are connected through three joints, namely the lateral swing joint, hip joint, and knee joint. The entire robot is composed of twelve joints. The specific structure is shown in Fig. 1.

3 Kinematic Analysis

3.1 Establishment of the D–H Coordinate System

Establish the D–H coordinate system for the right front leg of a quadruped robot as shown in Fig. 2, and the coordinate system parameters are shown in Table 1. Establish the base coordinate system on the lateral pendulum joint, with the vertical paper facing inward as the X-axis of the base coordinate system, and the direction of the lateral pendulum joint axis as the Z-axis of the base coordinate system. Determine the Y-axis direction according to the right-hand rule and fix the first coordinate system on this base coordinate system.

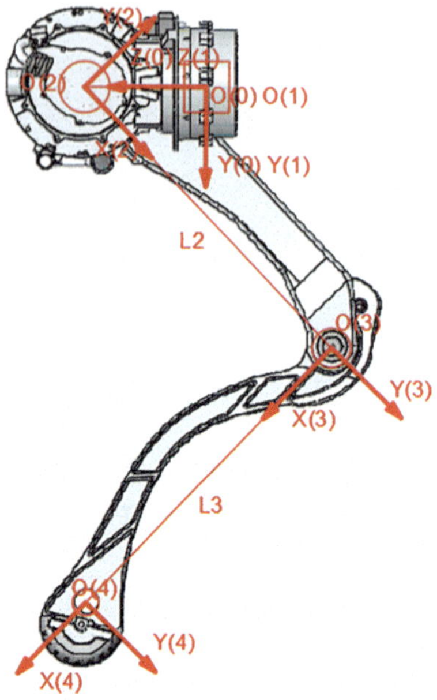

Fig. 2 Single-leg D–H coordinate system of the robot

Table 1 Single-leg D–H parameters of the robot

Connecting rod serial number	Rod length ai/mm	Torsional angle $\alpha i/(°)$	Joint angle $\theta i/(°)$	Distance d/mm
1	L_1	0	$\theta 1$	0
2	L_2	90	$\theta 2$	0
3	L_3	0	$\theta 3$	0
4	0	0	0	0

3.2 *Forward Kinematics Solution*

According to the solution process of forward kinematics, by solving each translation matrix and rotation matrix, the general solution of the homogeneous transformation matrix between adjacent coordinate systems can be obtained, as shown in Eq. (1)

$$
{}_{i-1}^{\ \ i}A = \begin{pmatrix} \cos\theta_i & -\sin\theta_i & 0 & L_i\cos\theta_i \\ \sin\theta_i & \cos\theta_i & 0 & L_i\sin\theta_i \\ 0 & 0 & 1 & 0 \\ 0 & 0 & 0 & 1 \end{pmatrix} \tag{1}
$$

By substituting the data from Table 1 into the general solution of the homogeneous transformation matrix, we can obtain ${}_0^1A$, ${}_1^2A$, ${}_2^3A$, ${}_3^4A$. Multiplying the homogeneous transformation matrices yields the homogeneous transformation matrix from the base coordinate system to the fourth coordinate system, as shown in Eq. (2).

$$
{}_0^4T = {}_0^1A\,{}_1^2A\,{}_2^3A\,{}_3^4A
= \begin{pmatrix} \cos(\theta_1+\theta_2+\theta_3) & -\sin(\theta_1+\theta_2+\theta_3) & 0 & L_1\cos\theta_1 + L_2\cos(\theta_1+\theta_2) + L_3\cos(\theta_1+\theta_2+\theta_3) \\ \sin(\theta_1+\theta_2+\theta_3) & \cos(\theta_1+\theta_2+\theta_3) & 0 & L_1\sin\theta_1 + L_2\sin(\theta_1+\theta_2) + L_3\sin(\theta_1+\theta_2+\theta_3) \\ 0 & 0 & 1 & 0 \\ 0 & 0 & 0 & 1 \end{pmatrix} \tag{2}
$$

3.3 *Inverse Kinematics Solution*

The expression for the robot foot-end coordinate $p_0 = \left(p_x\ p_y\ p_z\right)^T$, p_x, p_y, p_z is shown in Eqs. (3 and 4).

$$p_x = L_1\sin\theta_1 + L_2\sin(\theta_1+\theta_2) + L_3\sin(\theta_1+\theta_2+\theta_3) \tag{3}$$

$$p_{y=}L_1\cos\theta_1 + L_2\cos(\theta_1+\theta_2) + L_3\cos(\theta_1+\theta_2+\theta_3) \tag{4}$$

$$p_z = 0 \tag{5}$$

By calculation, the inverse kinematics formula can be obtained as shown in Eqs. (6–8).

$$\theta_1 = \arctan\frac{p_y}{p_x} \tag{6}$$

$$\theta_2 = \arctan\frac{p_y - L_1\cos\theta_1}{p_x - L_1\sin\theta_1} \tag{7}$$

$$\theta_3 = \arctan\frac{(p_y - L_1\cos\theta_1) - L_2\cos(\theta_1+\theta_2)}{(p_x - L_1\sin\theta_1) - L_2\sin(\theta_1+\theta_2)} \tag{8}$$

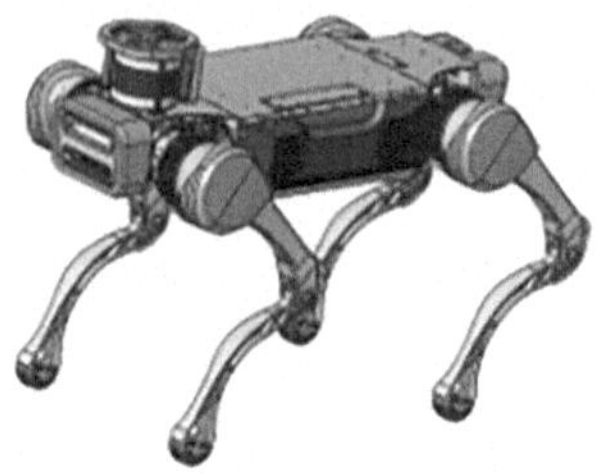

Fig. 3 Overall model of the robot

4 Virtual Prototype and Simulation Analysis

4.1 *Virtual Prototype*

The quadruped robot forms the overall framework of the robot through the connection of four legs and the body, with each leg consisting of three joints: the lateral swing joint, hip joint, and knee joint. The quadruped robot discussed in this article has a total of 12 joints, each with 1° of freedom for a total of 12° of freedom. As shown in Fig. 3.

4.2 *Simulation Analysis*

Establish a quadruped robot model in ADAMS, create a plane below the robot, and add gravity to the simulation to fix the plane to the ground. It is necessary to ensure that the bottom of the foot of the quadruped robot model coincides with the plane when it is stationary, to prevent errors caused by penetrating the model or slipping during the simulation process. Add angular displacement drive on the rotating pair. Ignoring the lateral joint, set the simulation time (End time) to 5 s and the simulation step size (step) to 500.

The displacement curves of the robot's center of mass in the three directions of XYZ are shown in Fig. 4. Figure 4a and b show the displacement curves of the robot's center of mass 1 and 2, respectively. The X-direction is the forward direction of the robot, and the displacement of the center of mass in the Y-direction (vertical direction) is basically 0, which can be ignored. This indicates that the quadruped robot maintains stability during motion; the displacement of the center of mass in the Z-direction varies slightly. By comparing the displacement of the centroids of robots 1 and 2, it can be found that Robot 1 is more stable in the X-direction of motion than Robot B, and the left and right swing amplitude in the Z-direction can be ignored, while Robot 2 has a more significant left and right swing amplitude in the Z-direction. This indicates that keeping the robot's body level can make the displacement curve of the robot more stable.

Fig. 4 Displacement of robot mass center in XYZ-direction

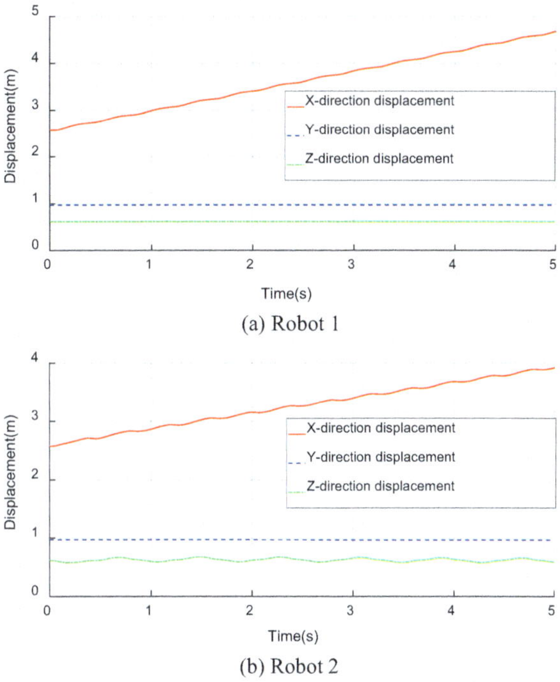

The velocity curve of the quadruped robot's center of mass in the XYZ-direction is shown in Fig. 5. Robot 1 moves in the X-direction, and the velocity curve is closer to a sine curve, with zero velocity in the YZ-direction. It can be clearly observed that Robot 2, which has not maintained its body level, has velocities in both the X- and YZ-directions, and the velocity curves in all three directions are irregular.

The contact force curve of the quadruped robot is shown in Fig. 6, which includes friction in the X-direction and gravity in the Y-direction. Compared to Robot 1, Robot 2 not only has friction in the X-direction and gravity in the Y-direction but also has friction in the Z-direction. This indicates that the motion direction of Robot 2 does not completely follow the set X-direction and still swings left and right in the Z-direction. Comparing the values of the friction curves in the two graphs, the X-direction friction force of Robot 1 remains around 200 N, while the friction force of Robot 2 is less than 200 N, slightly smaller than Robot 1. This results in a lower probability of slipping during the walking process of Robot 1 compared to Robot 2, indicating good operational stability.

Figures 7 and 8 show the joint torque curves of the hip and knee joints of a quadruped robot. The joint torque curve is a periodic curve with a period of approximately 0.8 s. The hip joint torque curve of Robot 1 first decreases and then increases in the X-direction within one cycle, then decreases and finally increases. After the curve stabilizes, the maximum value is 2 Nm and the minimum value is about 0.9 Nm. The maximum value of the knee joint torque curve after stabilization is 0.35 Nm, and

Fig. 5 Velocity of robot mass center in XYZ-direction

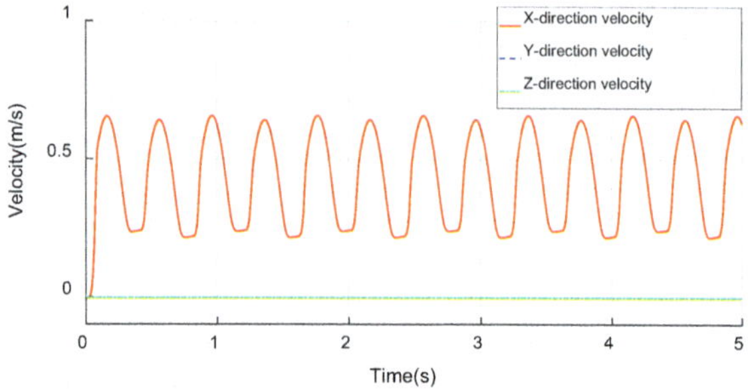

(a) Robot 1

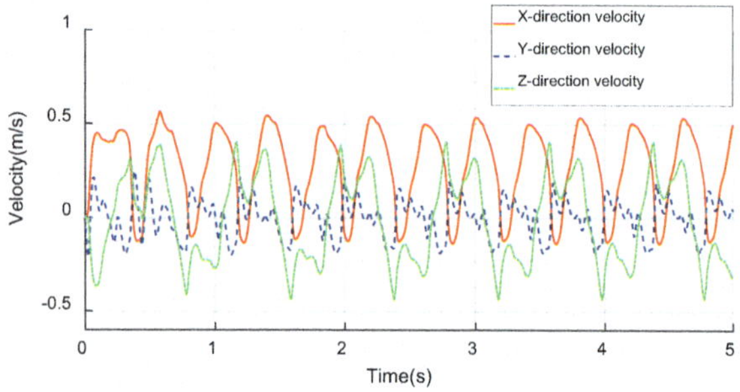

(b) Robot 2

Fig. 6 Contact force of the robot in XYZ-direction

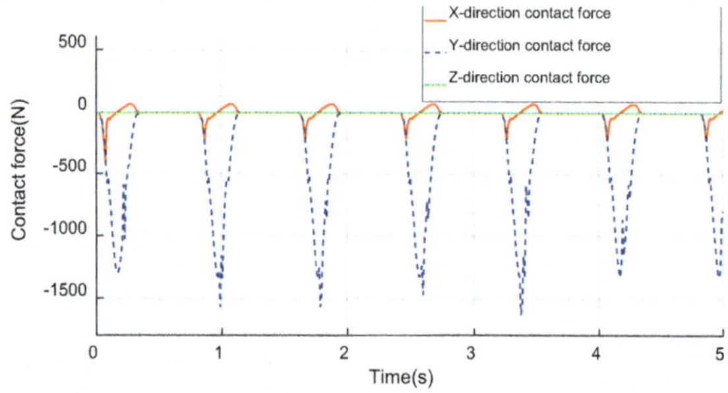

(a) Robot 1

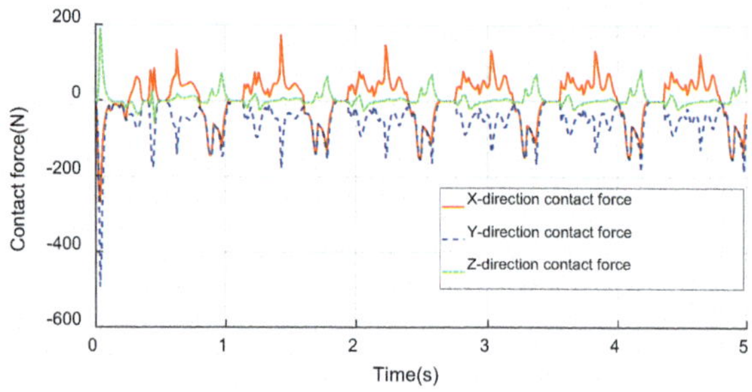

(b) Robot 2

Fig. 7 Joint moment of robot hip joint in XYZ-direction

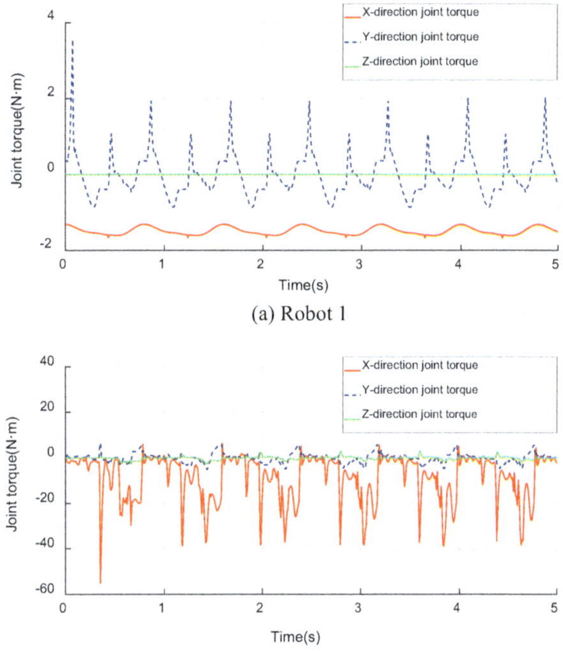

(a) Robot 1

(b) Robot 2

the minimum value is −0.17 Nm. The maximum value of the hip joint torque curve of Robot 2 after stabilization is 7 Nm, and the minimum value is about −39 Nm. The maximum value of the knee joint torque curve after stabilization is 32 Nm, and the minimum value is −3.5 Nm.

By comparison, Robot 2 has increased the joint torque in the Z-direction, and the absolute value of the joint torque is much greater than Robot 1. It can be seen that when the robot body is not kept horizontal, the robot will tilt, and the wear between joints is more severe, which reduces the service life of the robot.

The driving torque of the hip and knee joints of a quadruped robot is shown in Figs. 9 and 10. The robot moves in the X-direction, and according to the right-hand rule, the direction of the driving torque can be determined to be the Z-direction. The maximum value of the stable hip joint driving torque curve of Robot 1 is about 110 Nm, and the maximum value of the stable knee joint driving torque curve is about 305 Nm. The maximum value of the stable hip joint driving torque curve of Robot 2 is about 75 Nm, and the maximum value of the stable knee joint driving torque curve is about 68 Nm. The driving torque of Robot 1 is greater than that of Robot 2.

The power consumption curve of the quadruped robot is shown in Fig. 11. It can be clearly seen from the figure that the power consumption of Robot 1 is lower than that of Robot 2. When the body of the quadruped robot is kept horizontal, the

Fig. 8 Joint moment of the robot knee joint in XYZ-direction

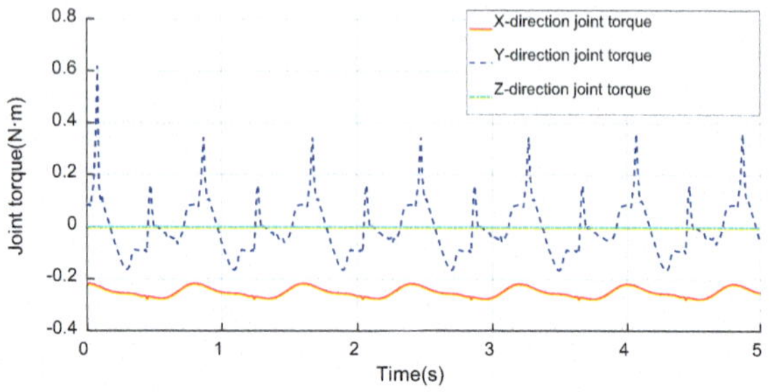

(a) Robot 1

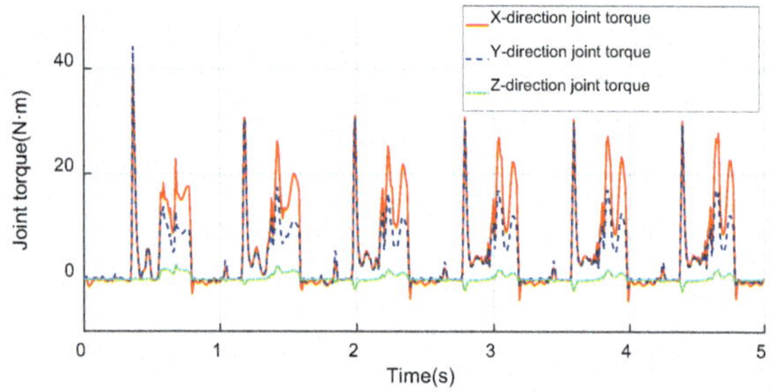

(b) Robot 2

Fig. 9 Driving torque of robot hip joint in XYZ-direction

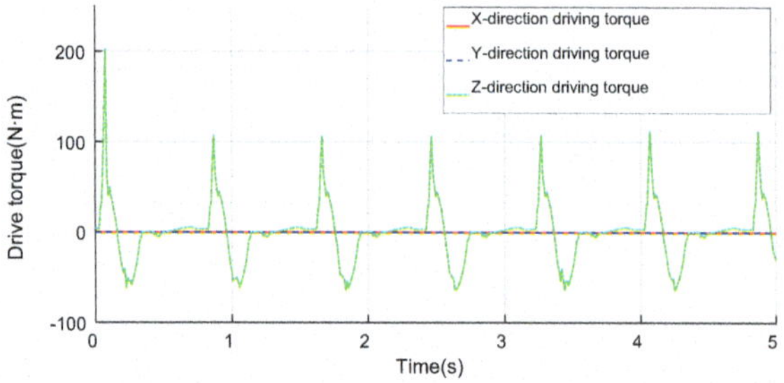

(a) Robot 1

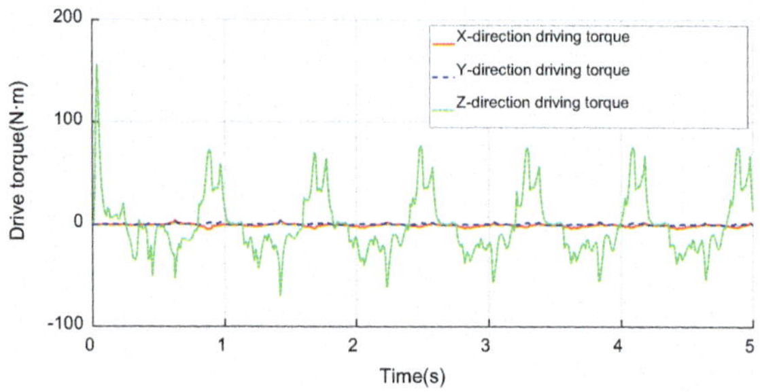

(b) Robot 2

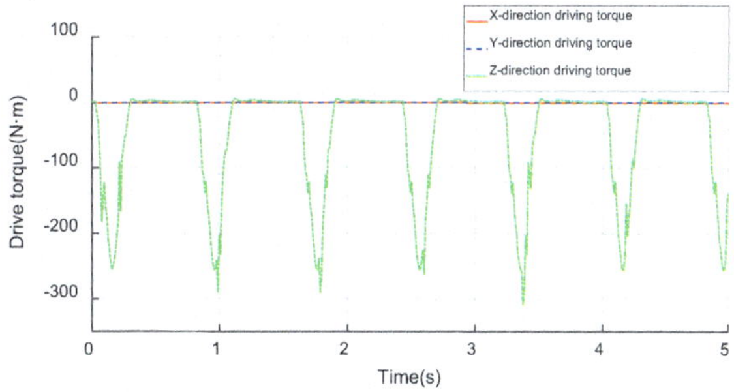

(a) Robot 1

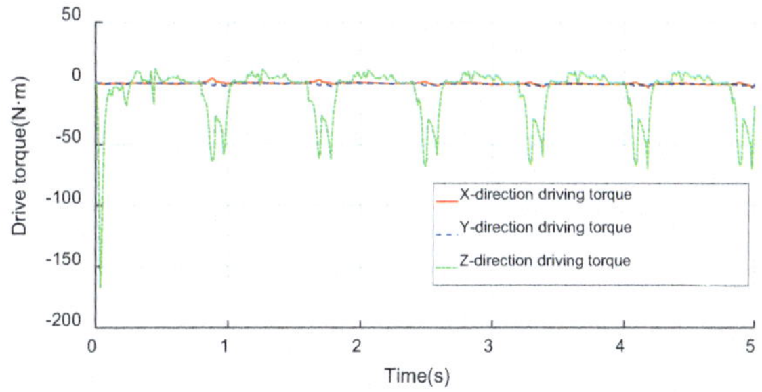

(b) Robot 2

Fig. 10 Driving torque of robot knee joint in XYZ-direction

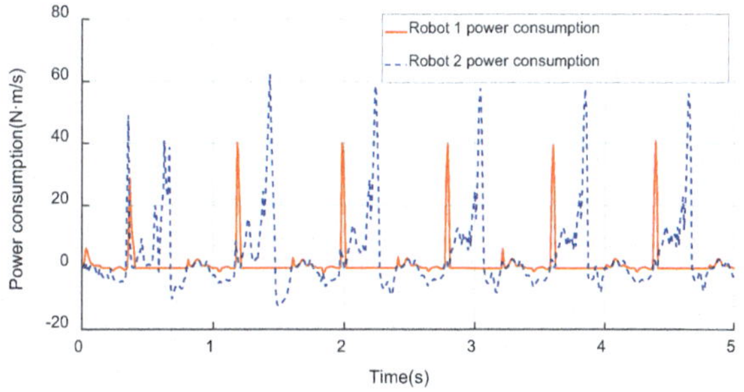

Fig. 11 Robot power consumption comparison

power consumption of the robot is smaller, which increases the endurance time of the quadruped robot and significantly improves its service life.

When the robot body is kept horizontal, the displacement and velocity curves of the robot's center of mass are smoother, and the components in other directions are almost zero, resulting in good motion stability; the robot has greater contact force, effectively increasing its grip ability and reducing the probability of slipping; The joint torque has no Z-direction component, which can prevent the robot from tilting and reduce wear between joints; the robot has a greater driving torque, which enables it to have better acceleration performance and performs better when carrying heavy objects or climbing slopes; robots have lower power consumption and longer endurance under the same conditions. To ensure better motion stability of quadruped robots under complex road conditions in substations, it is advisable to keep the robot body in a horizontal position as much as possible.

5 Physical Prototype and Its Applications

To verify the effect of keeping the body level of a quadruped robot on its motion performance, a substation safety production supervision robot was developed and applied in a certain substation. The actual safety production supervision image is shown in Fig. 12.

During the safety production supervision process, the robot experienced a series of challenges such as obstacle crossing, climbing stairs, and climbing gravel roads. The measured results of the displacement sensor installed on site are shown in Fig. 13. When the robot is moving, this almost zero left and right swing and relatively small up and down swing amplitude indicate that the robot has excellent stability and control ability. It can effectively suppress unnecessary swaying and bumps, improving the smoothness and reliability of the robot during driving. Enable it to better adapt to various situations and terrains, improve work efficiency, and reduce potential risks. The robot can smoothly climb a 45° slope and maintain stable driving on gravel roads. Robots perform well in a series of challenges such as driving on smooth roads, climbing stairs, and climbing gravel roads. This means that the robot can maintain a relatively stable driving posture and move forward smoothly.

Fig. 12 Quadrupedal robot substation inspection diagram

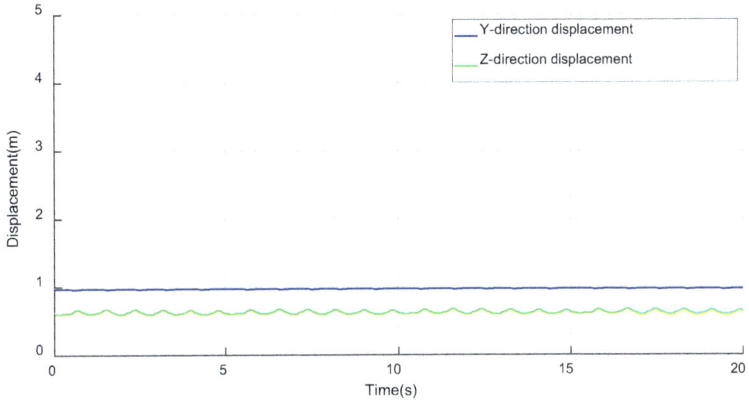

Fig. 13 The displacement curve of the robot's center of mass in the YZ-direction

6 Conclusion

(1) A detailed analysis was conducted on the special application environment of safety production supervision in substations, as well as the walking characteristics of common wheeled robots, tracked robots, and rail-mounted robots. It was pointed out that legged robots can better adapt to the special road surface of substations, and the structural parameters of quadruped robots were designed.
(2) A virtual model of a quadruped robot for substation safety production supervision was established, and virtual simulation analysis was conducted using ADAMS software. The displacement, velocity, contact force, joint torque, driving torque, and power consumption curves of the center-of-mass motion were obtained. The simulation results show that adding horizontal constraints to the robot body will help improve the motion performance of quadruped robots.
(3) A substation safety production supervision robot has been developed and applied in a certain substation. Practical applications have shown that quadruped robots with horizontal constraints on the robot body have good abilities in obstacle crossing, climbing, and walking on gravel roads, as well as improved load-bearing capacity and endurance time.

Acknowledgements Fund project: science and technology project of State Grid Zhejiang Electric Power Co., Ltd. (5211DS23000Q).

References

1. Jin D, Liu XG, Shi G, Song RP, Zu MX (2023) 3D point cloud registration method for substation robot patrol tracks. Infrared Technol 45:678–683
2. Huang S, Liu XF, Xie T (2023) Application of intelligent patrol robot in substation. Appl IC 40:156–157
3. Wei XD, Li Y, Li Q, Mei J, Li QM (2020) Development of climbing inspection robot for power towers. Electr Power 53:162–167
4. Ma AM (2023) Application of intelligent safety inspection robots in equipment operation and maintenance. Equipment Manage Maintenance 8:156–159
5. Li FP, Zeng HX, Wang X (2023) Obstacle avoidance path planning method of mechanical arm of wheeled inspection robot in substation. Mach Electron 41:32–36
6. Xiao X, Li XH, Luo HX, Zhou H, Liang JM, Xia M (2022) An identification method of the plate in the blind zone of substation inspection robots and its application. Guangxi Electr Power 45:36–41
7. Li J, Chen XM. Discussion on the technical specifications for inspection of wheeled inspection robots in substations. Automa Appl 2:60–62
8. Jin YC (2021) Research on four-wheel inspection robot for substation. Nanjing Forestry University, Nanjing
9. Xu SB, Li SC, Liang QH (2017) System development and analysis of the live-line maintenance tracked mobile robot navigation system. Mach Des Res 33:26–29
10. Zhang CH, Lu ZH, Liu XC (2021) Joint inspection technology and its application in a smart substation. Power Syst Prot Control 49:158–164
11. Chen ZY, Tu QZ, Zhang XP, Pan M, Jiang CM, Huang H (2020) Review of multi-legged crawling robot. J Ordnance Equipment Eng 41:1–12
12. Guo J, Zhao Y, Chen JC (2021) Kinematics analysis and simulation of quadruped robot based on gait planning. Modular Mach Tool Autom Manuf Tech 7:57–61+65
13. Han XJ, Shang LY, Yang Y (2015) Centroid position planning and stability analysis of a quadruped robot. Comput Integr Manuf Syst 32:308–313+359
14. Ma GY, Wang GM, Liu RC, Yao YL (2021) Structure design and kinematics analysis on the serial-parallel leg of quadruped robots. J S China Univ Technol Nat Sci Ed 49:103–112
15. Deep in the Cloud (2021) Deep in the cloud releases kung fu version of robot dog. Robot Technol Appl 02:15
16. Deep in the Cloud (2022) Endless shadow X20 breaking record for extreme operating load. Robot Technol Appl 02:4

Cross-Modal Active Speaker Detection Algorithm in Video and End-To-End Landing Solution

Yatao Yang and Siyu Yan

Abstract This research paper delves into the realm of active speaker detection (ASD) in video content with the primary objective of accurately identifying one or more active speakers. In contrast to traditional speech recognition, ASD focuses on verifying the sources of sound rather than the specific content of speech. Unlike speaker diarization, which categorizes speech segments by distinct speakers, ASD precisely identifies the active speaker in scenarios involving multiple speakers. The proposed OneStage architecture adopts a self-attention-based network to dynamically integrate cross-modal information. Notably, the incorporation of differential images serves to diminish the reliance on visual information. To alleviate the burden of annotation, a novel random ratio mask annotation and audio-visual asynchronous data augmentation are introduced, significantly reducing the annotation workload and enhancing the generalization of the model. Furthermore, the paper introduces the OnePass architecture tailored for lightweight real-time speaker detection on small terminal devices. With a modest 5 million parameters, 0.05 GFLOPs computation, and an impressive accuracy rate of 89.5%, the OnePass architecture achieves real-time speaker detection. Meanwhile, the OneStage architecture demonstrates exceptional performance with a 95.98% area under the curve (AUC) on the test set. These findings showcase the efficacy of the proposed architectures in advancing the field of active speaker detection in videos.

1 Introduction

Active speaker detection (ASD) aims to identify individuals speaking in a visual scene from a video, distinguishing it from speech recognition. Unlike speech recognition, ASD focuses on recognizing and confirming the speaker's identity rather than transcribing the spoken content. It assigns labels of "speaking" or "not speaking" to

---

Y. Yang · S. Yan (✉)
Shenzhen University, Shenzhen 518060, China
e-mail: yansiyu2021@email.szu.edu.cn

S. Yadav et al. (eds.), *Proceedings of 3rd International Conference on Artificial Intelligence, Robotics, and Communication*, Lecture Notes in Electrical Engineering 1172, https://doi.org/10.1007/978-981-97-2200-6_29

each person in the video clip. Furthermore, ASD is distinct from speaker diarization, where the former specifically identifies the active speaker in a multi-speaker environment, while the latter concentrates on differentiating between speakers and identifying their respective speech segments.

Currently, there are three primary methods for active speaker detection: visual-only, audio-only, and audio-visual approaches. Visual-only approaches struggle to differentiate between speaking and non-speaking lip movements (e.g., eating, yawning) and can be influenced by irrelevant facial and lip motions. On the other hand, audio-only approaches are susceptible to interference from background sounds or noise.

Mainstream frameworks, employing cross-modal methods based on audio and video, typically use the two-stream architecture. In this paper, we introduce a novel cross-modal fusion structure tailored for ASD. Additionally, we present an end-to-end solution designed for implementation on small terminals or edge computing devices. This approach not only enhances the accuracy of active speaker detection but also ensures its applicability in resource-constrained environments.

Our contributions can be summarized as follows:

(1) Proposed a cross-modal fusion network architecture with self-attention to integrate information from different modalities, modeling temporal sequences, and dynamically adjusting attention weights.
(2) Introduced ResMap as a replacement for RGB images in video modality, reducing reliance on visual info and focusing on mouth region for essential motion info.
(3) Introduced random ratio masking annotation method for weak labeling and mask learning, reducing annotation workload while maintaining learning effectiveness.
(4) Utilized asynchronous audio-visual data augmentation techniques to generate time slices and mix them, enhancing model's generalization capabilities.
(5) Proposed a lightweight speaker detection network architecture for small-scale terminals or edge computing devices, achieving real-time effectiveness for active speaker detection solutions in resource-constrained environments.

2 Related Work

Visual-Only Approaches. Visual-only methods not only include face detection and recognition, which are used for speaker identification through facial recognition technology, but also involve more complex techniques. Pose estimation, for instance, infers speaker identity through the analysis of body posture, such as head orientation and movements. Additionally, deep learning methods, using models like convolutional neural networks (CNN) [1, 2], recurrent neural networks (RNN), or transformers [3], are employed to learn speaker features from video data.

Audio-Only Approaches. Audio methods encompass speech activity detection, determining speaker presence through the analysis of energy or spectral features in audio signals. Speaker recognition utilizes speech recognition technology to identify the current speaker. Voiceprint recognition involves recognizing speakers by analyzing voice characteristics like tone, pitch, and speech rate. To enhance the accuracy and efficiency of these methods, deep learning models are frequently employed to extract advanced features from audio, aiding in both speaker detection and recognition tasks.

Audio-Visual Approaches. Studies [4] concatenate audio and visual features without considering temporal info. [5] proposes an integrated network using 3D CNN, time convolution, and LSTM for ASD prediction, showing improvement. ASDNet [6] encodes visual info using 3D-ResNet101, audio using SincNet [7], and models relationships using MLP. TalkNet [8] explores long-term features in active speaker detection tasks. UniCon [9] aims to improve audio-visual active speaker detection by unifying contextual info. SPELL [10] introduces spatiotemporal graph learning framework, encoding each person into frame nodes. Edges represent spatial or temporal interactions, simplifying ASD into node classification task and reducing memory consumption and computational complexity.

Audio Feature Extraction. MFCCs are a crucial method in speech signal processing and recognition. Introduced in 1976 by Markel and Gray [11], MFCCs transform speech signals into Mel-frequency scales and capture features like resonant peaks, pitch, and speech articulation using cepstral analysis. Ongoing research has refined MFCC calculation processes, making it a standard feature extraction method.

Cross-modal Fusion Methods. Cross-modal fusion methods can be categorized into model-agnostic and model-based methods. Model-agnostic methods include early fusion (feature-based), late fusion (decision-based), and hybrid fusion. Early fusion extracts features separately and merges them at the feature level. Late fusion trains models separately and merges results from multiple outputs. Hybrid fusion combines advantages of both but increases complexity and training difficulty. Model-based methods address cross-modal fusion from a technical and model perspective using methods like MKL [12], GM [13], NN [14], etc.

3 Method

ASD's ability to identify speakers lies in the fundamental combination of information from the visual and audio branches. The visual branch supplies lip motion information, while the audio branch provides audio-vibration information. These two sources of information are then integrated to make a determination (see Fig. 1).

3.1 Architecture

The entire network architecture is illustrated in the diagram (see Fig. 2), referred to as the OneStage framework. The visual modality takes the facial difference map

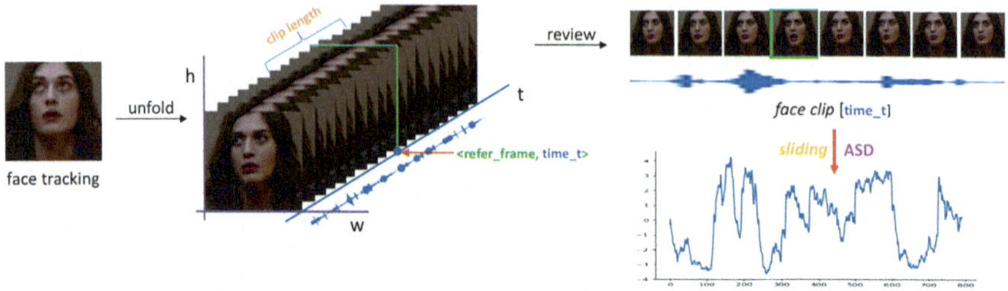

Fig. 1 Binary classification of facial slices with accompanying audio: determining whether the person in the current face clip is speaking or not

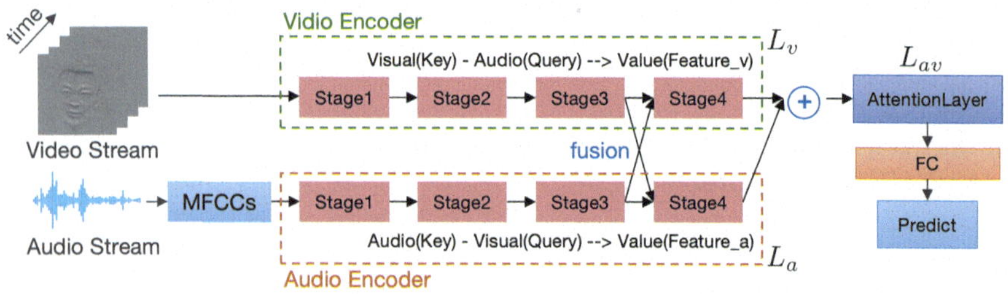

Fig. 2 The network architecture of OneStage

(ResMap) sequence as input, while the audio modality takes the Mel-frequency cepstral coefficients (MFCCs) of the audio. After partial fusion through two separate encoders, the data is fed into a self-attention layer. By introducing an attention mechanism, the model can adaptively focus on the most crucial modality or features for the current task. Additionally, this approach allows for the removal of the explicit temporal modeling module, reducing the overall model complexity.

Audio-Visual Encoder. Only two levels of The Video Encoder utilizes ResNet18, while the Audio Encoder employs ResNet34, primarily to maintain consistency with mainstream ASD network architectures for ease of comparison. In the Encoder, after encoding through the initial three stages, the first fusion of audio and video information takes place. At this stage, the model has already learned a substantial amount of feature information from both modalities, facilitating the retention of unique information from each modality. Simultaneously, it allows the network to learn the correlated information between the two modalities.

During the fusion process, a self-attention mechanism is employed to handle sequence information. For the fourth stage of the Video Encoder, the input involves using the video modality information from the first three stages as Key and the audio information as Query for computation. Similarly, for the fourth stage of the Audio Encoder, the input involves using the video modality information from the first three stages as Key and the audio information as Query for computation.

Cross-modal Fusion. In the backbone network of active speaker detection algorithm models, an attention mechanism is introduced using a transformer-based self-attention network structure. The attention mechanism allows the model to dynamically adjust its focus on each modality when processing inputs from different modalities. For instance, in the fusion of video and audio modalities, when the model processes video images, the attention mechanism can prioritize regions in the image that are relevant to the audio.

Loss Function. We propose an innovative strategy involving weak annotation and mask learning, which significantly alleviates the burdensome task of frame-by-frame labeling, offering a more flexible and efficient solution for data annotation. Specifically, our approach randomly erases segments of random sequence lengths based on a specified ratio. The erased portions no longer contribute to loss calculation during the training process. This measure not only reduces the complexity of annotation but also greatly saves time and human resources in the labeling task.

The weighted cross-entropy loss allocates different weights to samples of different classes, directing the model's attention toward the accurate classification of key categories. The calculation is as follows:

$$L(y, \hat{y}) = -\sum_{i} w_i \cdot y_i \cdot \log(\hat{y}_i)$$

In this context, w_i represents the weight assigned to the i-th class. The default weight distribution for positive and negative samples in the cross-entropy loss is 1:1. However, due to the approximate 1:3 ratio between positive and negative samples (speaking: non-speaking) in the entire dataset, the weight distribution in the cross-entropy loss is adjusted to 3:1. This modification aims to focus the loss more on positive examples. For categories with a larger number of samples, a smaller weight penalty is applied, while categories with fewer samples receive a larger weight penalty. This approach penalizes the network more heavily for misclassifying labels in categories with fewer samples, alleviating the issue of sample imbalance. The total loss includes the loss of the audio part (L_a), the loss of the video part (L_v), and the loss of the fused part (L_{av}):

$$Loss = 0.4 \times L_a + 0.4 \times L_v + L_{av}$$

3.2 *End-To-End Landing Solution*

As illustrated in the figure below (see Fig. 3), we have designed a simpler, faster, and more effective end-to-end pipeline for an active speaker detection solution, enabling real-time efficiency. The pipeline begins by performing scene detection on the video frames, segmenting them into different scene slices. Simultaneously, features are extracted from the audio of the input video. Next, face detection and tracking are

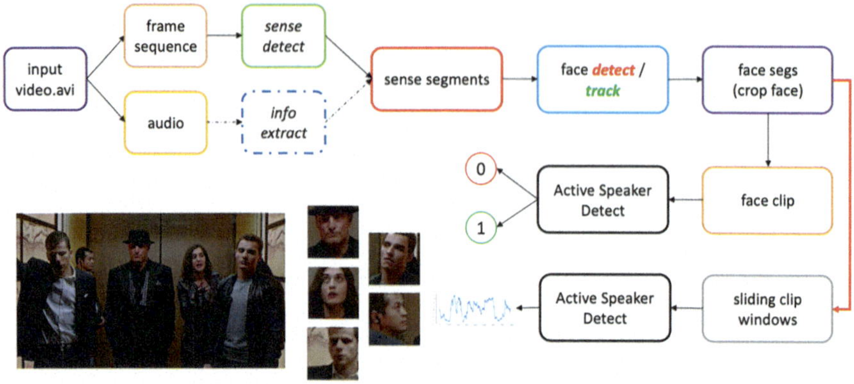

Fig. 3 The pipeline of OnePass

applied to each scene's video frames, producing a sequence of facial slices. After merging the visual features of the facial slices from the video frames with their corresponding audio features, temporal modeling is performed, ultimately yielding the detection results. We refer to this end-to-end solution as the OnePass architecture, which is lighter in comparison to the aforementioned OneStage architecture. The OnePass architecture is well-suited for deployment on edge devices with limited computational power, maintaining high accuracy and performance.

Audio Feature Extraction. To digitize analog audio signals, spectrograms are used for conversion into images. MFCCs are derived by applying a cosine transform to the Mel spectrogram, retaining elements 1–13. This 13-element vector serves as the MFCC representation, used as the audio feature input for the neural network.

Scene Detection. Automated segmentation of video shots into scenes is achieved by detecting transitions between shots. In this approach, content-based scene detection is used to segment video scenes, sequentially comparing frames to identify content changes and aid in rapid cut detection.

Face Detection. Face detection is performed for each video frame using the lightweight and fast FaceBoxes [15] model. Detected faces are cropped based on bounding box coordinates, and the resulting sequence of face images is encoded and fed into the network. FaceBoxes operates at 20 frames per second on a CPU for real-time processing and can reach 125 frames per second on a GPU, ensuring high precision with low computational load.

Network Architecture. Refining the algorithm for cross-sectional comparison involves initially using a voluminous backbone. However, for edge deployment with maintained performance, downsizing efforts focus on the encoder. The network structure is similar to OneStage, but the encoder backbone adopts the lightweight MobileNet [16], and adjustments to channels and network layers enable rapid downsizing.

Implemented Logical Architecture. In edge scenarios, a dual-thread approach is used to process video and audio data while handling model inference (see Fig. 4).

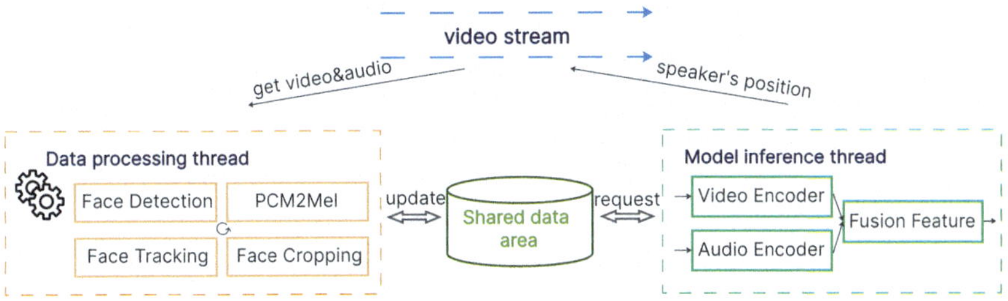

Fig. 4 Implemented logical architecture

The data-processing module processes video frames sequentially, including face detection, tracking, and cropping. Audio data are converted to MFCCs. The data capture module stores data in a shared queue to meet algorithm requirements within limited display memory. The model inference module uses data from the shared area for inference.

Results. The model boasts a parameter count of 5 M and a computational load of 0.05 GFLOPs. On our validation set, it achieves an accuracy (ACC) of 88.4%, showcasing its lightweight nature coupled with commendable performance. Notably, this model achieves real-time detection on GPU, demonstrating its efficacy in resource-intensive scenarios.

4 Experiments

4.1 Benchmark

Our dataset is sourced from the AVA—-Active Speaker Dataset [17], Columbia Active Speaker Dataset [18], and our own annotated data. The AVA—ActiveSpeaker Dataset comprises 38,500 face trajectories, 3.65 million face images, and corresponding audio. The Columbia Active Speaker Dataset consists of an 87-min group discussion video involving five group members engaged in interactive discourse.

4.2 Implementation Details

We resize each face to 112 × 112, and the MFCC dimension is set to 13. The final architecture is implemented using PyTorch [19], and all experiments are conducted on six NVIDIA RTX 3090 GPUs. The models employ the Adam optimizer across 25 training epochs, with a learning rate set at 0.0001 and a decay rate of 0.05 per epoch.

Data Augmentation. Video data augmentation includes common image techniques like flipping, rotation, and cropping on frames. For audio, augmentation introduces AVA dataset audio as background noise. Asynchronous techniques for audio-visual disparities diversify the dataset: (1) Time Warping: Stretching or compressing both audio and video for varying speech speeds. (2) Frame Dropout: Randomly masking video frames for simulated information loss. (3) Time Slicing: Segmenting audio or video signals, shuffling fragments. (4) Mixing: Blending audio or video segments for scenarios with multiple sources.

Evaluation Metric. The evaluation metric for the AVA—ActiveSpeaker dataset is mean Average Precision (mAP), while the Columbia Active Speaker Dataset utilizes the F1 score for assessment. Due to the imbalanced 1:3 speaker-to-non-speaker ratio, the primary evaluation metric is the area under the curve (AUC). AUC, derived from the receiver operating characteristic (ROC) curve, remains invariant despite changes in sample distribution.

4.3 Results

On the AVA dataset, our approach demonstrates superior performance compared to other state-of-the-art methods, as illustrated in Table 1. The data marked in bold in the table represents the best-performing experimental metrics. Specifically, our OnePass architecture achieves optimal performance on mobile networks, with minimal parameter count (0.05GFLOPs) and a remarkable mAP of 90.6%. Additionally, our OneStage architecture attains the best performance on ResNet-18/34, with lower model complexity compared to other methods.

Furthermore, our method exhibits excellent performance on our dataset, as shown in Table 2. The OnePass architecture achieves an accuracy of 88.4% and an AUC of 93.4%, while the OneStage architecture achieves an accuracy of 90.9% and an AUC of 96.0%.

Table 1 Performance comparisons with other state-of-the-art methods on the AVA dataset [17]

Method	E2E?	Feature encoding network	Params (M)	FLOPs (G)	mAP (%)
ASC [20]	×	ResNet-18 [21]	23.5	1.8	87.1
UniCon [9]	×	ResNet-18	>22.4	>1.8	92.2
TalkNet [8]	√	ResNet-18/34	15.7	1.5	92.3
ASDNet [6]	×	ResNeXt-18 [22] + SincDSNet [7]	51.3	14.9	93.5
SPELL [10]	×	ResNet-18-TSM [23]	22.5	2.6	**94.2**
Our-OnePass (small)	√	MobileNet [16]	**5**	**0.05**	90.6
Our-OneStage (big)	√	ResNet-18/34	**12.3**	**1.2**	**93.8**

Table 2 Performance comparisons with ASDNet and TalkNet on our dataset

Method	Face size	ACC (%)	AUC (%)	AP (%)	F1 (%)
ASDNet [6]	160*160	85.2	89.9	75.5	66.7
TalkNet [8]	112*112	88.7	94.5	87.5	78.8
Our-OnePass(small)	112*112	88.4	93.4	85.8	75.2
Our-OneStage(big)	112*112	**90.9**	**96.0**	**90.8**	**82.0**

As shown in Fig. 5, there are two hyperparameters that affect the experimental indicators. (a) indicates the relationship between the threshold for detecting whether the active speaker is speaking and the experimental indicators. From the figure, it can be seen that the experimental effect is best when the threshold is set to 0.03. (b) indicates the relationship between the length of the image sequence input to the network and the experimental indicators. The best effect is achieved when the image sequence length is 15 images.

Table 3 presents the experimental indicators for different masking ratios, with the bold ones representing the best-performing experimental metrics. From the table, it

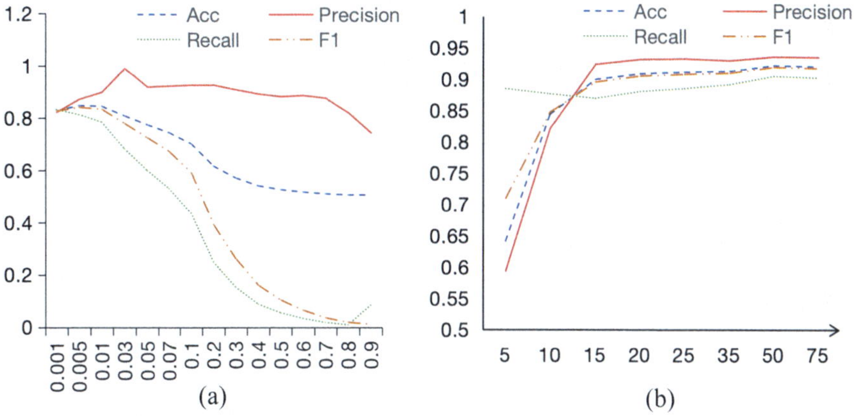

Fig. 5 Study on the impact of two hyperparameters, the horizontal axis of (**a**) represents the threshold for detection, while the horizontal axis of (**b**) corresponds to the length of the input image sequence. In both graphs, the vertical axis denotes the experimental metric

Table 3 Experimental indicators for various masking ratios on the AVA dataset [17]

Masking radio	ACC (%)	AUC (%)	AP (%)	F1 (%)
0.9	90.4	94.7	90.2	79.2
0.7	90.8	95.4	90.8	80.2
0.5	**91.0**	**95.7**	**90.8**	**81.4**
0.3	90.5	95.5	90.2	80.0
0.1	90.6	94.7	90.4	80.8

can be seen that the experimental effect is best when the ratio is set to 0.5, which randomly erases 50% of the sequence length.

5 Conclusion

Active speaker detection (ASD) plays a crucial role in enhancing applications such as speech interaction, audio conferences, and intelligent voice assistants. The advancements proposed in this paper contribute to the field by introducing the OneStage architecture for dynamic cross-modal fusion, leveraging self-attention networks and differential images to reduce visual information reliance. The random ratio mask annotation strategy and audio-visual asynchronous data augmentation significantly alleviate the annotation burden, enhancing model generalization. The lightweight and real-time OnePass architecture designed for small terminal devices further broadens the practical applicability of ASD.

References

1. He K, Zhang X, Ren S, Sun J (2016) Deep residual learning for image recognition. In: Proceedings of the IEEE conference on computer vision and pattern recognition, pp 770–778. https://doi.org/10.1109/cvpr.2016.90
2. LeCun Y, Bengio Y, Hinton G (2015) Deep learning. Nature 521(7553):436–444. https://doi.org/10.1142/9789814618045_0006
3. Vaswani A, Shazeer N, Parmar N, Uszkoreit J, Jones L, Gomez AN, Polosukhin I (2017) Attention is all you need. Adv Neural Inform Process Syst. 30. https://doi.org/10.5555/3295222.3295349
4. Afouras T, Owens A, Chung JS, Zisserman A (2020) Self-supervised learning of audio-visual objects from video. In: Computer vision—ECCV Springer international publishing, pp 208–224. https://doi.org/10.1007/978-3-030-58523-5_13
5. Chung JS (2019) Naver at Activitynet challenge 2019—task b active speaker detection (AVA). arXiv preprint arXiv:1906.10555. https://doi.org/10.48550/ARXIV. 1906.10555
6. Köpüklü O, Taseska M, Rigoll G (2021) How to design a three-stage architecture for audio-visual active speaker detection in the wild. In: Proceedings of the IEEE/CVF international conference on computer vision, pp 1193–1203. https://doi.org/10.1109/iccv48922.2021.00123

7. Ravanelli M, Bengio Y (2018, December) Speaker recognition from raw waveform with sincnet. In: 2018 IEEE spoken language technology workshop (SLT), pp 1021–1028. https://doi.org/10.1109/slt.2018.8639585
8. Tao R, Pan Z, Das RK, Qian X, Shou MX, Li H (2021) Is someone speaking? exploring long-term temporal features for audio-visual active speaker detection. arXiv preprint arXiv:2107.06592.https://doi.org/10.48550/ARXIV.2107.06592
9. Zhang Y, Liang S, Yang S, Liu X, Wu Z, Shan S, Chen X (2021, October) Unicon: Unified context network for robust active speaker detection. In: Proceedings of the 29th ACM international conference on multimedia, pp 3964–3972. https://doi.org/10.1145/3474085.3475275
10. Min K, Roy S, Tripathi S, Guha T, Majumdar S (2022, October) Learning long-term spatial-temporal graphs for active speaker detection. In: European conference on computer vision. Springer Nature Switzerland, Cham, pp 371–387. https://doi.org/10.1007/978-3-031-19833-5_22
11. Gray A, Markel J (1976) Distance measures for speech processing. IEEE Trans Acoust Speech Sign Process 24(5):380–391. https://doi.org/10.1109/tassp.1976.1162849
12. Yeh YR, Lin TC, Chung YY, Wang YCF (2012) A novel multiple kernel learning framework for heterogeneous feature fusion and variable selection. IEEE Trans Multimedia 14(3):563–574. https://doi.org/10.1109/tmm.2012.2188783
13. Getoor L, Taskar B (2007) An introduction to conditional random fields for relational learning. In: Introduction to statistical relational learning, MIT Press, pp 93–127. https://ieeexplore.ieee.org/document/6278196
14. Ngiam J, Khosla A, Kim M, Nam J, Lee H, Ng AY (2011) Multimodal deep learning. In: Proceedings of the 28th international conference on machine learning (ICML-11), pp 689–696. https://doi.org/10.5555/3104482.3104569
15. Zhang S, Zhu X, Lei Z, Shi H, Wang X, Li SZ (2017) Faceboxes: a CPU real-time face detector with high accuracy. In: 2017 IEEE international joint conference on biometrics (IJCB), pp 1–9. https://doi.org/10.1109/btas.2017.8272675
16. Howard AG, Zhu M, Chen B, Kalenichenko D, Wang W, Weyand T, Adam H (2017) Mobilenets efficient convolutional neural networks for mobile vision applications. arXiv preprint
17. Roth J, Chaudhuri S, Klejch O, Marvin R, Gallagher A, Kaver L, ... Pantofaru C (2020) Ava active speaker: an audio-visual dataset for active speaker detection. In: ICASSP 2020–2020 IEEE international conference on acoustics, speech and signal processing (ICASSP), IEEE, pp 4492–4496
18. Chakravarty P, Tuytelaars T (2016) Cross-modal supervision for learning active speaker detection in video. In: Computer Vision–ECCV 2016: 14th European conference, Amsterdam, The Netherlands, October 11–14, 2016, Proceedings, Part V 14.Springer International Publishing, pp 285–301. https://doi.org/10.1007/978-3-319-46454-1_18
19. Paszke A, Gross S, Massa F, Lerer A, Bradbury J, Chanan G, ... Chintala S (2019, December) PyTorch: an imperative style, high-performance deep learning library. In: Proceedings of the 33rd international conference on neural information processing systems, pp 8026–8037. https://doi.org/10.5555/3454287.3455008
20. Alcázar JL, Caba F, Mai L, Perazzi F, Lee JY, Arbeláez P, Ghanem B (2020) Active speakers in context. In: Proceedings of the IEEE/CVF conference on computer vision and pattern recognition, pp 12465–12474
21. He K, Zhang X, Ren S, Sun J (2016) Deep residual learning for image recognition. In: Proceedings of the IEEE conference on computer vision and pattern recognition, pp 770–778. https://doi.org/10.1109/cvpr.2016.90
22. Kopuklu O, Kose N, Gunduz A, Rigoll G (2019) Resource efficient 3d convolutional neural networks. In: Proceedings of the IEEE/CVF international conference on computer vision workshops, pp 0–0. https://doi.org/10.1109/iccvw.2019.00240
23. Lin J, Gan C, Han S (2019) Tsm: Temporal shift module for efficient video understanding. In: Proceedings of the IEEE/CVF international conference on computer vision, pp 7083–7093. https://doi.org/10.1109/iccv.2019.00718

Simulation of Inter-Satellite Terahertz Communication Channel Based on Parallel Ray Tracing Algorithm

Yuanzhi He and Chensheng Ma

Abstract With the development of the new generation of satellite communication technology, inter-satellite high-speed communication through satellite networking has become the demand of the development of human society. The terahertz band is an attractive solution for high-speed inter-satellite communication due to its extremely large bandwidth and the absence of extremely high atmospheric absorption losses in space. Due to the short wavelength of terahertz waves, the scattering mechanisms of satellite platforms in the terahertz band are complex and difficult to estimate. In order to quickly and accurately estimate the propagation characteristics of terahertz signals between satellites, a channel simulation method based on parallel ray tracing algorithm is proposed in this paper. According to the characteristics of space communication environment, the propagation process and multi-dimensional characteristics of terahertz waves from satellite to satellite are analyzed. A terahertz communication channel model was established by using ray tracing method and parallel algorithm, and the channel characteristics of terahertz inter-satellite links were analyzed.

1 Introduction

At present, satellite communication systems are faced with two key problems: the increasing shortage of traditional spectrum resources and the rapid growth of satellite network traffic demand [1]. This has forced the satellite communications industry to urgently develop new spectrum resources to meet the urgent needs of users. Thus, the frontier of satellite communications research lies in frequencies above 100 GHz [2, 3], often referred to as the "terahertz range." Looking forward to 2030 and beyond, the terahertz band (0.1 ~ 10 THz), as one of the core technologies expected to meet the development requirements of future satellite networks, will give full play to its advantages of ultra-large bandwidth and ultra-high communication rate, supporting massive device connections and ultra-high transmission rates of Tbit/

Y. He (✉) · C. Ma
Academy of Military Sciences, Institute of Systems Engineering, Beijing 100141, China
e-mail: he_yuanzhi@126.com

S. Yadav et al. (eds.), *Proceedings of 3rd International Conference on Artificial Intelligence, Robotics, and Communication*, Lecture Notes in Electrical Engineering 1172, https://doi.org/10.1007/978-981-97-2200-6_30

s [4]. However, harnessing this relatively unexplored area of the electromagnetic spectrum will involve a number of challenging issues. One of the problems is the inter-satellite terahertz communication channel model in space applications [5].

The working environment of satellites in orbit puts forward high requirements for temperature control technology, and covering the surface of satellites with multi-layer heat-insulating materials is a commonly used and effective temperature control method. The temperature control materials used in all countries are generally polyimide films (Kapton) or fluorinated ethylene propylene films (FEP), which are usually strong mirror counter materials. Affected by factors such as satellite manufacturing process and in-orbit working space environment, the temperature control materials will form irregular folds on the surface of the satellite, and the height of the folds can be up to 10 mm. Due to the short wavelength of the terahertz frequency band, most targets that can be regarded as smooth surfaces in the microwave frequency band can only be regarded as non-metallic rough objects in the terahertz frequency band, which exhibit different scattering characteristics from smooth conductor targets [6]. Therefore, the scattering of terahertz waves on the satellite coating material is an important component of the modeling of inter-satellite terahertz communication channel. Experimental method and theoretical simulation are two important ways to study terahertz scattering characteristics. In the experimental aspect, the main research method is to use the contraction field to measure the contraction. However, for large size targets in real space, it is difficult to measure them with existing experimental systems due to complex system design and time consuming, so only theoretical simulation technology can be used to study the terahertz scattering characteristics of electrically large size rough medium targets. A variety of numerical methods, such as ray tracing high-frequency electromagnetic technique [7], Kirchhoff approximation [8], finite-difference Time-domain method [9], perturbation method, radiative transfer theory [10], and Monte Carlo [11], have been applied to the simulation calculation of scattering characteristics of THz rough targets. A lot of research on terahertz multipath channels, especially indoor channels, has been done [12, 13]. But research on multipath channels in satellite communication environment is just starting. At the same time, the actual space target size is much larger than the incident wavelength, and it has electric and super-electric large size, which makes the calculation of massive data difficult for the traditional electromagnetic algorithm. Therefore, how to quickly and accurately model the interplanetary terahertz communication channel is a difficult problem.

With the rapid development of GPU technology, GPU + CPU heterogeneous computing mode has been widely used in various scientific research fields. For ray tracing algorithms with high computational complexity and parallelization, GPU can also be used for parallel acceleration optimization [14, 15]. Therefore, it is of great significance to study the terahertz scattering characteristics of the folded and rough surfaces of electrically large-sized satellite cladding materials using parallel algorithms. In this paper, the ray tracing method combined with CUDA parallel acceleration algorithm is used to quickly and accurately analyze the channel characteristics of inter-satellite links in terahertz band.

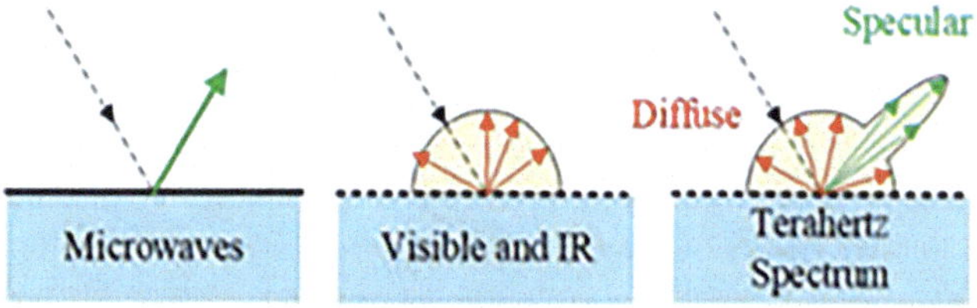

Fig. 1 Propagation changes of electromagnetic waves of different frequencies on the same surface

2 Scattering Model of Satellite Surface Cladding Materials

2.1 Surface Roughness Criterion

Rayleigh criterion [16] is commonly used to determine whether a surface is rough, and the Rayleigh criterion formula is as follows:

$$h_c = \frac{\lambda}{8\cos\theta_i} = \frac{c}{8f \times \cos\theta_i}, \tag{1}$$

where λ is the wavelength, f is the frequency, and θ_i is the incidence Angle. We define h_0 as the maximum height difference of a rough surface. If h_0 is greater than h_c, the surface is considered rough; otherwise, it can be considered smooth. According to Rayleigh criterion, the frequency is proportional to the roughness when the difference between the incident angle and the maximum height of the rough surface is the same. As shown in Fig. 1, as the wavelength decreases, scattering will play a more important role in propagation. It is important to study the scattering characteristics of terahertz waves on rough surfaces for the accuracy of the terahertz channel model.

2.2 Rough Surface Characterization Parameters

To study the effect of roughness on scattering, it is necessary to determine the statistical parameters of rough surface first. At present, relevant theories have introduced the statistics of height distribution of rough surface [17]. The root means square height of random rough surface height fluctuation δ and surface correlation length l are two basic statistical parameters to describe the height distribution characteristics of rough surface. The determination of the maximum height difference δ provides a basis for estimating the independence of two points on a rough surface. For smooth surfaces, this means δ is zero, l approaching infinity. The following is a brief introduction to these parameters.

Root means square height of the rough surface δ reflects the roughness of the rough surface, which is defined as

$$\delta = \sqrt{E\left[f^2(x, y)\right] - E[f(x, y)]^2}, \tag{2}$$

where $f(x, y)$ is the height fluctuation function of the rough surface. For a rough surface with a specific distribution, a single root mean square height cannot uniquely describe the characteristics of the rough surface, so the autocorrelation function is introduced to show the degree of correlation between any two points on the random surface, so as to describe the characteristics of the rough surface more accurately. The autocorrelation function is defined as

$$G\left(d_x, d_y\right) = E\left[f(x, y) f\left(x + d_x, y + d_y\right)\right], \tag{3}$$

where d_x and d_y are the distances between two points on a rough surface in directions x and y, respectively. δ^2 is the height variance of the surface of the rough surface, so the autocorrelation function is normalized and the correlation coefficient is defined as

$$\rho(R) = \frac{G(R)}{\delta^2}. \tag{4}$$

The surface correlation length l is a basis for estimating whether any 2 points on a surface are independent of each other, that is, if the distance between 2 points on a surface is greater than 1 in the horizontal distance, the height values of the 2 points can be considered statistically approximately independent of each other. According to the above definition, the value of l when the correlation coefficient $\rho(R)$ is reduced to $1/e$ is called the correlation length, which is $\rho(l) = 1/e$.

2.3 Rough Surface Modeling Based on Monte Carlo Method

The Monte Carlo method is often used to model rough surfaces. Monte Carlo method, also known as linear filtering method [18], filters it with power spectrum in the frequency domain, and then does inverse fast Fourier transform on the result to obtain the height fluctuation of the rough surface. Since the rough surface is thought to be composed of a large number of harmonics superimposed, the amplitude of the harmonics is an independent Gaussian random variable whose variance is proportional to the power spectral density $S(k)$ of a particular wave number. Assuming that the length of the two-dimensional random rough surface to be generated in the direction x and y is L_x and L_y, respectively, the number of equally spaced discrete points is M and N respectively, and the distance between adjacent two points is Δx and Δy, respectively, the height fluctuation function of the rough surface can be expressed as

$$f(x_m, y_n) = \frac{1}{L_x L_y} \sum_{m_k=-\frac{M}{2}+1}^{\frac{M}{2}} \sum_{n_k=-\frac{N}{2}+1}^{\frac{N}{2}} F(k_{mk}, k_{nk}) e^{i(k_{mk} x_m + k_{nk} y_n)}, \tag{5}$$

where the discrete wave number $k_{mk} = \frac{2\pi m_k}{L_x}$, $k_{nk} = \frac{2\pi n_k}{L_y}$,

$$F(k_{mk}, k_{nk}) = 2\pi \sqrt{l_x l_y S(k_{mk}, k_{nk})} \cdot \begin{cases} \frac{[N(0,1)+iN(0,1)]}{\sqrt{2}} & , m_k \neq 0, \frac{M}{2} \text{and } n_k \neq 0, \frac{N}{2} \\ [N(0,1) + iN(0,1)] & , m_k = 0, \frac{M}{2} \text{and } n_k = 0, \frac{N}{2} \end{cases} \tag{6}$$

where $N(0, 1)$ represents a normally distributed random number with a mean of 0 and a variance of 1. $S(k_{mk}, k_{nk})$ is the power spectrum function of the rough surface obeying the Gaussian distribution.

The main covering material of the satellite capsule body is golden polyimide film, and the solar wings are mainly composed of panels. We assume that the root-mean-square roughness of the golden polyimide film and the panels are 1 mm and 0.031 mm, respectively, and the relevant lengths are 10 mm and 1.8 mm, respectively. Golden polyimide films and panels were modeled by Monte Carlo method, and the resulting model is shown in Fig. 2.

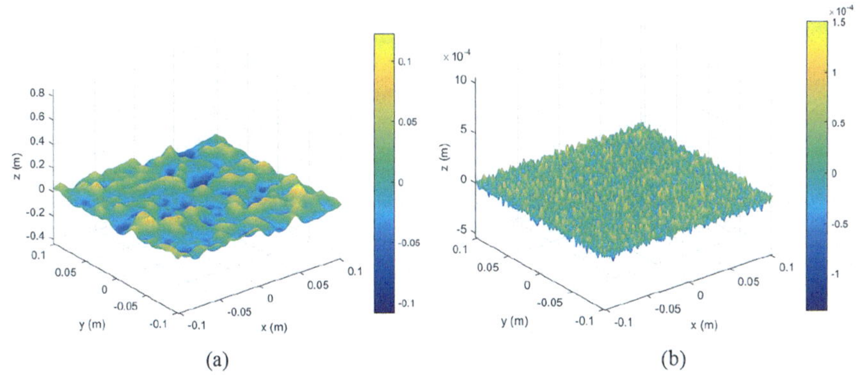

Fig. 2 Golden polyimide film (**a**) and panel (**b**) rough surface model

3 Channel Modeling Based on Parallel Ray Tracing Algorithm

3.1 Channel Modeling

Ray tracing method is a method to estimate the propagation characteristics of high-frequency electromagnetic wave based on geometric optics and geometric diffraction theory. The core idea is to use rays to simulate electromagnetic beams, treat the transmitting point as a point source, treat the emitted electromagnetic wave as a ray transmitted in all directions, and then calculate each path that can reach the field point from the source point. In this paper, an interplanetary terahertz communication channel model based on ray tracing method is developed. For rough surfaces, reflection consists of specular reflection and diffuse reflection. Therefore, the received signal in the model is represented as the superposition of direct signal, specular reflection signal and diffuse reflection signal, and the superposition of specular reflection signal and diffuse reflection signal is represented as reflected signal.

The signal transfer function can be expressed as

$$\begin{aligned} H(f) &= H_{LoS}(f) + \sum_{i=1}^{I_{\mathrm{Re}f}} H_{\mathrm{Re}f,i}(f) \\ &= H_{LoS}(f) + \sum_{p=1}^{P_{\mathrm{Re}f}} H_{sp,p}(f) + \sum_{q=1}^{Q_{\mathrm{Re}f}} H_{sc,q}(f) \end{aligned} \tag{7}$$

where H_{LoS}, $H_{\mathrm{Re}f}(f)$, $H_{sp}(f)$, and $H_{sc}(f)$ are the transmission equations of direct, reflective, specular, and diffuse reflection paths, respectively.

Specifically, for direct path, since there is no atmosphere in space, path loss consists only of free space loss. Therefore, the transfer function of a direct path can be expressed as

$$H_{LoS}(f) = \left(\frac{c}{4\pi f d}\right) e^{-j2\pi f \tau_{LoS}}, \tag{8}$$

where c is the speed of light; f is the carrier frequency; d is the distance between the transmitting end and receiving end, and $\tau_{LoS} = d/c$ is the time of the direct path.

For the reflection path, the reflectivity of specular reflection and diffuse reflection superimposed by electromagnetic wave incident is

$$R = R_0 \cdot e^{-\left(\frac{4\pi\delta\cos\theta_i}{\lambda}\right)^2} + R_0 \frac{2^5\pi^4}{S^2}\left(\frac{\delta}{\lambda}\right)^4 \cos^3\theta_i (\Delta\theta)^2, \tag{9}$$

where R_0 denotes the reflectivity of the same material under a completely smooth surface, where $S = \sqrt{2}\delta/l$ is the root mean square slope, λ is the wavelength, and θ_i

is the incidence Angle. Reflectivity is the square of the modulus value of the reflection system. According to the rough surface reflectivity model of the terahertz frequency band above, similar terms of specular reflection and diffuse reflection coefficients can be combined to obtain the modified expression of the rough surface reflection coefficient:

$$R(f) = R_0(f) \cdot e^{-8\left(\frac{\pi\delta\cos\theta_i}{\lambda}\right)^2}, \tag{10}$$

where $R_0(f)$ represents the reflection coefficient of a smooth surface, which can be given by Fresnel equations for horizontal and vertical polarization:

$$R_0(f)_{TE} = \frac{Z\cos\theta_i - Z_0\cos\theta_t}{Z\cos\theta_i + Z_0\cos\theta_t}, \tag{11}$$

$$R_0(f)_{TM} = \frac{Z\cos\theta_t - Z_0\cos\theta_i}{Z\cos\theta_t + Z_0\cos\theta_i}, \tag{12}$$

where μ_0 and ε_0 are free space permeability and permittivity. In the terahertz band, it can be accurately represented by the frequency-dependent complex refractive index $n_t = n + i\alpha$, where the real part n is the refractive index and the imaginary part α is the absorption coefficient.

The transport equations of specular reflection and diffuse reflection can be expressed as

$$H_{sp}(f) = \left(\frac{c}{4\pi f(r_1 + r_2)}\right)e^{-j2\pi f\tau_{sp}} \cdot R(f), \tag{13}$$

$$H_{sc}(f) = \left(\frac{c}{4\pi f(s_1 + s_2)}\right)e^{-j2\pi f\tau_{sc}} \cdot R(f), \tag{14}$$

where r_1 and r_2 are the distance from the transmitting point and receiving point to the specular reflection point, τ_{sp} is the time for the specular reflection ray to reach the receiving point, s_1 and s_2 are the distance from the transmitting point and receiving point to the specular reflection point, and τ_{sc} is the time for the specular reflection ray to reach the receiving point.

3.2 *Parallel Optimization of Ray Tracing Algorithm*

In the ray tracing algorithm, each ray has a high degree of independence, that is, the rays do not interfere with each other, and the calculation result of each ray will not affect the calculation of other rays. Therefore, this paper uses parallel algorithm to optimize parallel acceleration, that is, two-dimensional thread blocks and grid blocks are used on GPU to correspond to a ray, and each thread independently completes

Table 1 Simulation configuration

Source	Font size and style
Frequency	210–230 GHz
Polarization	Linear polarization
Gain	24 dB
Power	10 dBm
Distance	1000 km

all the computing tasks of a ray. Among the algorithms implemented in parallel, the CUDA programming model can be mainly divided into two parts. On the one hand, the logic control part of the whole program is run at the host end of the CPU, which mainly includes reading the satellite model data file, initializing various electromagnetic parameter information, generating rays, copying data to the GPU, reading and retrieving data from the device GPU, calculating path loss according to the effective ray path calculated by the GPU, and so on. On the other hand, the core part of the program runs in the kernel function of the GPU side, which is mainly responsible for the parallelization of tracing rays and the acceleration optimization process of the algorithm.

4 Design of Inter-Satellite Communication Link

4.1 Design of the Link

This section simulated inter-satellite terahertz communication between two geosynchronous orbit satellites, and the simulation parameters were configured according to the simulation scenario, as given in Table 1. In order to balance simulation accuracy and computational complexity, the propagation mechanism is set as direct radiation, highest secondary reflection, and secondary diffuse reflection.

4.2 Satellite Model

According to the known Tiantong-1 satellite model, its data is imported into the modeling tool Sketch Up to complete the three-dimensional geometric modeling of the satellite platform. During the modeling process, solid structures such as the satellite's cabin body, connecting structure, and solar panels are built with columns and geometric planes. The terahertz antenna is constructed on the hull, all surfaces characterized by different materials. In order to construct the satellite model with cladding materials, the original satellite geometry is modified by mapping on the

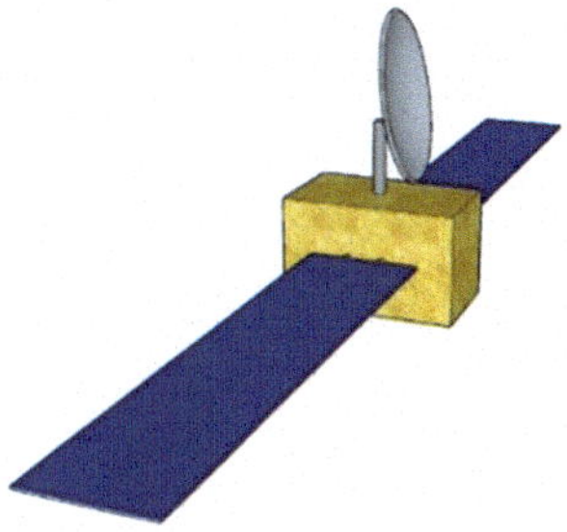

Fig. 3 Schematic of the satellite model

capsule that requires cladding materials for thermal control, and the satellite model is obtained as shown in Fig. 3.

5 Simulation Results

On the basis of selecting the 210–230 GHz band as the simulation frequency, the channel model based on parallel ray tracing algorithm was used to extract the multipath parameters of the communication scene in Sect. 4, and the impact response function (CIR) of the channel in the scenario with and without cladding materials was obtained, as shown in Fig. 4. The results show that in terahertz communication between satellites, the direct path is slightly delayed and is the main contributor to the received signal. The signal power of the reflected paths generated by the cladding materials is small. And the delay of the reflected paths is larger than direct path. It is worth noting that the delay paths caused by the coated materials are more than that of the direct signal, and the delay is more concentrated.

The Power Delay Profile (PDP) can be calculated by using the multipath impulse response function CIR. The power delay spectrum is an important parameter of multipath channel, which characterizes the diffusivity of the channel in time delay.

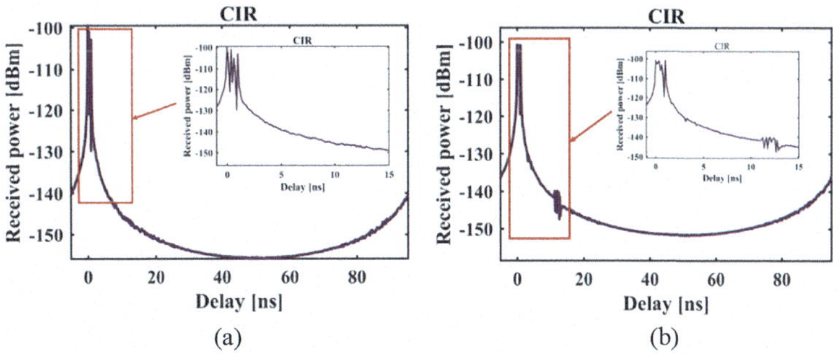

Fig. 4 Channel impulse response function curve (**a**) no cladding materials and (**b**) covered by cladding materials

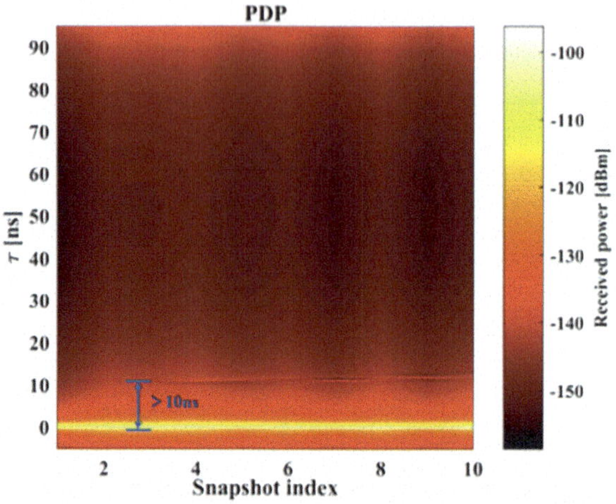

Fig. 5 Power delay spectrum

Figure 5 shows the power time delay spectrum during communication between satellites with cladding materials, where 10 snapshots are set up. Observation of Fig. 5 shows that the reflected paths are more delayed and contribute less to the received power compared to the direct path. It can also be seen that the delay of the reflected paths is at least 10 ns different from the delay of the direct path.

6 Conclusion

We propose a channel simulation method based on parallel ray tracing algorithm. According to the propagation process of terahertz waves between satellites, the ray tracing method and CUDA parallel acceleration algorithm are used to establish an inter-satellite terahertz communication channel model. We find that the folded surface of the satellite will cause multiple delay paths that are centralized. In the future, the terahertz signal of these delay paths can be used to enhance the received signal power.

Acknowledgements The national key research and development program sponsored this work under grant number 2019YFB1803201.

References

1. Rappaport TS et al (2019) Wireless communications and applications above 100 GHz: opportunities and challenges for 6G and beyond. IEEE Access 7:78729–78757
2. Akyildiz IF, Jornet JM, Han C (2014) Terahertz band: next frontier for wireless communications. Phys Commun 12:16–32

3. Bariah L et al (2020) A prospective look: key enabling technologies, applications and open research topics in 6G networks. IEEE Access 8:174792–174820
4. He Y, Ma C (2022) Research on frequency selection method for spatial terahertz communication under multiple constraints. In: 2022 IEEE 22nd International conference on communication technology (ICCT). Nanjing, pp 774–781
5. He Y, Ma C (2022) Analysis of the effect of antenna pointing error caused by satellite perturbation on space terahertz communication. Appl Sci 12(21):10772
6. Khamse AM, Dong X, Ferdinand N (2023) The scattering channel model for terahertz wireless communications. IEEE Open J Commun Soc 4:810–822
7. Li Z et al (2009) Electromagnetic scattering characteristics of PEC targets in the terahertz regime. IEEE Antennas Propag Mag 51(1):39–50
8. Christian Jansen SM et al (2011) Diffuse scattering from rough surfaces in THz communication channels. IEEE Trans Terahertz Sci Technol 1(2):462–472
9. Sundberg G, Zurk LM, Schecklman S, Henry S (2010) Modeling rough-surface and granular scattering at terahertz frequencies using the finite-difference time-domain method. IEEE Trans Geosci Remote Sens 48(10):3709–3719
10. Nam KM, Zurk LM, Schecklman S (2012) Modeling terahertz diffuse scattering from granular media using radiative transfer theory. Prog Electromagnet Res B 38:205–223
11. Kallel A, Gastellu-Etchegorry JP (2017) Canopy polarized BRDF simulation based on non-stationary Monte Carlo 3-D vector RT modeling. J Quant Spectrosc Radiat Transfer 189:149–167
12. Porretta M, Zanier F, Garcia-Rodriguez A, Mongrard O, Powe M (2014) Use of a ray-tracing tool to predict the international space station multipath and masking environment. In: 2014 7th ESA workshop on satellite navigation technologies and European workshop on GNSS signals and signal processing (NAVITEC), pp 1–13
13. Guan K, Yi H, He D, Ai B, Zhong Z (2021) Towards 6G: paradigm of realistic terahertz channel modeling. China Commun 18(5):1–18
14. Epstein BR, Rhodes DL (2010) GPU-accelerated ray tracing for electromagnetic propagation analysis. In: IEEE international conference on wireless information technology and systems. Honolulu, pp 1–4
15. Tan J, Su Z, Long Y (2015) A full 3-D GPU-based beam-tracing method for complex indoor environments propagation modeling. IEEE Trans Antennas Propag 63(6):2705–2718
16. Rappaport TS (2022) Wireless communications: principles and practice. Prentice hall PTR
17. Xie P et al (2022) Terahertz wave propagation characteristics on rough surfaces based on full-wave simulations. Radio Sci 57(6):1–16
18. Kuga Y, Phu P (1996) Experimental studies of millimeter-wave scattering in discrete random media and from rough surfaces–Summary. J Electromagn Waves Appl 10(3):451–453

Optimization of Milling Robot Trajectory Based on Whale Optimization Algorithm

Yang Zhou, Qingxin Li, and Yong Sun

Abstract In response to the problem of vibration and low work efficiency during robot milling, taking into account the work indicators of milling time and stability, combined with the kinematic constraints of the robotic arm, the time impact optimization problem is transformed into a multi-objective optimization problem. The paper proposes an improved whale optimization algorithm, which introduces a differential mutation strategy to improve the problem of traditional whale algorithms easily getting stuck in local optima. Compared with traditional whale optimization algorithms, the improved whale algorithm significantly improves the convergence speed and accuracy during the milling process. Finally, it is verified by a solid robot. The results showed that the pulse value of the planned trajectory was reduced by up to 99.76%, and the efficiency was improved by 47.62%, effectively avoiding robot vibration during the milling process.

Y. Zhou
Software College Shenyang University of Technology, Shenyang, China
e-mail: zhouyang6@smail.sut.edu.cn

Y. Zhou · Q. Li (✉) · Y. Sun
Key Laboratory of Networked Control Systems, Chinese Academy of Sciences, Shenyang 110016, China
e-mail: liqingxin@sia.cn

Shenyang Institute of Automation, Chinese Academy of Sciences, Shenyang 110016, China

Y. Sun
e-mail: sunyong1@sia.cn

Q. Li · Y. Sun
Institutes for Robotics and Intelligent Manufacturing, Chinese Academy of Sciences, Shenyang 110169, China

S. Yadav et al. (eds.), *Proceedings of 3rd International Conference on Artificial Intelligence, Robotics, and Communication*, Lecture Notes in Electrical Engineering 1172, https://doi.org/10.1007/978-981-97-2200-6_31

1 Introduction

Joint robots, with their excellent spatial flexibility, high efficiency, and precision, are widely used in modern industrial automation production, such as handling, palletizing, welding, and cutting. In some machining situations with low precision requirements and soft cutting materials, CNC machine tools have gradually been replaced by industrial robots. The application research of robots in milling processing has gradually become a hot direction. Many practical applications have shown that in addition to focusing on motion effects and task requirements, planning the motion trajectory of robots also needs to consider factors such as work efficiency and motion stability, in order to achieve optimal trajectory planning [1, 2]. Qianqian and Shaojie proposed a trajectory planning method based on an improved hybrid particle swarm optimization algorithm to address the issues of robot work efficiency and operational stability. Establish a new simulated annealing mechanism and combine it with particle swarm optimization algorithm for multi-objective optimization; by using a nonlinear inertia weight reduction strategy and dynamic learning factors to balance local and global search capabilities, the operating time of the robotic arm is shortened and the pulsating impact of the robotic arm is effectively reduced [3]. Mingyue Z et al. used the 3–5-3 polynomial interpolation algorithm combined with the butterfly optimization algorithm that introduced Levy flight and sine cosine algorithm to optimize the motion trajectory in order to solve the optimization problem of the speed and acceleration of each joint of the robotic arm, greatly reducing the motion time of the robotic arm [4]; Paulo Salgado, Get ú Lio Igrejas, and others proposed a new particle swarm optimization algorithm to address the issue of stability during robot operation. The particles were replaced by spline functions to generate smooth motion trajectories with two consecutive differentiable curvatures, reducing the impact of vibration on the robot during operation [5]; Yi et al. proposed a multi-objective genetic algorithm to optimize the joint space trajectory of drilling robots in order to improve their operational efficiency and reduce impact vibration during the operation process. The NSGA-II algorithm with constraint processing was used to optimize the interpolation trajectory of B-splines. Finally, a trajectory scheme selection standard of "comprehensive comparison operator" was designed to obtain the comprehensive optimal solution. The results show that the optimal solution obtained by using the comprehensive comparison operator is superior in both time performance and impact performance [6]; Jie et al. focused on the issues of low convergence accuracy and susceptibility to local optima in whale optimization algorithm (WOA) for the core unit 6R robot in intelligent manufacturing production lines. They proposed a nonlinear convergence factor and introduced nonlinear inertia weights to improve the basic WOA, while optimizing the time. The experimental results show that the complete loading and unloading process of the robot after motion optimization in this research scenario is shortened by 20–50% compared to before optimization, indicating that the improved WOA is effective, reliable, and has practical significance in related field applications [7]; in summary, a large number of researchers have studied the trajectory planning of robots. However, some algorithms only optimize for time or vibration, while others have strong optimization ability and fast convergence speed. However, there is also an imbalance between global exploration and local development capabilities, which can easily lead to local optima. Some algorithms also do not consider smooth running trajectories. Aiming at the above shortcomings, a whale optimization algorithm combining differential strategy is proposed for joint space trajectory planning of milling robots. This algorithm enhances population diversity, accelerates convergence speed, and avoids getting stuck in local optima too early. By comparing the application results of traditional whale optimization algorithms, the superiority of IWOA is verified. The simulation results verified the effectiveness of the improved algorithm.

2 Objective Function Construction

Ensuring smooth operation and trajectory during robot milling can reduce joint vibration and improve milling quality [8]. In order to effectively mill the surface of the work piece and achieve consistency in material removal, each point on the milling path needs to be planned. During the robot machining process, the vibration of the robot body directly affects the milling trajectory. In order to ensure its stability and prevent overshoot and chattering, it is necessary to maintain continuous speed and acceleration, but this will lead to a significant increase in execution time. Therefore, it is necessary to strike a balance between minimizing time and minimizing pulse. Introducing proportional parameters ω and v, weight the time and pulse terms to establish an objective function, where the pulse is represented by joint angle plus acceleration. Given the kinematic constraints such as velocity of the UR5 robot, an improved whale optimization algorithm is used for trajectory planning, and cubic splines are used to interpolate the working trajectory. Firstly, the milling path is defined in Cartesian space and discretized into multiple path points. Then, the angle change curves of the six joints of UR5 are obtained by solving inverse kinematics. Considering the constraints of velocity, acceleration, and pulse, establish the objective function:

$$\mathbf{Function} = \boldsymbol{\omega} \sum_{\mathbf{i}=1}^{\mathbf{n}-1} \mathbf{l_i} + \mathbf{v} \sum_{\mathbf{J}=1}^{6} (\mathbf{maxP_j'''(t)}). \tag{1}$$

Among them:

$$\begin{cases} \left|\mathbf{P_j''}(t)\right| \leq \mathbf{V_{M_j}} \\ \left|\mathbf{P_j''}(t)\right| \leq \mathbf{A_{M_j}} \\ \left|\mathbf{P_j'''(t)}\right| \leq \mathbf{J_{M_j}}. \end{cases} \tag{2}$$

$j = 1, 2, 3 \ldots 6$, represents the six joints of the UR5 robotic arm, $i = 1, 2, \ldots n$ represents the points on the milling path. The number of interpolation points in the curve directly affects the fitting degree of the workpiece contour. The more interpolation points there are, the higher the fitting degree. However, having too many path points can reduce the efficiency of the robot. Therefore, determining the appropriate number of path points is crucial for balancing accuracy and efficiency. ω and v are weight factors. Within the range of [0,1], multiple experiments were conducted on the values of ω and v, and it was determined that the optimal optimization effect was achieved when the final values were 0.45 and 0.55, respectively. l_i is the calculation time interval between two consecutive points on the joint path $\left|P_j'(t)\right|$, $\left|P_j''(t)\right|$, and $\left|P_j'''(t)\right|$ represent the velocity, acceleration, and pulse value of the second joint, respectively. V_{M_j}, A_{M_j}, and J_{M_j} are the upper bounds of the velocity, acceleration, and pulse of joint i, respectively.

3 Improved Whale Optimization Algorithm

The whale optimization algorithm is a novel heuristic optimization algorithm proposed by Mir Jalili et al. from Griffith University in Australia [9]. The whale optimization algorithm simulates the feeding behavior of whales through bubble nets and divides the algorithm into three stages: surround predation, spiral updating, and searching for prey [10]. The selection of circular contraction mechanism and spiral ascent mechanism follows the following rules: humpback whales surround the school of fish in a circular contraction manner, while swimming upward in a spiral ascent manner to prey. During the optimization process, the algorithm searches for individuals with a 50% probability to choose to attack or surround prey. The mathematical model is as follows:

$$\overrightarrow{\mathbf{X}}(t+1)=\begin{cases}\overrightarrow{\mathbf{X}^*}(t)-\vec{\mathbf{A}}\cdot\vec{\mathbf{D}} & p<0.5\\ \overrightarrow{\mathbf{D}'}\cdot\mathbf{e}^{\mathbf{bl}}\cdot cos(2\pi\mathbf{l})+\overrightarrow{\mathbf{X}^*}(\mathbf{t}) & p\geq 0.5.\end{cases} \tag{3}$$

In Eq. (3), p is a random number between [0,1], t represents the number of iterations, $\vec{A}$ and $\vec{D}$ are vector factors, $\overrightarrow{X^*}$ is the current optimal position vector, $\vec{X}$ is the vector representation of whale position, and X^* is the current optimal solution position.

$\overrightarrow{D'}=\left|\overrightarrow{X^*}(t)-\vec{X}(t)\right|$ is the linear distance between the retrieved prey and the target prey, b is the shape parameter of the logarithmic spiral, and l is a random number between $[-1, 1]$. After initialization, the search agent will update their current position toward the best selected location. This behavior can be represented by the following equation:

$$\vec{D}=\left|\vec{C}\cdot\overrightarrow{X^*}(t)-\vec{X}(t)\right|, \tag{4}$$

$$\vec{X}(t+1) = \overrightarrow{X^*}(t) - \vec{A} \cdot \vec{D}. \quad (5)$$

The coefficient vector $\vec{A}$ and $\vec{C}$ can be calculated using the following equation:

$$\vec{A} = 2\vec{a} \cdot \vec{r} - \vec{a}, \quad (6)$$

$$\vec{C} = 2 \cdot \vec{r}. \quad (7)$$

$\vec{a}$ linearly decreases from 2 to 0, and $\vec{r}$ is a random vector between [0,1]. The update of the position of the humpback whale is based on a random search of the positions between each whale. Therefore, a random value greater than 1 or less than −1, i.e., $|\vec{A}|>1$, is chosen to keep the search agent away from the reference individual. Update whale position based on arbitrarily selected search agents. The mathematical model is as follows:

$$\vec{D} = \left|\vec{C} \cdot \overrightarrow{X_{rand}} - \cdot\vec{X}\right|, \quad (8)$$

$$\vec{X}(t+1) = \vec{X}(t+1) = \overrightarrow{X_{rand}} - \vec{A} \cdot \vec{D}. \quad (9)$$

$\overrightarrow{X_{rand}}$ is a random position vector selected from the current generation.

Compared with other intelligent optimization algorithms, the whale optimization algorithm has the characteristics of simple structure and easy implementation, but there is a problem of easily falling into local optima. Therefore, traditional whale optimization algorithms are improved. Introducing differential mutation strategy into whale optimization algorithm can increase the exploration ability of the algorithm, avoid getting stuck in local optimal solutions, and improve the convergence speed and accuracy of the algorithm. Specifically, the differential mutation strategy can perturb the current optimal solution during the search process, thereby expanding the search range and discovering more possible solutions. In the whale optimization algorithm, differential mutation strategy can be applied to individuals during the search process. By performing differential mutation on individuals, new individuals are generated, thereby increasing the diversity of the population. The differential mutation strategy is as follows:

$$\mathbf{DE/rand/1/bin} : \mathbf{V_{i,G}} = \mathbf{X_{r1,G}} + \mathbf{F} * \left(\mathbf{X_{r2,G}} - \mathbf{X_{r3,G}}\right). \quad (10)$$

Among them, $r1$ and $r2$ are any two positive integers within the interval [1, N] that are different from each other and different from i, F is the variation factor, and is one Constants greater than 0 but less than 2 directly affect the degree of expansion or contraction of the difference vector in Eq. (10). In each iteration after initializing the population, perform mutation operations on all target vectors to obtain the corresponding mutation vector $V_{i,G}$.

In terms of specific implementation, the differential mutation strategy can adopt the following steps: Select two individuals in the current population for differential operation to obtain the differential vector; scaling the difference vector to obtain a mutation vector; hybridize the mutation vector with a third individual to obtain the attempted individual; compare the fitness of the attempted individual with the parent individual and retain the better individual; repeat the above steps until the termination condition is reached. By introducing differential mutation strategy, whale optimization algorithm can better balance global search and local search during the search process, improving the efficiency and accuracy of the algorithm.

Steps to improve whale optimization algorithm:

Improving Whale Optimization Algorithm

Input: the time period during which the robot runs between adjacent interpolation points

output: optimized adjacent interpolation point time period

initialize whale population X_i (i=1,2,...,n)

complete the mutation operation according to equations (10)

update whale population X_i'

calculate the fitness value of each search agent

X^*=Best Search Agent

while (t<maximum iteration algebra)

for each search agent

update parameters a, A, C, I, p

if 1 (p<0.5)

if 2 (|A|<1)

update the position of the current search agent according to equation (5)

else if 2 ((|A|≥ 1)

choose a random search agent X_{rand}

update the position of the current search agent according to equation (9)

end if 2

else if 1 (p≥0.5)

update the position of the current search agent according to equation (3)

end if 1
end for
check if any search agents have exceeded the search space, and if so, correct it
calculate the fitness value of each search agent and perform selection operations
if there is a better solution, update the position of X^*
let t=t+1
end while
return the value of X *

4 Experimental Analysis

This experiment takes the UR5 six degree of freedom robotic arm as the research object, with a circular grain as the milling model. The real experimental scene is shown in Fig. 1. The spatial coordinate points at the edge of the grain are collected to represent the milling trajectory, which is first converted into Cartesian coordinate path points at the end of the robot. In this paper, four path points in Cartesian coordinate systems are selected as given in Table 1; then, using the inverse kinematics of the six degree of freedom robotic arm, calculate the joint angle interpolation points corresponding to the four path points in the Cartesian coordinate system, as given in Table 2.

Using cubic spline interpolation to fit joint space trajectory points, following the steps of the improved whale optimization algorithm and combining kinematic constraints, all angle interpolation points are substituted into the algorithm for optimization, resulting in a fitness curve graph, as shown in Fig. 2. Figures 3 and 4 show the optimized velocity and acceleration curves. Using cubic spline trajectory

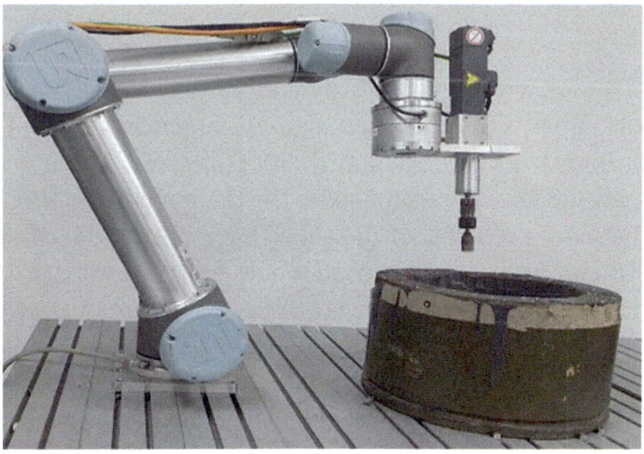

Fig. 1 Experimental scene diagram

Table 1 Cartesian space path points of robotic arm

Interpolation points	Position vector/mm			Attitude vector/rad		
1	−29.69	−283.22	408.83	0.399	3.101	0.193
2	−38.22	−284.75	409.32	0.399	3.105	0.187
3	−43.63	−285.84	409.29	0.399	3.104	0.170
4	−52.20	−290.08	409.33	0.399	3.105	0.170

Table 2 Interpolation points in the joint space of the robotic arm

Joint i	M_0/rad	M_1/rad	M_2/rad	M_3/rad
1	−1.245	−1.254	−1.263	−1.271
2	−0.921	−0.922	−0.922	−0.923
3	−2.085	−2.086	−2.090	−2.086
4	−1.706	−1.704	−1.704	−1.704
5	1.570	1.570	1.570	1.570
6	0.326	0.317	0.308	0.299

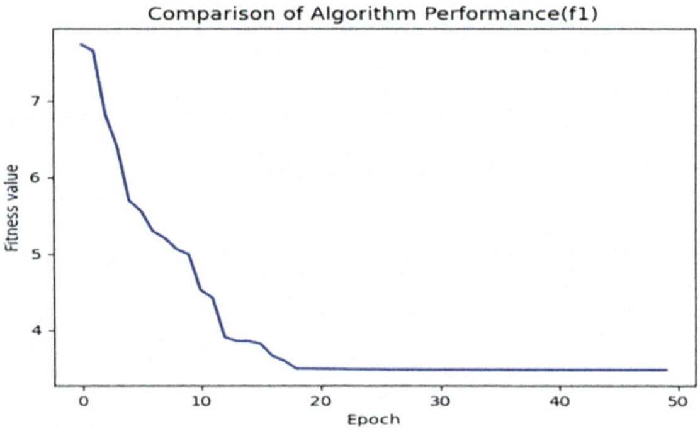

Fig. 2 Fitness curve

planning without optimization algorithm, traditional whale optimization algorithm, and improved whale optimization algorithm for optimization solution, the robot milling time was calculated and compared. Table 3 shows the robot milling time and comparison results under the three methods. From the table, it can be seen that the working hours of the robot have been reduced after using the three methods. Among them, the improved whale optimization algorithm has the shortest running time, with a reduction rate of 47.62%. Figure 5 shows the pulse comparison of the six joints under the three methods. The maximum pulse values of each joint before optimization are 27.0°/s^3, 2.50°/s^3, 1.25°/s^3, 1.48°/s^3, 0.0022°/s^3, and 17.20°/s^3, respectively. The maximum pulse values of each joint after optimization are 0.253°/s^3, 0.482°/s^3, 0.247°/s^3, 0.239°/s^3, 0.0007°/s^3, and 0.042°/s^3, the maximum pulse

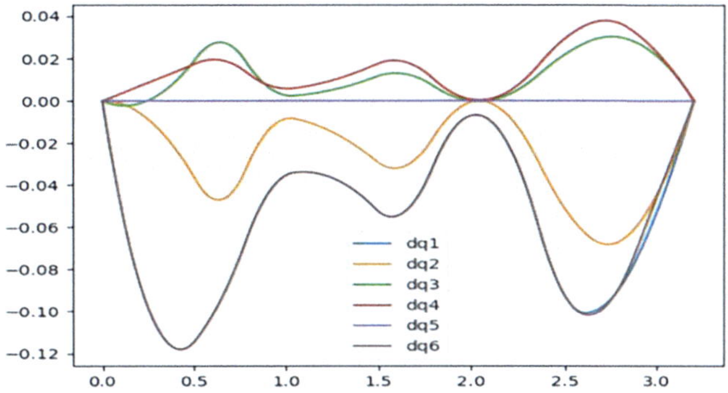

Fig. 3 Optimized speed curve

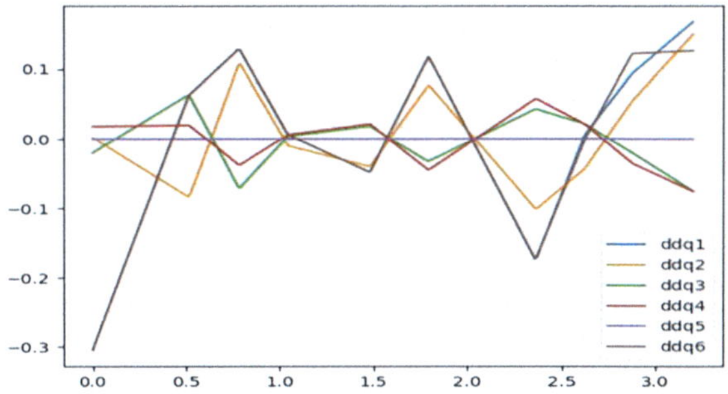

Fig. 4 Optimized acceleration curve

Table 3 Comparison of milling time under three meth

Algorithm	Run time before optimization (s)	Optimized running time (s)	Reduction rate (%)
Cubic spline	18.52	15.63	15.60
WOA	18.52	11.35	38.71
IWOA	18.52	9.70	47.62

values of each joint were reduced by 99.06%, 80.72%, 80.24%, 83.85%, 68.18%, and 99.76%, respectively. Before optimization, the fluctuation range of joint 1 was the largest, but after optimization, its pulse values were significantly reduced. By reducing joint pulses, suppressing joint vibration, improving the accuracy of robot end trajectory, and ensuring milling quality.

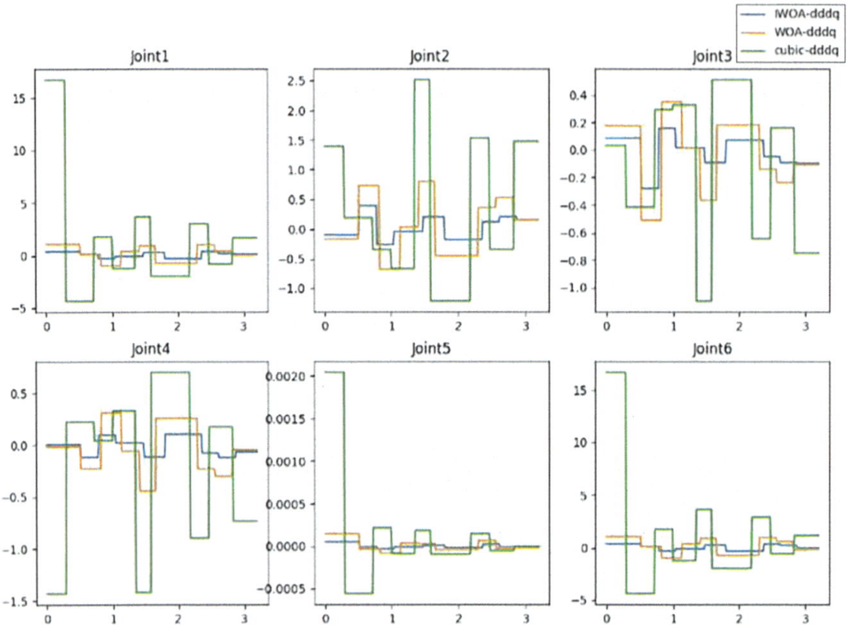

Fig. 5 Comparison of robot milling pulses under three methods

5 Conclusion

This article proposes an improved whale optimization algorithm to optimize the milling trajectory of a six degree of freedom UR5 robotic arm. In order to solve the problem of instability caused by vibration of various joints in the milling process of the robotic arm and improve work efficiency, the improved whale optimization algorithm is adopted to optimize each trajectory segment of cubic spline interpolation. Under the premise of satisfying the kinematic constraints of the robotic arm, the time required for the robotic arm to complete the entire trajectory was reduced from 18.52 s to 9.7 s. The overall time was shortened by nearly 50% compared to before using the algorithm. The maximum pulse values of each joint were reduced by 99.06%, 80.72%, 80.24%, 83.85%, 68.18%, and 99.76%, respectively. This not only shortened the operating time of the robotic arm but also improved its stability during operation. Therefore, it has been verified that the improved whale optimization algorithm is effective in optimizing milling trajectory time and pulse.

Acknowledgements This work was supported by the National Natural Science Foundation of China under Grant 92267301.

References

1. Chao H, Jian M, Li M, Chaoxing X, Chen W, Dawen R (2020) Time optimal robotic arm trajectory planning based on improved particle swarm optimization algorithm. J Shanghai Univer Eng Technol 34(03):238–246
2. Research on Huang Chao's Multi degree of Freedom Collaborative Robot Visual Assembly Method, Shanghai University of Engineering and Technology (2021). https://doi.org/10.27715/dcnki.gshgj.2021.000005
3. Qianqian D, Shaojie X (2022) Robot trajectory planning based on improved hybrid particle swarm optimization algorithm. Mach Tool and Hydraulic 50(17):50–56
4. Mingyue Z, Mingwei Z, Guiqi L et al (2023) Time optimal trajectory planning for robotic arms based on improved butterfly algorithm. Comput Sci 50(S2):119–126
5. Salgado P, Igrejas G, Afonso P (2017) Hybrid PSO-cubic spline for autonomous robots optimal trajectory planning. In: 2017 IEEE 21st international conference on intelligent engineering systems (INES), Larnaca, Cyprus, pp 000131–000136. https://doi.org/10.1109/INES.2017.8118542
6. Ni Yi Y, Jianfeng JY et al (2023) Trajectory planning of elevator shaft drilling robots based on multi-objective genetic algorithm. Mech Electron 41(11):49–56
7. Jie L, Xinyu B, Xiaoyong W, Juan L (2023) Robot time optimal trajectory planning with improved whale optimization algorithm. Mech Design and Manuf 02:270–275
8. Xinggui W, Chaowan Y, Xingrui M et al (1993) Optimal joint trajectory planning for robotic arms based on dynamic constraints. Robotics 05:20–26. https://doi.org/10.13973/j.cnki.robot.1993.05.005
9. Ting W (2021) Optimization and control of grinding robot trajectory based on improved whale optimization algorithm. North Central University. https://doi.org/10.27470/d.cnki.ghbgc.2021-0000004
10. Zhiqiang Z, Zhihong Z, Xing Z et al. (2021) Efficient time optimal trajectory planning for robots with full dynamics constraints. Inform Control 50(06):701–708+721. https://doi.org/10.13976/j.cnki.xk.2021.0546

Research on Chinese Word Segmentation Algorithm in the Tobacco Field Based on the BERT-BiLSTM-CRF Model

Yongsheng Wang, Weihua Feng, Xiaona Zhang, Hao Zhou, Hongxia Wang, Yinghao Zhang, and Yali Liu

Abstract This study addresses the issue of poor performance in segmenting specialized terms in Chinese texts within the tobacco domain, using traditional segmentation methods. A Chinese word segmentation method in the tobacco field has been proposed with combining bidirectional encoder representations from transformers and bidirectional long short-term memory-conditional random field (BERT-BiLSTM-CRF). Initially, a tobacco domain text annotation corpus is constructed, integrating general and specialized tobacco dictionaries to enhance the BERT model training. This approach takes advantage of BERT's understanding of language structures and semantic relationships, allowing more effective use of external knowledge. Then, the BiLSTM model serializes the text and further extracts features in context. Finally, the CRF model is utilized for learning word position labels, thus achieving an optimal word segmentation sequence. Experiments on tobacco domain texts and MSRA generic corpora demonstrate F_1-scores of 97.18% and 98.46%, respectively, indicating superior segmentation performance and generalization capabilities of this method compared to common approaches.

1 Introduction

In recent years, a plethora of textual data has been generated in the tobacco field, reaching millions of records including scientific papers, patents, standards, and technological achievements [1]. This data, rich in knowledge, encompasses fundamental details and potential insights into academic structures, research trends, and collaborative relationships. Efficiently leveraging this information can avert redundant research, reduce costs, and bolster knowledge discovery, thereby revealing the academic worth of scientific documents and the patterns of knowledge evolution [2].

Y. Wang · W. Feng · Y. Liu (✉)
Zheng Zhou Tobacco Research Institute of CNTC, Zhengzhou 450000, China
e-mail: lyl200681@163.com

X. Zhang · H. Zhou · H. Wang · Y. Zhang
China Tobacco Henan Industrial Co., Ltd, Zhengzhou 450000, China

S. Yadav et al. (eds.), *Proceedings of 3rd International Conference on Artificial Intelligence, Robotics, and Communication*, Lecture Notes in Electrical Engineering 1172, https://doi.org/10.1007/978-981-97-2200-6_32

The rapid growth of tobacco-specific texts poses significant challenges for information retrieval to meet the detailed knowledge service demands of users. Therefore, applying natural language processing (NLP) to analyze this text data is essential [3], with word segmentation being a foundational task.

Chinese word segmentation is the process of segmenting continuous character sequences into word sequences [4, 5]. It is a critical first step in text mining and impacts subsequent tasks like entity recognition and semantic analysis. While specialized segmentation research has been conducted in areas like medicine and the military [6, 7], there's a lack of research on segmentation algorithms for the tobacco field, where the presence of specialized terminology and unregistered words is substantial.

Traditional segmentation methods, both rule-based and machine learning-based, have limitations [8]. Rule-based methods suffer from ambiguity and rely heavily on comprehensive dictionaries [9]. Machine learning methods, while accurate, depend on the design of feature templates and require extensive annotated data. Deep learning, however, has made rapid strides in NLP [10, 11], decreasing reliance on domain dictionaries and feature templates, and enabling the automatic learning of feature hierarchies to exploit textual context. Thus, this paper proposes a BERT-BiLSTM-CRF deep learning algorithm tailored for tobacco text segmentation, demonstrating that this model achieves improved segmentation with minimal domain-specific annotated data.

2 Related Work

2.1 Construction of Annotated Tobacco Text Corpus

The neural network segmentation methods have significantly advanced cross-domain segmentation but still face limitations due to the lack of specialized domain-annotated data. To address the scarcity of publicly available annotated tobacco text corpora and the high costs of manual annotation [12], we developed text annotation standards tailored for the tobacco field and constructed a dedicated corpus. This corpus, a semantic-rich source describing tobacco entities and related texts, utilizes interactive iterative learning for semantic labeling. We collected over 76,000 domain-specific entries and 86,000 general entries from various disciplines related to tobacco over the past decade, ensuring representativeness and comprehensive coverage. Figure 1 shows the construction process of a text annotation corpus in the tobacco field. Among them, the tobacco text corpus collection stage is mainly responsible for the collection and integration of tobacco papers, patents, books, and other text corpora. The corpus preprocessing stage is mainly responsible for data cleaning, deduplication, denoising, and other text normalization processing of the collected raw corpus. At

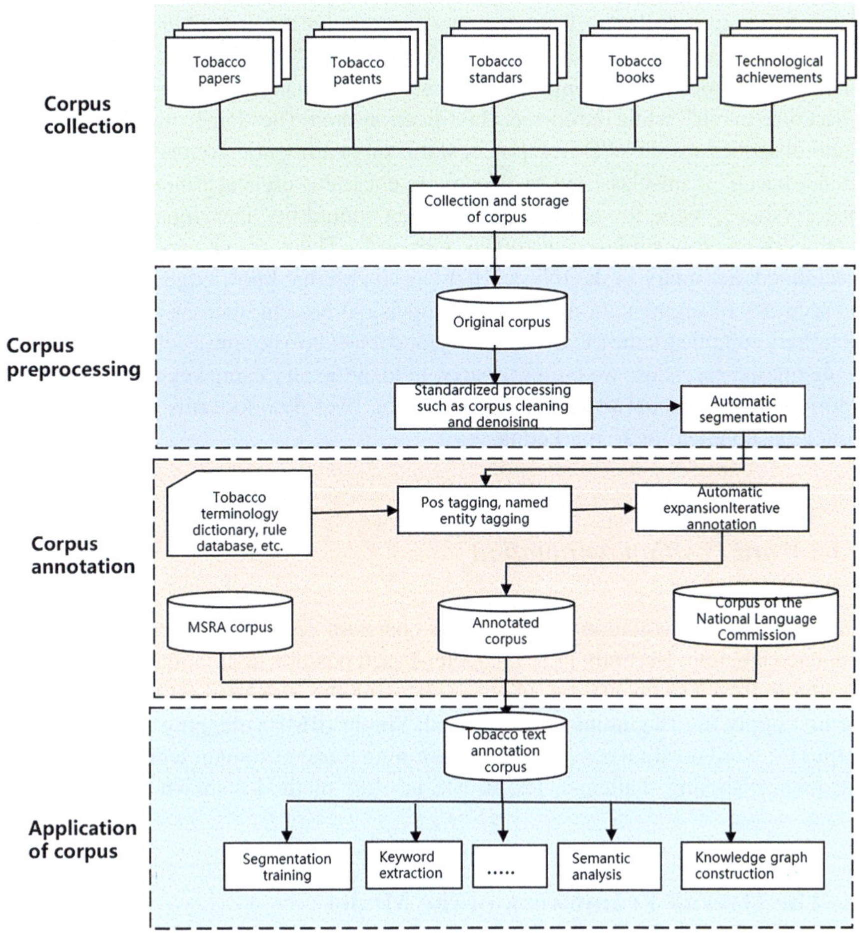

Fig. 1 Construction process of a text annotation corpus in the tobacco field

the same time, Chinese word segmentation algorithms are used to segment the standardized corpus. The corpus annotation process annotates and recognizes the preprocessed corpus based on the established annotation standards, ultimately generating a unified tobacco text annotation corpus, further supporting the training of tobacco segmentation algorithm models and research on domain knowledge discovery.

2.2 Construction of Tobacco Field Dictionary

As an important component of word segmentation models, dictionaries directly guide the model in breaking down domain-specific nouns [13, 14]. Tobacco science research involves multiple disciplines and contains a large number of proprietary terms, which are often misclassified. For example, "Research on paper-making reconstituted tobacco" is misclassified as "Paper-making method/reconstituted tobacco/research", while the correct classification should be "Paper-making reconstituted tobacco/research". "Development status on heating non-combustible reconstituted tobacco" is misclassified as "Development status on/heated/non-combustible/tobacco/slice", while the correct classification should be "Development status on/heating non-combustible/reconstituted tobacco". There is an urgent need for a specialized dictionary in the tobacco field as an external knowledge aid to improve the accuracy of segmentation algorithm models. At present, there is no professional dictionary suitable for the tobacco industry, and it needs to be constructed. To enhance segmentation precision, we built a tobacco field dictionary using keyword extraction algorithms and manual selection, incorporating over 59,000 terms covering areas from tobacco breeding to marketing.

2.3 *Word Position Annotation*

Chinese word segmentation views texts as character sequences segmented through sequence labeling. Recently [15], character-based position annotation methods have greatly improved segmentation. In our study using the BERT-BiLSTM-CRF model, we first apply the Beginning, Middle, End, Single (BMES) tagging method to the corpus for word position annotation, transforming tobacco domain segmentation into a sequence labeling challenge. The BEMS labeling method is shown in Table 1.

3 The Overall Framework of the Model

When using existing Chinese word segmentation methods for tobacco text processing, there are problems such as unregistered words and low accuracy in identifying ambiguous words, which cannot meet the needs of tobacco text analysis. How to use limited annotation training corpora and domain dictionaries to improve tobacco segmentation performance and provide more diverse segmentation results for downstream tasks under the condition of limited domain annotation data? Based on the characteristics of tobacco text data, combined with tobacco domain dictionaries,

Table 1 BEMS annotation

Annotation name	Annotation description
B	The beginning character of the word
M	The middle character of the word
E	The ending word of the word
S	The single word

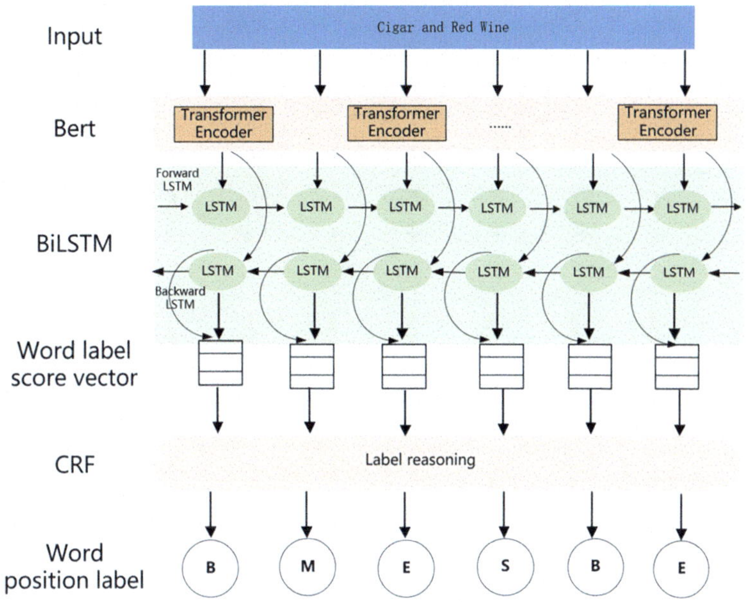

Fig. 2 Overall framework of the BERT-BiLSTM-CRF model

and drawing on the basic ideas of transfer learning, this project adopts the BERT-BiLSTM-CRF deep learning algorithm, A word segmentation algorithm model suitable for the tobacco field has been studied and constructed. The model framework consists of three layers: the BERT character embedding layer, the BiLSTM feature extraction layer, and the CRF label prediction layer. The framework is shown in Fig. 2. Starting with BERT's pre-trained model to process tobacco text, it captures semantically rich character vector representations. These vectors are then input into the BiLSTM layer to obtain a feature representation of the text sequence, which predicts a tagging sequence. The CRF layer fully utilizes the adjacent label information to calculate and output the most probable label sequence, thus optimizing the entity classification and completing the word segmentation process.

3.1 The BERT Model

The key advantage of the BERT-BiLSTM-CRF model for tobacco domain segmentation is the application of BERT, a pre-trained language representation model based on transformer architecture [16]. It offers bidirectional representations and transforms input characters into word vectors by consulting a vocabulary table. The model outputs vectors enriched with the full semantic context of the input text. This integration means that for tobacco domain segmentation, the embedding layer doesn't

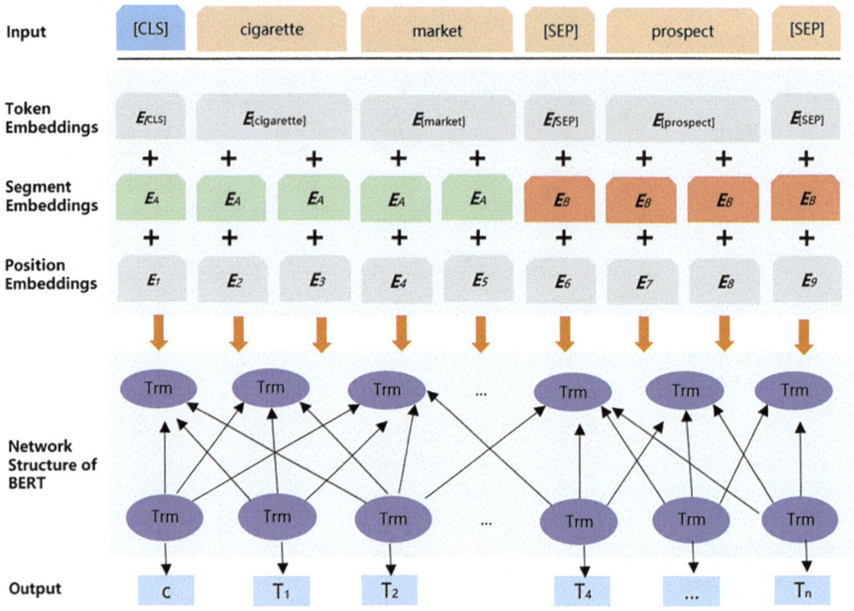

Fig. 3 Overall structure of the BERT model

require pre-training of large-scale characters or word vectors. Input characters are directly fed into BERT, automatically extracting rich lexical, syntactic, and semantic features, it is shown in Fig. 3. The BERT model outputs word vectors comprised of Token, Segment, and Position Embeddings. Additionally, BERT introduces special [CLS] and [SEP] tokens to indicate the beginning and separation of text sequences, respectively.

3.2 *The BiLSTM Model*

The long short-term memory (LSTM) model [17], a specific type of recurrent neural network (RNN), is engineered to resolve the challenges of gradient vanishing and exploding in lengthy sequences [18]. It does this through a series of gates—the input, forget, and output gates—along with a memory cell that collectively filters out irrelevant information while preserving what's essential for future predictions. However, a unidirectional LSTM can only assimilate information in one direction, lacking the ability to fully grasp the entire context of a sentence. To overcome this, the bidirectional long short-term memory (BiLSTM) model processes each word sequence in both forward and backward directions, merging the outputs at each time step to retain context from both past and future states. This dual-structure allows for a complete contextual understanding of the sentence or text. The structure of the

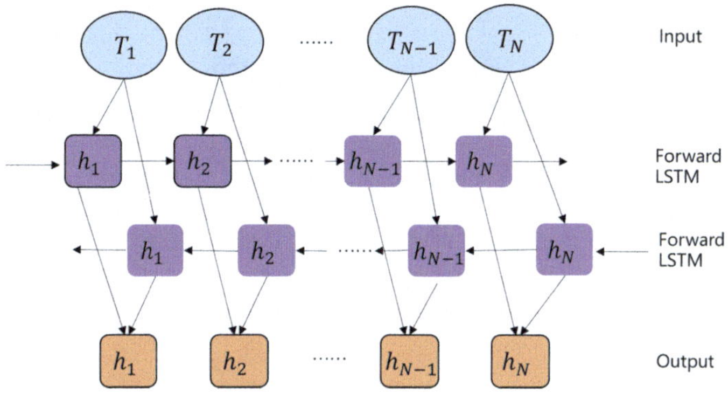

Fig. 4 Overall structure of the BiLSTM model

BiLSTM model is shown in Fig. 4 and is crucial for the feature extraction layer within the BERT-BiLSTM-CRF framework for tobacco text segmentation. Here, character sequences from tobacco texts are inputted, with the BiLSTM modeling the contextual information, leading to a contextual representation vector sequence for the tobacco domain. These vectors are then used to calculate specific scores for each tag before being sent to the CRF prediction layer.

3.3 The CRF Model

The CRF model is an algorithm that determines the conditional probability distribution of output random variables given specified input random variables [19, 20]. It is commonly used in NLP tasks like part-of-speech tagging and named entity recognition. The model's input comes from the BiLSTM feature extraction layer, which might not account for dependencies between characters and labels, leading to possible inaccuracies. The CRF model's strength lies in learning hidden structural rules within sentences, modeling the label sequences to yield the most optimal global sequence, and enhancing the accuracy of segmentation. The CRF model structure is shown in Fig. 5. Among them, Y_n represents all possible label sequences for a given sentence.

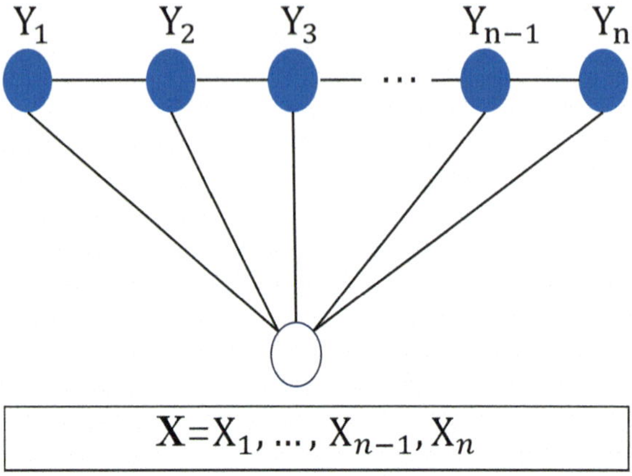

Fig. 5 Overall structure of the CRF model

4 Experiment

4.1 Experimental Data and Evaluation Indicators

Currently, there is no publicly available annotated corpus for tobacco text, which is essential for verifying the effectiveness of the proposed BERT-BiLSTM-CRF segmentation model for the tobacco domain. Therefore, for experimental validation, the corpus constructed in Sect. 2.1 is utilized. In the experiment, 80% of the sentences were randomly selected to form the training set, 10% for validation, and the remaining 10% for testing. The MSRA (Microsoft Research Asia) corpus was also chosen for a comparative study [21].

For the evaluation of the segmentation model, the metrics used are precision (*P*), recall (*R*), and F1-score, which are standard measures of model performance [22]. The formulas for these metrics are as follows:

$$P = \mathrm{TP}/(\mathrm{TP} + \mathrm{FP})^{*}100\% \tag{1}$$

$$R = \mathrm{TP}/(\mathrm{TP} + \mathrm{FN})^{*}100\% \tag{2}$$

$$F_1 = 2^{*}P^{*}R/(P + R)^{*}100\% \tag{3}$$

Here, TP (True Positives) represents the number of words correctly segmented; TP + FP (False Positives) is the total number of words segmented by the model; TP + FN (False Negatives) is the number of words in the standard segmentation set.

Table 2 Experimental environment and parameters

Parameters	Value
CPU	Intel (R) Xeon (R) Gold 5220R CPU @ 220 GHz 219 GHz
GPU	GeForce RTX3090
Operating system	Ubuntu 20.04
Programming language	Python 3.9
Deep learning framework	PyTorch1.7.1

4.2 Experimental Environment and Parameters

The experimental environment used in the study is shown in Table 2. For the construction of the BERT-BiLSTM-CRF model specifically for tobacco domain text segmentation, the PyTorch framework was used. The model's parameters were set as follows: a transformer with 12 layers, 768-dimensional hidden units, 12 self-attention heads, and an embedding dimension of 128. The batch size for the training set was set at 64 with a learning rate of 5 × 10–5. The model underwent 100 epochs of training, incorporating a dropout rate of 0.5 to mitigate overfitting.

4.3 Experimental Results and Analysis

To evaluate the performance of the proposed BERT-BiLSTM-CRF model for tobacco domain text segmentation, we conducted experiments on both a tobacco text dataset and the MSRA dataset, comparing it with other models like Bert-CRF, BiLSTM-CRF, CRF model, and the Jieba segmentation algorithm. The outcomes of these experiments are presented in Tables 3 and 4.

Table 3 Word segmentation effect of different models in datasets of tobacco

Word segmentation model	*P* (%)	*R* (%)	F_1 (%)
Jieba	78.83	76.85	77.83
CRF	91.02	85.27	88.05
BiLSTM-CRF	96.01	94.70	95.35
BERT-BiLSTM-CRF	**97.05**	**97.31**	**97.18**

Table 4 Word segmentation effect of different models in datasets of MSRA

Word segmentation model	*P* (%)	*R* (%)	F_1 (%)
Jieba	90.24	82.22	86.04
CRF	91.18	87.69	89.40
BiLSTM-CRF	96.49	95.59	96.04
BERT-BiLSTM-CRF	**98.35**	**98.58**	**98.46**

The results, particularly the F1-score of the BERT-BiLSTM-CRF model reaching 97.18% on the tobacco corpus, significantly outperformed the other models, highlighting the model's advantage in segmenting tobacco domain texts. This success is attributed to the model's integration of domain-specific terminological features and the transfer learning from the pre-trained BERT model, which effectively captures character-level, word-level, sentence-level, and even inter-sentence features, thus enhancing segmentation performance in the tobacco domain.

A comparison between the results on the general domain dataset (MSRA) and the tobacco domain dataset reveals that all four segmentation algorithms performed better on the general domain dataset. Specifically, the BERT-BiLSTM-CRF and BiLSTM-CRF models showed an increase of 1.28% and 0.69%, respectively, in the general domain compared to the tobacco domain. This could be due to the abundance of specialized terminology and a large number of out-of-vocabulary words in the tobacco texts, which leads to incomplete training of the models and thus does not achieve the desired optimal segmentation outcome. To improve the segmentation results in the tobacco dataset, it is necessary to augment the training data within this domain and further optimize and train the model to better understand Chinese texts related to tobacco, thereby maximizing the segmentation capabilities of the model.

5 Conclusion

This paper conducted research on Chinese word segmentation in the tobacco industry and proposed a tobacco domain word segmentation algorithm based on the BERT-BiLSTM-CRF model. By integrating a tobacco domain dictionary and incorporating tobacco-specific features into the neural network training process, the algorithm improved its adaptability within the tobacco domain. Leveraging the powerful semantic representation capabilities of the BERT model further enhanced the extraction of character vector features, consequently boosting the overall performance of domain-specific word segmentation. Experimental comparisons were conducted on tobacco domain text corpora and the MSRA general corpora. The results demonstrated that the tobacco domain word segmentation algorithm based on the BERT-BiLSTM-CRF model achieved F1-scores of 97.18 and 98.46% on these respective datasets. These findings indicate that, compared to commonly used word segmentation methods such as BERT-CRF, BiLSTM-CRF, CRF, our approach exhibits superior word segmentation performance.

Acknowledgements This work is supported by the Key Science and Technology Program of Henan Province. China (No. 232102210073) and the Planning Project for the Scientific Research and Technological Development of China Tobacco Henan Industrial Co., Ltd. (AW202169), as well as by the Science and Technology Major Program of China National Tobacco Corporation (N0.110202101031(SJ-02), No. 110202101081(SJ-05)). We would like to express our sincere thanks to the anonymous reviewers and editors for their valuable comments and suggestions for this paper.

References

1. Wang YS, Liu YL, Jia N, et al. (2022) Design and implementation of knowledge retrieval service platform for tobacco literatures. Tobacco Sci Technol 55(3):107-112. https://doi.org/10.16135/j.issn1002-0861.2022.0021
2. Wang WJ, Li N, Zheng XZ, et al. (2021) Key technologies of the service platform of scientific research knowledge graph in tobacco field. Acta Tabacaria Sinica 27(4):83–91. https://doi.org/10.16472/j.chinatobacco.2020.256
3. Zwilling M (2023) Big Data challenges in social sciences: an NLP analysis. J Comput Inf Syst 63(3):537–554. https://doi.org/10.1080/08874417.2022.2085211
4. Wen Ph, Fen Lh, Zhang T (2022) A hybrid Chinese word segmentation model for quality management-related texts based on transfer learning. PloSone 17(10):e0270154. https://doi.org/10.1371/JOURNAL.PONE.0270154
5. Qiu Q, Xie Z, Wu L (2018) A cyclic self-learning Chinese word segmentation for the geoscience domain. Geomatica 72(1):16–26. https://doi.org/10.1139/geomat-2018-0007
6. Li S, Li MZ, Xu YJ, et al. (2018) Capsules based Chinese word segmentation for ancient Chinese medical books. IEEE Access 670874–70883. https://doi.org/10.1109/access.2018.2881280
7. Li JL, Wang PQ, Han QY (2018) Research on Chinese word segmentation technology for military field. Comput Modernization 1006–2475. https://doi.org/10.3969/j.issn.1006-2475.2018.11.020
8. Zhou W, Wang A, Shu H et al. (2018) Word segmentation by alternating colors facilitates eye guidance in Chinese reading. Mem Cogn 46:729–740. https://doi.org/10.3758/s13421-018-0797-5
9. Yang S (2021) Investigating word segmentation of Chinese second language learners. Read Writ 34:1273–1293. https://doi.org/10.1007/s11145-020-10113-6
10. Yuan L (2022) A joint method for Chinese word segmentation and part-of-speech labeling based on deep neural network. Soft Comput 26:5607–5616. https://doi.org/10.1007/s00500-022-07093-w
11. Yu J, Xie L, Xiao X et al (2017) A hybrid neural network hidden Markov model approach for automatic story segmentation. J Ambient Intell Human Comput 8:925–936. https://doi.org/10.1007/s12652-017-0501-9
12. Pan J, Yan M, Richter EM et al (2022) The Beijing sentence corpus: a Chinese sentence corpus with eye movement data and predictability norms. Behav Res 54:1989–2000. https://doi.org/10.3758/s13428-021-01730-2
13. Islam A, Inkpen D, Kiringa I (2008) Applications of corpus-based semantic similarity and word segmentation to database schema matching. VLDB J 17:1293–1320. https://doi.org/10.1007/s00778-007-0067-9
14. Zhang JQ (2023) A Levenshtein distance-based method for word segmentation in corpus augmentation of geoscience texts. Annals of GIS 29.2:293–306. https://doi.org/10.1080/19475683.2023.2165543
15. Qun, N, Yan H, Qiu, XP et al. (2020) Chinese word segmentation via BiLSTM+Semi-CRF with relay node. J Comput Sci Technol 35:1115–1126. https://doi.org/10.1007/s11390-020-9576-4
16. Ma K et al. (2023) Ontology-based BERT model for automated information extraction from geological hazard reports. J Earth Sci 34.05:1390–1405. https://doi.org/10.1007/S12583-022-1724-Z
17. Sahbi B, Mohamed B, Adel M, et al. (2021) Deep neural networks for predicting solar radiation at hail Region, Saudi Arabia. IEEE Access 936719–36729. https://doi.org/10.1109/ACCESS.2021.3062205
18. Prem Anand M, Anand M, Adams Joe M et al (2023) Lightweight Bi-LSTM method for the prediction of mechanical properties of concrete. Multimed Tools Appl. https://doi.org/10.1007/s11042-023-17796-3
19. Ijazul H et al. (2023) Correction of whitespace and word segmentation in noisy Pashto text using CRF. Speech Commun 153. https://doi.org/10.1016/J.SPECOM.2023.102970

20. Nuo Q, Hang Y, Peng XQ et al (2020) Chinese word segmentation via BiLSTM+Semi-CRF with relay node. J Comput Sci Technol 35(5):1115–1126. https://doi.org/10.1007/s11390-020-9576-4
21. Hongshuai L, Ge J, Yuanyuan Z (2021) Chinese named entity recognition model based on BERT. In: MATEC web of conferences, pp 33606021–06021. https://doi.org/10.1051/MATECCONF/202133606021
22. Hao FW, Meili L, Suiping W (2023) Perceptual intake explains variability in statistical word segmentation. Cognition 241105612–105612. https://doi.org/10.1016/J.COGNITION.2023.105612

A MPGNN-Based Unsupervised Learning Framework for Power Control in D2D Networks

Yan Zhang, Yucheng He, and Yu Zhang

Abstract The Internet of Things (IoT) allows physical devices to be connected over wireless networks, and while device-to-device (D2D) communication has become a promising technology for IoT, traditional solutions for D2D resource allocation are often computationally complex and time-consuming. In order to improve the computational efficiency and improve the generalization performance of the algorithm, a large-scale power control unsupervised learning framework algorithm based on Message Passing Graph Neural Network (MPGNN) was proposed, which aims to maximize the weighting and rate of all D2D links. To achieve the above goals, we first model the topology of the wireless network as a graph model, define the channel matrix as the graph structure, and then model the objective function of the system as a loss function and train the model parameters in the MPGNN with the help of the unsupervised learning model. After offline training, each D2D link can obtain the optimal power control strategy in a distributed manner according to the channel state information (CSI). Simulation shows that compared with the classic algorithm based on optimization theory, Weighted Minimum Mean Square Error (WMMSE), the computational efficiency of the proposed algorithm is increased by more than 50 times when the number of users is 100 pairs. In addition, compared with the traditional neural network algorithm, deep neural network (DNN), the proposed algorithm improves the generalization ability of the system under the same parameter settings.

Y. Zhang (✉) · Y. He · Y. Zhang
Xiamen Key Laboratory of Mobile Multimedia Communications, Huaqiao University, Xiamen 362021, China
e-mail: 1059291489@qq.com

Y. He
e-mail: yucheng.he@hqu.edu.cn

Y. Zhang
e-mail: zydbh120@163.com

S. Yadav et al. (eds.), *Proceedings of 3rd International Conference on Artificial Intelligence, Robotics, and Communication*, Lecture Notes in Electrical Engineering 1172, https://doi.org/10.1007/978-981-97-2200-6_33

1 Introduction

With the rapid popularization of IoT technology in the fields of industry and people's livelihood, IoT communication or device-to-device communication has become an important application scenario in the next five generations of mobile networks. In this context, device-to-device (D2D) communication has received extensive attention from academia and industry. Compared to traditional cellular communications, D2D communication no longer relies on various base stations or relay nodes, but allows data transmission directly from neighboring devices [1]. By taking advantage of the good wireless channel environment between neighboring devices and sharing the scarce spectrum resources in traditional cellular networks, D2D communication can effectively improve spectrum efficiency, reduce power overhead, shorten communication delay, and offload base station load. However, in the future dense mobile network architecture, the coupling relationship between links will become more complex, resulting in the non-convex characteristics of the power control problem [2], and the global optimal solution cannot be guaranteed in polynomial time. Therefore, most of the existing studies focus on the design of suboptimal algorithms with low complexity, aiming to balance optimization performance and computational complexity. Traditional algorithms based on optimization theory, such as geometric programming (GP) [3] and weighted minimum mean square error (WMMSE) [4], have excellent optimization performance but high computational complexity, and with the increase of network scale, it is difficult to meet the requirements of low latency in communication. In recent years, deep learning (DL) has proven to be a competitive candidate for solving non-convex resource allocation problems, and its basic idea is to achieve fast power control by learning the mapping relationship of power allocation functions, when the model is trained, it can be directly applied to new communication scenarios. H. Sun et al. originally proposed a "learning optimization" method [5] that uses a deep neural network (DNN) to simulate the input–output mapping of WMMSE in a supervised manner. The pre-trained neural network model requires much less computation than WMMSE, but its main drawbacks are the computational burden of generating labels from WMMSE and the poor generalization ability to large problem sizes. Unsupervised learning DNNs are used in Ref. [6, 7] to obtain the power control strategy by directly maximizing the target utility function, which improves the computational burden problem, without needs to generate labels, but they still have the problem of generalization difficulties. Based on the above limitations, this paper proposes an unsupervised learning framework based on Message Passing Graph Neural Network (MPGNN) [8], which uses the underlying topology of the wireless network to model it into a graph model, specifically, a pair of D2D is regarded as a node, in which the channel gain of the direct link is modeled as node features, the channel gain of the interference link is modeled as the edge feature, and the MPGNN is used to learn the complex relationship between the node and the edge. It can be naturally generalized to large-scale wireless networks and adapt to the dynamic characteristics of the wireless transmission environment. At the same time, this calculation process only needs to input the channel matrix, and we can get the best power control scheme according to the trained MPGNN model, without the need to solve any optimization problems, only involving simple algebraic operation, which greatly improves the calculation speed compared with traditional algorithms, such as WMMSE.

The rest of this article is structured as follows. In the second section, we introduce our proposed system model, and in the third section, we describe the proposed large-scale power control algorithm based on the MPGNN unsupervised learning framework, and in the fourth section, we evaluate the performance of the proposed algorithm and compare it with other algorithms. Finally, in the section five, we summarize our observations and propose directions for future work.

2 System Model

As shown in Fig. 1, the D2D communication system considered in this article consists of K D2D communication pairs $N = \{1, 2, \ldots, k, j, \ldots K\}(k \in N, k \neq j)$, where N represents the set of D2D communication pairs, k and j represent the index values of the receiving user and the sending user, and each D2D user pair contains a sending user and a receiving user. For the sake of subsequent description, the identity K represents the K-th D2D communication pair. In order to focus on the design of the power control algorithm and maximize the system capacity, this paper considers the single-channel scenario, so it is assumed that all D2D communication pairs reuse the same channel resources (time–frequency resource blocks). In the single-antenna scenario, each network node is equipped with a single omnidirectional antenna at the transmitter and receiver ends. The channel gain of the target link between the transmitting user and the receiving user (as shown in the solid red arrow in Fig. 1) is h_{kk}, , which includes large-scale path loss, shadow fading, and small-scale channel fading. Similarly, the channel gain of the interfering link (as shown in the green solid arrow in Fig. 1) between the transmitting user and the receiving user is h_{kj}. In addition, the variance of the received user-side additive Gaussian noise is recorded as σ_k^2. . Based on the above definition, the signal-to-noise ratio (SINR) of the target link can be expressed as:

$$\mathrm{SINR} = \frac{|h_{kk}|^2 p_k}{\sum\limits_{j \neq k} |h_{kj}|^2 p_j + \sigma_k^2}, \tag{1}$$

where p_k indicates the transmit power of the transmitting user side.

The goal of this paper is to find the optimal power distribution, so that the weighted rate and maximum, the formula of the problem can be written as:

$$\begin{aligned} &\underset{p}{\text{maxmize}} \ \sum_k w_k \log_2(1 + \mathrm{SINR}_k) \\ &\text{subject to } 0 \leq p_k \leq P_{\max}, \ \forall k, \end{aligned} \tag{2}$$

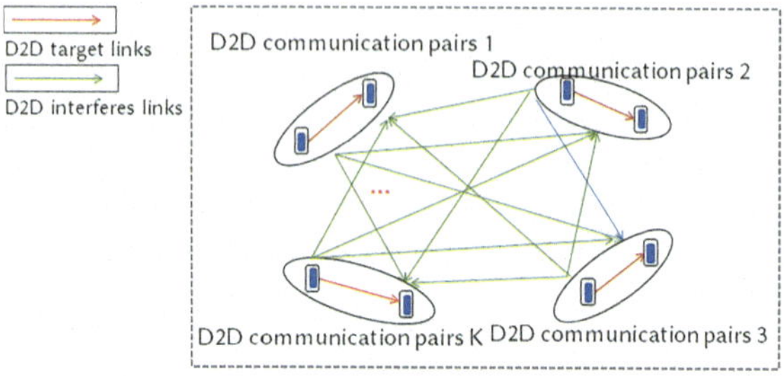

Fig. 1 D2D communication system

where w_k is the weight of user k. The priority of different users is reflected, and the higher the weight, the higher the priority of the user.

3 A Scalable Power Control Policy for MPGNN-Based Unsupervised Learning Framework

From Formulas (1) and (2), it can be seen that the objective function (2) is a summation form of a fractional function, so it is a non-concave function and cannot be equivalently transformed into a concave function through variable transformation [9]. Therefore, (2) belongs to the classic non-deterministic polynomial time difficulty problem in optimization theory [10]. Although traditional model-driven optimization algorithms can obtain an approximate solution in polynomial time, their computational complexity will increase exponentially as the number of links increases. In addition, due to the fading characteristics of the channel, the DNN models trained during the training phase may experience a sharp decline in optimization performance during the testing phase, accompanied by changes in network size or deployment parameters, resulting in difficulty in generalization [11]. In order to improve the computational speed of the algorithm while maintaining its generalization ability for D2D communication networks, we propose a large-scale power control algorithm based on the unsupervised learning framework of MPGNN.

3.1 Graph Model

The network topology of the D2D communication system (target link and interference link in Fig. 1) is a natural graph model, and the Message Passing Graph Neural Network (MPGNN), as a deep learning architecture for graph model data, exhibits excellent performance in extracting and characterizing the connection relationships and logical structures in the graph model. Compared to traditional neural network models (DNN), MPGNN has the following advantages: it can handle non-Euclidean spatial data, better mine hidden information in graph models by learning

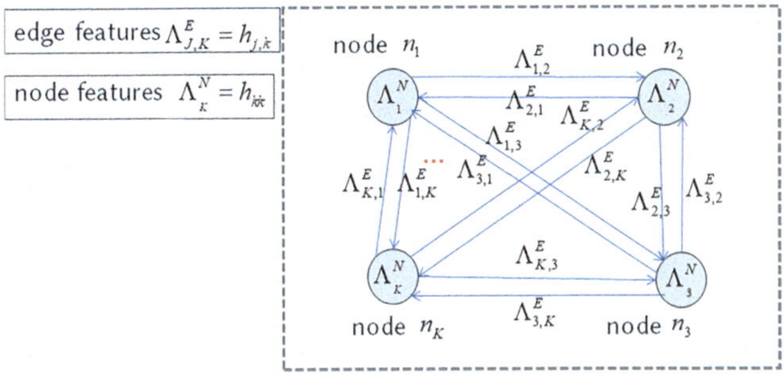

Fig. 2 Graph model of D2D communication system

complex relationships between nodes and edges, and performs well in processing large-scale complex graph data. In our task, CSI data is non-Euclidean spatial data, so GNN theory has the potential to solve complex power control problems in D2D communication systems.

As shown in Fig. 2, the data structure in the graph model can be represented as a binary $G = (N, E, \Lambda)$, where N represents the set of nodes (as shown in the blue node in Fig. 2), $N = \{n_1, n_2, n_3, ..., n_K\}$, and E represents the set of edges (as shown in the green directed segment in Fig. 2), $E = \{[n_1, n_2], [n_2, n_1], ..., [n_K, n_J]\}$. Λ represents the eigenvalues of the graph structure, which are node features and edge features, where in the channel gain value of the direct link of each D2D communication pair in the objective function is taken as the node feature, and the channel gain value of the interfering link is taken as the edge feature, specifically, the feature of the K-node are: $\Lambda^N_K = h_{kk}$, and the feature of the edge from the J-node to the K-node is $\Lambda^E_{J,K} = h_{j,k}$ [12].

3.2 *Our Algorithm Based on MPGNN Frameworks*

In this section, we introduce our algorithm, first of all, the preprocessing of data, followed by the design of the graph neural network framework, and finally the loss function has been updated by the unsupervised paradigm implementation for reverse gradient calculation to obtain the optimal power allocation scheme. The specific steps are as follows:

1. Construct graph data: After we generate CSI data, we first do a normalization process, and then treat a pair of D2D communication pairs as a node, the features of the k-th node pair are written as: $\Lambda^N_k = (p_k, h_{kk})$, and the edge feature of the j-th node pair connecting the k-th node pair is written as: $\Lambda^E_{j,k} = h_{j,k}$, so the input graph is $G = (N, E, \Lambda) = (\Lambda^N, \Lambda^E)$
2. Initialization: The power allocation value p_k is initialized as an all-one vector, and the MPGNN network parameters are initialized randomly

3. Node Message Passing Process: As the core operation of MPGNN, the main function of message passing is to share information between nodes through graph convolution operations, so as to extract the feature information of nodes. It consists of the following two stages: 1) Aggregation: In an arbitrary graph convolutional layer, each node collects the implicit state of all neighboring nodes and connected edges in the previous layer according to the connection relationship between nodes. 2) Combination: Each node updates its own hidden state based on the hidden state collected during the aggregation process. Specifically, the aggregation phase: $a_k^{(t)} = \mathrm{MAX}\left(\mathrm{MLP1}\left(\Lambda_j^{(t-1)} \middle\| \Lambda_{j,k}^E\right)\right)$; Combination stage: $d_k^{(t)} = \mathrm{MLP2}(a_k^{(t)} \| \Lambda_k^{(t-1)})$, through message passing, each node collects the information of neighbor nodes and connected edges, and in the process of continuous iteration of message passing, gradually learning the relationship between nodes and global characteristics will help nodes perceive and understand the entire graph model from a global perspective. This global information fusion capability and feature mining capability make MPGNN more efficient and accurate when processing graph models with complex topologies.
4. Output optimal power control: After the node is updated, the final embedding feature is the optimal power allocation $p_k = d_k^{(t)}$.
5. Backpropagation: Adam algorithm is used to optimize the model parameters W_1, W_2 in MPGNN in the way of backpropagation.

We named the proposed algorithm MPGNN-PC, and the pseudocode looks like this:

ALGORITHM 1: MPGNN-PC
generate channel state information and convert the channel state information into the graph model described above, and the node features are as follows: $\Lambda_k^N = (p_k, h_{kk})$, edge features denote as: $\Lambda_{j,k}^E = h_{j,k}$.
Initialize to p_k all 1 vectors, and initialize the MPGNN network parameters randomly.
for t: = 1 to T do.
for k: = 1 to K do
$d_{N(k)}^{(t)} \leftarrow \mathrm{MAX}(\mathrm{MLP1}(\mathrm{W}_1[\Lambda_j^{(t-1)} \| \Lambda_{j,k}^E]))$
$d_k^{(t)} \leftarrow \mathrm{MLP2}(\mathrm{W}_2[d_{N(k)}^{(t)} \| \Lambda_k^{(t-1)}])$
end
If $t \neq T$ then
$d_k^{(t)} \leftarrow d_k^{(t)} \| d_k^{(t-1)}$
end

(continued)

(continued)

loss function $= -\sum_{k} w_k \log_2\left(1 + \frac{|h_{kk}|^2 p_k^{(t)}}{\sum_{j \neq k} |h_{kj}|^2 p_j + \sigma_k^2}\right)$, backpropagation optimization using Adam's algorithm W_1, W_2.

4 Numerical Evaluation

4.1 Simulation Settings

In this paper, a square single-cell system is considered and the cell area is: 1000 $m \times 1000$ m. In this system, the transmitting users in the D2D communication pair are randomly distributed in the cell range, and the receiving users are randomly distributed in the area from 2 to 65 m away from the corresponding transmitting users, the range of transmitting power of the D2D communication pair is 0 ~ 1 W, and the sum weight of the target link is a random value between 0.5 ~ 1. In the channel transmission model, the path loss is $148.1 + 37.6\ \log_2(d_{kj})$dB, which d_{kj} represents the distance between the transmitting user and the receiving user, the shadow fading obeys a lognormal distribution with a mean value of 0 and a variance of 8 dB, the small-scale fading obeys the Rayleigh distribution with a scale parameter of 1, and the power of the receiving user-side additive Gaussian white noise in the D2D communication pair is $\sigma_k^2 = -102$ dBm.Tables 1 and 2 are used to set the specific parameters of system and hyper-parameters. In the algorithm implementation, simulation platforms and test tools based on Python 3.7.13, PyTorch 1.10.1, and PYG 1.0.2 are used.

Table 1 System parameters

Parameters	Numeric value
The area of the residential area	1000 $m \times 1000$ m
D2D communication to distance	$2 \sim 65$ m
D2D communication pair transmit power	$0 \sim 1$ W
Path loss	$148.1 + 37.6\ \log_2(d_{kj})$dB
Shadow fading	log normal(0, 8)
Small-scale decline	Rayleigh(1)
Gaussian white noise	– 102 dBm
Sum weights	$0.5 \sim 1$

Table 2 MPGNN-PC decision model hyper-parameter settings

Parameters	Numeric value
The number of convolutional layers in the MPGNN graph	2
Pooling	MAX Pooling
MLP1 layers and neuronal parameters	(2, 2, 64, 128)
MLP2 layers and neuronal parameters	(2, 129, 64, 1)
MLP1 hidden layer activation function	Relu
MLP1 output layer activation function	Relu
MLP2 hidden layer activation function	Relu
MLP2 output layer activation function	Sigmoid
Optimizer	Adam
Learning rate	0.001

4.2 Algorithm Performance Comparison

In this section, we compare the traditional optimization algorithm (WMMSE), the traditional neural network algorithm (DNN), and our proposed algorithm (MPGNN-PC) in terms of computational efficiency and generalization performance, and prove that our algorithm still has good generalization ability while maintaining the speed of the algorithm.

Computational efficiency First of all, we compare the proposed algorithm with WMMSE. For the sake of fair competition, we are running MPGNN-PC and WMMSE on the same hardware, MPGNN-PC does not use GPU acceleration, and we randomly generate 1000 channel instances and calculate the average execution time. Table 3 shows that the average execution time of solving problems of different sizes increases dramatically with the increase of problem size, while the growth of MPGNN-PC is not significant. Compared to WMMSE, when the problem was extended to 100 user pairs, MPGNN-PC achieved more than 200x speedup.

Secondly, we evaluate the computational efficiency between our proposed algorithm and DNN via the sampling complexity. The smaller the number of samples, the higher the computational efficiency. We set the number of users be 10 and 20, and the signal-to-noise ratio be 10 dB, as shown in Fig. 3. For $K = 10$, MPGNN-PC trained with 20 and 40 training samples achieve similar performance to DNNs trained with 2 000 and 4 000 training samples, respectively. Notably, for $K = 20$,

Table 3 Comparison of the calculation time between WMMSE and MPGNN-PC algorithm units

Numbers of D2D pairs	$K = 10$	$K = 50$	$K = 100$	$K = 130$
WMMSE algorithm	0.0300	0.3048	2.2105	3.1456
MPGNN-PC algorithm	0.0070	0.0076	0.0080	0.0096

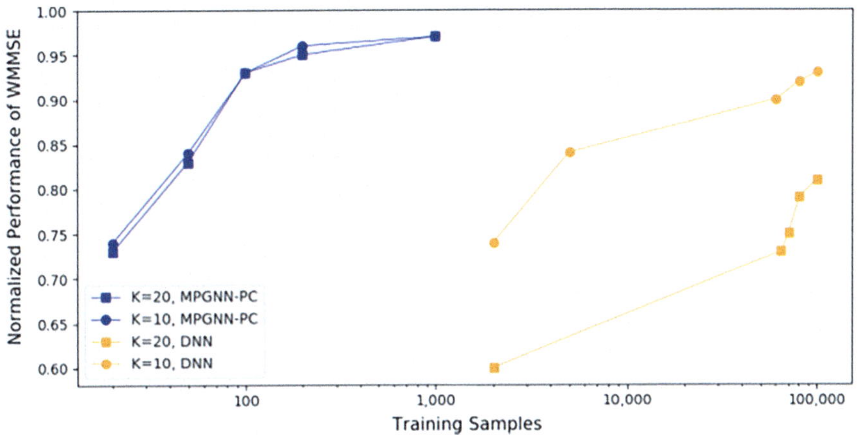

Fig. 3 Performance between MPGNN-PC and DNN when K = 10 and 20

MPGNN-PC trained with 20 and 40 training samples achieve similar performance to DNNs trained with 16 000 and 64 000 training examples, respectively. This proves that MPGNN-PC requires less sample complexity than DNN, which is about the order of magnitude $O(K^2)$.

Generalization performance We take the performance of WMMSE as the normalized reference benchmark, first train MPGNN-PC with ten user pairs in an area of 1000×1000 m, then change the number of users in the test set, and compare the performance of MPGNN-PC and DNN with the increase of the number of users, as shown in Fig. 4. As can be seen, the MPGNN-PC performance remains stable, indicating that our proposed algorithm can be well generalized to larger network scales, while the DNN performance decreases sharply, indicating generalization failure, which is consistent with our analysis.

5 Conclusions

In a spectrum-densely multiplexed D2D communication system, this paper aims to maximize the weighting and rate by optimizing the transmit power of the D2D communication pair. In order to improve the computing efficiency of the algorithm while maintaining the generalization performance by using the method based on deep learning, an unsupervised power control algorithm based on MPGNN was proposed, which was modeled into graph data according to CSI, and the power mapping function was approximated in the feature space, then the resulting node embedding was the power optimal control decision. There is no need to rely on optimization theory and complex mathematical algebraic calculations. The simulation results show that our proposed algorithm has excellent computational advantages, and the calculation is

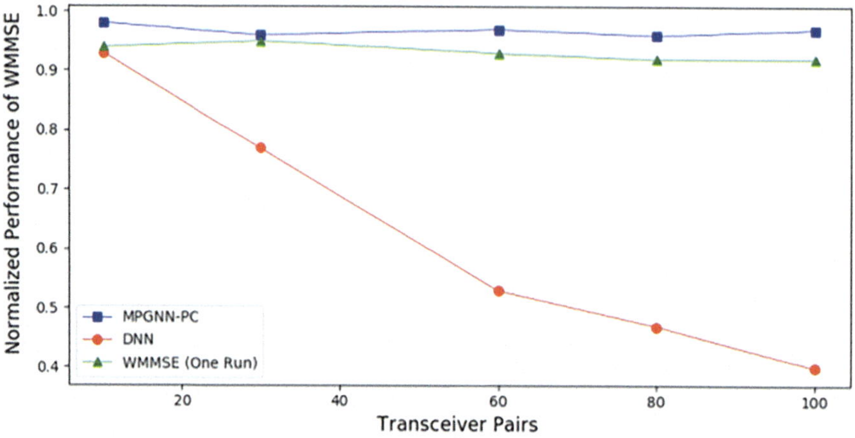

Fig. 4 Performance between MPGNN-PC, DNN, WMMSE when K increases from 10 to 100

improved by more than 50 times compared with the traditional optimization algorithm WMMSE when user pairs are 100. At the same time, our proposed algorithm has strong generalization ability in different system parameter settings compared with DNN.

References

1. Zhang S, Liu J, Guo H, Qi M, Kato N (2020) Envisioning device-to-device communications in 6G. IEEE Netw 34(3):86–91. https://doi.org/10.1109/MNET.001.1900652
2. Chiang M, Hande P, Lan T, Tan CW (2008) Power control in wireless cellular networks. Now
3. Tang R, Zhao J, Qu H, Zhang Z (2017) User-centric joint admission control and resource allocation for 5G D2D extreme mobile broadband: a sequential convex programming approach. IEEE Commun Lett 21(7):1641–1644. https://doi.org/10.1109/LCOMM.2017.2681664
4. Chiang M, Hande P, Lan T, Tan CW et al (2008) Power control in wireless cellular networks. Found Trends Networking 2(4):381–533
5. Sun H, Chen X, Shi Q, Hong M, Fu X, Sidiropoulos ND (2018) Learning to optimize: training deep neural networks for interference management. IEEE Trans Signal Process 66(20):5438–5453
6. Lee W, Kim M, Cho D (2018) Deep power control: transmit power control scheme based on convolutional neural network. IEEE Commun Lett 22(6):1276–1279
7. Liang F, Shen C, Yu W, Wu F (2020) Towards optimal power control via ensembling deep neural networks. IEEE Trans Commun 68(3):1760–1776
8. Abode D, Adeogun R, Berardinelli G (2023) Power control for 6G industrial wireless subnetworks: a graph neural network approach. In: 2023 IEEE wireless communications and networking conference (WCNC). Glasgow, United Kingdom, pp 1–6. https://doi.org/10.1109/WCNC55385.2023.10118984
9. Cui W, Shen K, Yu W (2019) Spatial deep learning for wireless scheduling. IEEE J Sel Areas Commun 37(6):1248–1261

10. Shen Y, Shi Y, Zhang J, Letaief KB (2021) Graph neural networks for scalable radio resource management: architecture design and theoretical analysis. IEEE J Sel Areas Commun 39(1):101–115. https://doi.org/10.1109/JSAC.2020.3036965
11. Eisen M, Zhang C, Chamon LFO, Lee DD, Ribeiro A (2019) Learning optimal resource allocations in wireless systems. IEEE Trans Signal Process 67(10):2775–2790
12. Zhang X, Zhao H, Xiong J, Liu X, Zhou L, Wei J (2021) Scalable power control/beamforming in heterogeneous wireless networks with graph neural networks. In: 2021 IEEE global communications conference (GLOBECOM). Madrid, Spain, pp 1–6. https://doi.org/10.1109/GLOBECOM46510.2021.9685457

GPSR Compliance

The European Union's (EU) General Product Safety Regulation (GPSR) is a set of rules that requires consumer products to be safe and our obligations to ensure this.

If you have any concerns about our products, you can contact us on ProductSafety@springernature.com

In case Publisher is established outside the EU, the EU authorized representative is:

Springer Nature Customer Service Center GmbH
Europaplatz 3
69115 Heidelberg, Germany

The manufacturer's authorised representative in the EU is Springer
Nature Customer Service Centre GmbH, Europaplatz 3, 69115 Heidelberg,
Germany. If you have any concerns regarding our products, please
contact ProductSafety@springernature.com

Printed and bound by CPI Group (UK) Ltd, Croydon, CR0 4YY
01/10/2026
02225855-0012